FUNCTIONAL ANALYSIS

N. Ya. Vilenkin, E. A. Gorin, A. G. Kostyuchenko, M. A. Krasnosel'skii, S. G. Krein (Editor), V. P. Maslov, B. S. Mityagin, Yu. I. Petunin, Ya. B. Rutitskii, V. I. Sobolev, V. Ya. Stetsenko, L. D. Faddeev, E. S. Tsitlanadze

FUNCTIONAL ANALYSIS

translated from the Russian by
RICHARD E. FLAHERTY

English edition edited by
GEORGE F. VOTRUBA
University of Montana,, Missoula, U.S.A.

with the collaboration of
LEO F. BORON
University of Idaho, Moscow, U.S.A.

IƵƳ∆NI

WOLTERS-NOORDHOFF PUBLISHING GRONINGEN
THE NETHERLANDS

ISBN: 90 01 90980 9

Library of Congress Catalog Card Number: 75-90855

Printed in the Netherlands

CONTENTS

Chapter I

Fundamental concepts of functional analysis

Chapter II

Linear operators in Hilbert space

Chapter III

Linear differential equations in a Banach space

Chapter IV

Nonlinear operator equations

Introductory remarks

a nonlinear operator 160.   Integration of abstract functions 162.
Urysohn operator in the spaces C and L_p. 164.   Operator f 167.
Hammerstein operator 167.   Derivatives of higher order 168.   Po-
tential operators 170.

Chapter V

Operators in spaces with a cone

Chapter VI

Commutative normed rings

Chapter VII

Operators of quantum mechanics

Chapter VIII

Generalized functions

EDITOR'S FOREWORD TO THE RUSSIAN EDITION

Functional analysis originated at the beginning of the present century and became an independent mathematical discipline during the third and fourth decades; it developed rapidly and continues to do so. After the appearance of the remarkable book by the Polish mathematician S. BANACH (Théorie des opérations linéaires, Warsaw, 1932) the ideas and language of functional analysis permeated the most diverse branches of mathematics and its applications. This process has now gone so far that it is sometimes difficult to distinguish functional analysis from those fields in which it is applied.

On the other hand, the discussion of certain problems of classical functional analysis turned out to be rather restricted and this led to an examination of its basic concepts, that is, to a detailed analysis of its axiomatics. This process occurred during the past decade and can not yet be considered completed. We recall that I. M. GELFAND began his talk on functional analysis at the Fourth All-Union Mathematical Conference with the pessimistic words: "We still do not have a good definition of *space*, nor do we yet have a good definition of *operator*."

The authors of the present book were confronted by two dangers: to become lost in the numerous logical and conceptual sources of functional analysis or to become dissipated among the infinite number of branches in the delta of functional analysis as it flows into the sea of mathematical sciences. In order to avoid these dangers the autors strove to keep close to the main channel – the theory of operators and operator equations. The main material of this book is devoted to this theory. An exception is the extensive last chapter "Generalized functions" by N. YA. VILENKIN which could also be part of a book on mathematical analysis since it contains the results of the influence of the ideas and methods of functional analysis on problems of mathematical analysis.

A priori restrictions on the size of the volume forced the authors to exclude entirely or partially the material of several large areas of functional analysis. It is obviously proper to indicate them.

The theory of linear topological spaces is not discussed. S. G. KREIN,

Yu. I. Petunin and E. S. Tsitlanadze included only some of the basic concepts of these spaces in Chapter I. To fill this gap it is recommended that the reader study the book "Functional analysis" by Kantorovich and Akilov or the book "Topological vector spaces" by N. Bourbaki.

Ergodic theory is related to the material of Chapter I; this theory is discussed in the book "Theory of operators" by Dunford and Schwartz. This interesting and methodologically elegant material is not included in this book.

Chapter II, written by A. G. Kostyuchenko, S. G. Krein and V. I. Sobolev, does not touch upon von Neumann's profound theory of rings of operators. There are given only illustrations of the numerous applications of the theory of operators in Hilbert space to partial differential equations.

In Chapter III, by S. G. Krein, there are introduced those facts from the theory of semigroups which it is natural to regard as theorems on differential equations in Banach space. Of course, this represents only an insignificant part of this theory, an exhaustive discussion of which in the book "Functional analysis and semigroups" by Hille and Phillips takes up more than 500 pages!

Chapter IV, by M. A. Krasnosel'skii and Ya. B. Rutitskii, omits a number of problems of nonlinear functional analysis: analytic methods of investigating nonlinear equations, nonlinear differential equations with bounded and unbounded operators, and others.

This book excludes completely Fantapie's theory of analytic functionals.

The general theory of partially ordered spaces is touched upon only lightly in Chapter V, written by M. A. Krasnosels'ki and V. Ya. Stetsenko. The reader to referred to the comprehensive monograph "Functional analysis in partially ordered spaces" by Vulikh, Kantorovich and Pinsker.

Chapter VI, by E. A. Gorin and B. S. Mityagin, contains only a discussion of the theory of commutative normed rings and its application to harmonic analysis. Chapter VII, by V. P. Maslov and L. D. Faddeev, contains basic material on the application of the theory of operators to the basic problems of quantum mechanical systems.

Finally, we have excluded such large areas as the theory of representations of groups and approximation methods in the solution of operator equations.

Despite these gaps in the book, we hope it will be useful to mathematicians, theoretical physicists and researchers in disciplines which make use of functional analysis.

In using this book the reader should bear in mind that there are no lemmas and theorems and the formulas are not numbered. The sections listed in the table of contents are to be regarded as the "elementary particles" of the material discussed. To find a topic it is recommended that the reader look for the section listed in the table of contents or in the index and then read the entire section; it usually contains a connected and relatively complete discussion of the chain of questions being considered.

There are no references in the book to papers or other works in which the various results were obtained. Moreover, as a rule, we do not indicate the authors of these results. An exception are only the theorems and concepts which are now firmly associated with the names of the mathematicians who discovered them. The bibliography consists of books in Russian, survey articles from "Uspekhi matematicheskikh nauk" and several works which contain material related to this book and which is not discussed in textbooks. Basically, these works refer to the material of Chapters III and VII.

Chapter VIII was written by N. YA. VILENKIN in constant contact with I. M. GELFAND who is in reality its author. Material from "Generalized functions" ([10]–[14]) from which the tables are adapted is used extensively in this chapter. The authors of Chapter VI frequently received aid from G. E. SHILOV. L. A. GUREVICH, M. KH. GOLDMAN, O. M. KOZLOV, M. C. KREIN, V. M. TIKHOMIROV gave the authors individual pieces of information which at times constituted entire pages of the book. We express our profound gratitude to all of them.

S. G. KREIN

CHAPTER I

FUNDAMENTAL CONCEPTS OF FUNCTIONAL ANALYSIS

§ 1. Linear Systems

1. *Concept of a linear system.* The concept of a linear system is one of basic importance in functional analysis.

A set E is called a *real (complex) linear system* if for every two of its elements x and y there is assigned an element $x+y$ of the set, called their *sum*, and if for any element x and any real (complex) number λ there is assigned an element λx of the set E, called their *product*, where these operations satisfy the following conditions (axioms):

1) $(x+y)+z=x+(y+z)$ (associativity of addition);
2) $x+y=y+x$ (commutativity of addition);
3) there exists an element 0 in E such that $0x=0$ for every $x \in E$;
4) $(\lambda+\mu)x=\lambda x+\mu x$ $\Big\}$ (distributivity);
5) $\lambda(x+y)=\lambda x+\lambda y$ $\Big\}$
6) $(\lambda\mu)x=\lambda(\mu x)$ (associativity of multiplication);
7) $1x=x$.

Thus a linear system is an algebraic structure which reflects properties connected with addition and multiplication by numbers of vectors in Euclidean spaces.

In a linear system E, an operation which is naturally called *subtraction* can be introduced: the *difference* $x-y$ is defined to be the element $x+(-1)y$ of E.

Examples of linear systems

a) Let E_n be the collection of all vectors of an n-dimensional Euclidean space. The operations of the addition of two vectors $x=\{\xi_1, \xi_2, ..., \xi_n\}$ and $y=\{\eta_1, \eta_2, ..., \eta_n\}$ of the set E_n and the multiplication of the vector $x=\{\xi_1, \xi_2, ..., \xi_n\}$ by the real number λ are introduced in a natural manner:

$$x + y = \{\xi_1 + \eta_1, \xi_2 + \eta_2, ..., \xi_n + \eta_n\},$$
$$\lambda x = \{\lambda\xi_1, \lambda\xi_2, ..., \lambda\xi_n\}.$$

The set E_n equipped with these operations becomes a real linear system.

b) The set Λ of all possible complex sequences $x = \{\xi_1, \xi_2, ..., \xi_n, ...\}$ is an example of a complex linear system in which the operations of the addition of the elements $x = \{\xi_1, \xi_2, ..., \xi_n, ...\}$ and $y = \{\eta_1, \eta_2, ..., \eta_n, ...\}$ and the multiplication of the element x by the complex number λ are defined analogously:

$$x + y = \{\xi_1 + \eta_1, \xi_2 + \eta_2, ..., \xi_n + \eta_n, ...\},$$
$$\lambda x = \{\lambda\xi_1, \lambda\xi_2, ..., \lambda\xi_n, ...\}.$$

c) The set $C(0, 1)$ consisting of all possible continuous functions defined on the interval $[0, 1]$ becomes a real linear system if the operations of addition of functions and multiplication of a function by a number are introduced in the usual way.

2. *Linear dependence and independence.*   A system of elements $x_1, x_2, ..., x_n$ is called *linearly independent* if a relation of the form $\sum_{k=1}^{n} \lambda_k x_k = \theta$ is possible only when $\lambda_1 = \lambda_2 = \cdots = \lambda_n = 0$. Otherwise, the elements $x_1, x_2, ..., x_n$ are called *linearly dependent*.

An infinite system of elements is called linearly independent if every finite collection of distinct elements of this system is linearly independent.

A linearly independent system $\{x_\alpha\}$ is called an *algebraic basis* of the linear system E if every element $x \in E$ is representable in the form of a linear combination of a finite number of elements of $\{x_\alpha\}$:

$$x = \sum_{i=1}^{n} \lambda_i x_{\alpha_i}.$$

Since an algebraic basis is a linearly independent system, the indicated representation of an element x is unique.

Every linear system has an algebraic basis. Any two algebraic bases of a linear system E have the same cardinal number χ. This cardinal number is called the *dimension of the linear system E.*

A linear system E is called *finite-dimensional* if its dimension is a natural number n. In this case an algebraic basis consists of n elements. In the case of an infinite χ the linear system E is called *infinite-dimensional*.

3. *Linear manifolds and convex sets.* A non-empty subset M of a linear system E is called a *linear manifold* if for any two elements x_1, x_2 of the set M all linear combinations $\lambda_1 x_1 + \lambda_2 x_2$ also belong to M.

Let S and T be two subsets of a linear system E. The *algebraic sum* $S + T$ of the sets S and T is understood to be the set consisting of all elements of the form $x + y$, where $x \in S$ and $y \in T$. Two linear manifolds M and N of E are called *algebraically complementary* if $M \cap N = \theta$ and $M + N = E$. For every linear manifold M of a linear system E there exists an algebraically complementary linear manifold N.

The *segment* defined by the elements x and y of a linear system is understood to be the collection of all elements of the form $\alpha x + (1 - \alpha) y$, where $0 \leqslant \alpha \leqslant 1$. A set S in a linear system E is called *convex* if it contains the segment defined by any two of its elements. The simplest example of a convex set is given by an arbitrary linear manifold $M \subset E$.

For any arbitrary set $S \subset E$ there exists a smallest convex set $\tilde{S}$ containing S, called the *convex hull of the set S*. The convex hull $\tilde{S}$ consists of all possible elements of the form

$$x = \sum_{k=1}^{n} \alpha_k x_k,$$

where $\alpha_k \geqslant 0$, $\sum_{k=1}^{n} \alpha_k = 1$, $x_k \in S$ and n is an arbitrary natural number.

§ 2. Linear topological, metric, normed and Banach spaces

1. *Linear topological space.* A linear topological space is a composite structure. Its structure is induced by a linear system and a topological space. The concept of a topological space reflects properties connected with the intuitive concepts of neighborhood, limit and continuity in ordinary Euclidean spaces. In a linear topological space both of these structures are interrelated. This interrelationship reflects the properties of the continuity of the algebraic operations on vectors in Euclidean spaces.

In functional analysis, infinite-dimensional linear topological spaces, which along with properties common to Euclidean spaces have several qualitatively new properties, are usually studied.

Let E be a linear system equipped with a separated (Hausdorff) topology defined by a system of neighborhoods $\{V_x\}$ (see [5]). The set E is called a *linear topological space* if the algebraic operations are continuous in the topology of E, that is:

1) For every pair of elements $x, y \in E$ and every neighborhood V_{x+y} of the element $x+y$ there exists a neighborhood V_x of the element x and a neighborhood V_y of the element y such that

$$V_x + V_y \subset V_{x+y};$$

2) For every element $x \in E$, every number λ and every neighborhood $V_{\lambda x}$ of the element λx, a neighborhood V_x of the element x and a number $\delta > 0$ can be found such that

$$\mu V_x \subset V_{\lambda x} \quad \text{for} \quad |\mu - \lambda| < \delta.$$

In a linear topological space, the system $\{V_\theta^{(\alpha)}\}$ of neighborhoods of zero defines completely the topology of this space: an arbitrary neighborhood $V_x^{(\alpha)}$ of the element x is obtained from some neighborhood of zero $V_\theta^{(\alpha)}$ by means of its displacement by the element x:

$$V_x^{(\alpha)} = x + V_\theta^{(\alpha)}.$$

The finite-dimensional Euclidean space R_n with the usual topology is the simplest example of a linear topological space.

Other examples of linear topological spaces

a) A topology is introduced in the linear system Λ of all complex sequences (see example b), §1) by means of a system of neighborhoods in the following manner: the collection of elements $x = \{\xi_1, \xi_2, \dots, \xi_n, \dots\}$ whose coordinates satisfy the condition

$$\sum_{n=1}^{\infty} \frac{1}{2^n} \frac{|\xi_n^0 - \xi_n|}{1 + |\xi_n^0 - \xi_n|} < \varepsilon,$$

where ε is some positive number, is called a neighborhood of the element $x_0 = \{\xi_1^0, \xi_2^0, \dots, \xi_n^0, \dots\}$. The linear system Λ equipped with this topology becomes a linear topological space which is denoted by s.

b) The linear system $C(0, 1)$ of real-valued functions continuous on the interval $[0, 1]$ (see example c), §1) can be transformed into a linear topological space by introducing a system of neighborhoods: the collection of all functions $x(t) \in C(0, 1)$ for which $|x_0(t) - x(t)| < \varepsilon$ for all $t \in [0, 1]$ and some number $\varepsilon > 0$ is called a neighborhood of the function $x_0(t) \in C(0, 1)$.

An element x_0 is called a *limit point* of a set S in a topological space E if every neighborhood of x_0 contains an element of the set S. The set of all limit points of the set S is called the *closure* of the set S and is denoted by $\bar{S}$.

A sequence $\{x_n\}$ of elements of a topological space E is said to *converge* to the element x_0 if for each neighborhood V_{x_0} of the element x_0 a natural number N can be found such that $x_n \in V_{x_0}$ for all $n > N$.

It must be noted that the closure of a set S in a linear topological space does not always coincide with the set of limits of all possible convergent sequences of elements of S. However, for several particular types of linear topological spaces, for example, for linear metric and normed spaces, the closure of a set coincides with the collection of limits of all convergent sequences of elements of the set.

If the closure of a set S coincides with the entire topological space E, then the set S is called *everywhere dense*.

2. *Locally convex space.* A set $\Gamma = \{V_x^{(\alpha)}\}$ of neighborhoods of the element x is called a *fundamental system of neighborhoods* of this element if every neighborhood of x contains a neighborhood of the set Γ.

A linear topological space E is called *locally convex* if it has a fundamental system of neighborhoods of zero, each of which is convex.

The spaces R_n, $C(0, 1)$ and s are locally convex.

A finite real-valued non-negative function $p(x)$, defined on a linear system E is called a *seminorm* if it has the following properties:

1) $p(\lambda x) = |\lambda| p(x)$ for all $x \in E$ and λ;

2) $p(x + y) \leqslant p(x) + p(y)$ for all $x, y \in E$.

If some collection of seminorms Γ is given on a linear system E, then a (not necessarily separable) locally convex topology can be defined on E by taking as a fundamental system of neighborhoods of zero the set of convex sets defined by relations of the form $p(x) < \varepsilon$ where p ranges over Γ and ε ranges over the set of all positive numbers. This topology will be separable if and only if a semi-norm $p \in \Gamma$ can be found for every element $x_0 \neq \theta (x \in E)$ such that $p(x_0) \neq 0$.

Every locally convex topology on a linear system E can be given in the way indicated by means of some collection of seminorms.

3. *Metric linear space.* A metric space is an important special case of a topological space. A set E is called a *metric space* if for every pair of its elements x, y there is a real number $\varrho(x, y)$, the *distance* between the elements x and y, which satisfies the following conditions (axioms):

1) $\varrho(x, y) \geqslant 0$; $\varrho(x, x) = 0$, and if $\varrho(x, y) = 0$, then $x = y$;

2) $\varrho(x, y) = \varrho(y, x)$ (axiom of symmetry);

3) $\varrho(x, y) \leqslant \varrho(x, z) + \varrho(z, y) (z \in E)$ (triangle inequality).

The elements of a metric space are also called points; if the set E is transformed into a metric space by the introduction of a distance, then one says that a *metric* is introduced in the set E or also that the set E is *metrized.*

If $x_n \in E$, $x \in E$ and $\varrho(x_n, x) \to 0$ as $n \to \infty$, then one says that x_n *converges* to x:

$$x_n \to x.$$

The $\inf\limits_{x \in A,\, y \in B} \varrho(x, y)$ is called the *distance* between the sets A and B of a metric space.

A linear system E is called a *metric linear space* if it is metrized and the algebraic operations are continuous in the metric of E, that is:

1) If $x_n \to x$ and $y_n \to y$, then $x_n + y_n \to x + y$;

2) If $x_n \to x$ and $\lambda_n \to \lambda$, then $\lambda_n x_n \to \lambda x$.

Example of a metric linear space. A distance between the elements $x = \{\xi_1, \xi_2, \ldots, \xi_n, \ldots\}$ and $y = \{\eta_1, \eta_2, \ldots, \eta_n, \ldots\}$ can be introduced in the linear topological space s (see example a), no. 1) by means of the formula

$$\varrho(x, y) = \sum_{n=1}^{\infty} \frac{1}{2^n} \frac{|\xi_n - \eta_n|}{1 + |\xi_n - \eta_n|}.$$

A metric on a set E *generates* (or *induces*) a topology in E in a natural manner: if we take for a neighborhood $V_{x_0}^{(\varepsilon)} (0 < \varepsilon < \infty)$ of the point x_0 the collection of all elements $y \in E$ such that $\varrho(x_0, y) < \varepsilon$ for a given $\varepsilon > 0$, then the set E becomes a topological space.

A topological space E is called *metrizable* if a metric can be introduced in it such that the topology which is induced by this metric in E coincides with the initial topology of the topological space E.

The following criterion for the metrizability of locally convex linear topological spaces is valid: *In order that a locally convex linear topological space be metrizable, it is necessary and sufficient that its topology be given by a denumerable set of seminorms.*

Example. The space D of all infinitely differentiable functions $x(t)$ vanishing outside of some interval (depending on the function) plays a large role in the theory of generalized functions. A locally convex topology is introduced in D by means of a system of seminorms:

$$p_n(x) = \sup_{-\infty < t < \infty} |x^{(n)}(t)|.$$

The space D is a metrizable linear topological space.

Important examples of non-metrizable linear topological spaces will be given below (see §4, no. 3).

4. *Normed linear space.* Along with the concepts of the algebraic operations, the concept of "length" or norm of a vector plays a large role in geometry, analysis and in several other fields of mathematics.

A linear system E is called a *normed linear space* if for each element $x \in E$ there corresponds a real number $\|x\| \geq 0$, called the *norm of the element x*, satisfying the following conditions (*axioms of a normed linear space*):

1) $\|x\| = 0$ if and only if $x = \theta$;
2) $\|\lambda x\| = |\lambda| \|x\|$ (homogeneity of the norm);
3) $\|x+y\| \leq \|x\| + \|y\|$ (triangle inequality).

We say that two normed linear spaces E and F are *isometric* if a one-to-one correspondence $x \leftrightarrow y$ exists between their elements x and y ($x \in E$, $y \in F$) satisfying the following conditions:

1) if $x_1 \leftrightarrow y_1$ and $x_2 \leftrightarrow y_2$, then

$$\lambda_1 x_1 + \lambda_2 x_2 \leftrightarrow \lambda_1 y_1 + \lambda_2 y_2$$

for arbitrary numbers λ_1, λ_2;

2) if $x \leftrightarrow y$, then $\|x\| = \|y\|$.

Two normed linear spaces are called *isomorphic* if a one-to-one correspondence $x \leftrightarrow y$ ($x \in E$, $y \in F$) exists between their elements such that

1) if $x_1 \leftrightarrow y_1$ and $x_2 \leftrightarrow y_2$, then $\lambda_1 x_1 + \lambda_2 x_2 \leftrightarrow \lambda_1 y_1 + \lambda_2 y_2$;
2) there exist constants $C_1 > 0$ and $C_2 > 0$, not depending on the

elements x and y, such that if $x \leftrightarrow y$ then

$$C_1 \|y\| \leqslant \|x\| \leqslant C_2 \|y\|.$$

With regard to isometry, all the properties of the space which are connected with the algebraic operations and the norm of elements are preserved, so that distinctions between isometric space concern only the nature of their elements. In abstract functional analysis, in the study of normed linear spaces, we are interested only in those properties of the space which are connected with the algebraic operations and the norms of this space, and we do not study the nature of the elements themselves. Therefore isometric normed linear spaces are simply identified. In applications of functional analysis, it is usual to have to deal with concrete normed linear spaces in which it is impossible to ignore the nature of the elements. The translation of all the obtained facts from the "language" of one space to the "language" of a space isometric to it is often practically unrealizable. In this case the identification of isometric spaces is not sensible.

The quantity

$$\varrho(x, y) = \|x - y\| = \|y - x\|$$

has all the properties of a metric; therefore normed linear spaces are, in particular, metric spaces and hence topological spaces. With regard to the isomorphism of two normed linear spaces, only those properties are preserved which depend on the nature of the algebraic operations and the topological structures of these spaces.

Since a normed linear space is a topological space, the concepts of the limit of a sequence, closure of set, neighborhood of a point, and so on, are defined. In view of the fact that the basic subject in the sequel will be normed linear spaces, all these concepts are formulated here directly in terms of normed linear spaces.

A sequence $\{x_n\}$ of elements of a normed linear space E is said to *converge to the element* x_0 if $\|x_0 - x_n\| \to 0$ as $n \to \infty$.

The set of all points $x \in E$ such that $\|x_0 - x\| < r$ ($\|x_0 - x\| \leqslant r$) is called an open (closed) *sphere* $S(x_0, r)$ of *radius* $r > 0$ with *center* at the point x_0.

A *neighborhood* of the point x_0 is understood to be any subset $V_{x_0} \subset E$ containing an open sphere of some radius with center at the point x_0.

An element x_0 is called a *limit point* of the set $T \subset E$ if every neighborhood of x_0 contains elements of the set T. In order for the element x_0 to be a limit point of the set T, it is necessary and sufficient that there exists a sequence $\{x_n\} \subset T$ which converges to x_0.

A set $T \subset E$ is called *closed* if all limit points of T belong to T.

A set $T \subset E$ is called *open* if every one of its points is *interior*, that is, contained in T together with some neighborhood. (In other words, for an arbitrary point $x_0 \in T$ and for some $r > 0$, $S(x_0, r) \subset T$.)

The sphere $S(x_0, r)$ of radius r with center at the point x_0 is a convex set; in addition, it follows from the definition of a neighborhood of the point x_0 that the spheres $S(x_0, r)$ form a fundamental system of neighborhoods of the point x_0 where r ranges over the set of all positive real numbers. Hence it follows that a normed linear space is a locally convex linear topological space.

The smallest closed convex set in E containing the set T is called the *closed convex hull of the set T* in the normed linear space E.

A linear topological space E is called *normable* if a norm can be introduced in E such that the topology induced by this norm on the set E coincides with the initial topology of the space E.

The question naturally arises: under what conditions is a linear topological space E normable? The answer is formulated with the aid of the concept of a bounded set in a linear topological space. A set $V \subset E$ is called *bounded* if for an arbitrary sequence of elements $x_n \in V$ and an arbitrary numerical sequence $\lambda_n \to 0$ the sequence of elements $\{\lambda_n x_n\}$ converges to zero.

In order that a linear topological space be normable, it is necessary and sufficient that there exist a bounded convex neighborhood of zero (A. N. Kolmogorov).

The spaces s and D are examples of non-normable linear topological spaces.

A closed linear manifold M in a normed linear space E is called a *linear subspace*. Every finite-dimensional linear manifold M is closed and is therefore a linear subspace.

Every linear manifold M in the normed linear space E is itself a normed linear space with respect to the norm of the space E.

We say that a space E is decomposed into a *direct sum* of its subspaces M_1 and M_2, $E = M_1 \oplus M_2$, if an arbitrary element $x \in E$ has a unique representation $x = x_1 + x_2$ where $x_1 \in M_1$, $x_2 \in M_2$.

5. *Examples of normed linear spaces.*

1. Euclidean space R_n. Let E_n be the linear system consisting of all possible n-dimensional vectors $x = \{\xi_1, \xi_2, ..., \xi_n\}$. A norm can be introduced in E_n by the formula

$$\|x\| = \sqrt{\sum_{i=1}^{n} \xi_i^2} \,.$$

The linear system E_n with this norm is called the *Euclidean space R_n*. The triangle inequality (axiom 3) follows from the well-known Minkowski inequality for finite sums

$$\left[\sum_{i=1}^{n} |\xi_i + \eta_i|^p\right]^{\frac{1}{p}} \leqslant \left[\sum_{i=1}^{n} |\xi_i|^p\right]^{\frac{1}{p}} + \left[\sum_{i=1}^{n} |\eta_i|^p\right]^{\frac{1}{p}} (p \geqslant 1)$$

by setting $p = 2$.

2. The space m_n. A norm can be defined in another manner for the vector $x = \{\xi_1, \xi_2, ..., \xi_n\}$ in E_n:

$$\|x\| = \max_{1 \leqslant i \leqslant n} |\xi_i| \,.$$

The linear system E_n with this norm is called the *space m_n.*

A norm can be introduced in the space E_n in various ways, but all the normed spaces which are obtained are isomorphic.

3. The spaces $l_p (p \geqslant 1)$. The elements of the space $l_p (p \geqslant 1)$ are the numerical sequences $x = \{\xi_1, \xi_2, ..., \xi_n, ...\}$ for which the series $\sum_{i=1}^{\infty} |\xi_i|^p$ is convergent.

It follows from the Minkowski inequality that l_p forms a linear system. After the introduction of the norm

$$\|x\| = \left\{\sum_{i=1}^{\infty} |\xi_i|^p\right\}^{\frac{1}{p}},$$

l_p becomes a normed linear space.

4. The space m. Let m be the set consisting of all possible bounded sequences $x = \{\xi_1, \xi_2, ..., \xi_n, ...\}$. If we set

$$\|x\| = \sup_i |\xi_i| \,,$$

then m becomes a normed linear space.

5. The space c. The space c, whose elements are convergent sequences, is extracted from the space m. The norm is defined in c as in m. The space c is a linear subspace of the space m.

6. The space c_0. The linear subspace of the space c which consists of all sequences convergent to zero is called the *space c_0*.

7. The space $L_p(0, 1)$. The analogue of the space l_p among the function spaces is the space $L_p(0, 1)(p \geqslant 1)$ consisting of all functions *) whose p-th power is summable on the segment $[0, 1]$, that is, measurable functions $x(t)$ such that

$$\int_0^1 |x(t)|^p \, dt < \infty.$$

It follows from the Minkowski inequality

$$\left[\int_0^1 |x(t) + y(t)|^p \, dt \right]^{\frac{1}{p}} \leqslant \left[\int_0^1 |x(t)|^p \, dt \right]^{\frac{1}{p}} + \left[\int_0^1 |y(t)|^p \, dt \right]^{\frac{1}{p}}$$

that $L_p(0, 1)$ becomes normed if we set

$$\|x\| = \left[\int_0^1 |x(t)|^p \, dt \right]^{\frac{1}{p}}.$$

Convergence in $L_p(0, 1)$ is mean convergence of order p. Namely, $x_n \to x_0$ means that

$$\int_0^1 |x_n(t) - x_0(t)|^p \, dt \to 0.$$

8. The space $C(0, 1)$. In the linear system $C(0, 1)$ (see §1, c)), consisting of all functions continuous on the interval $[0,1]$, the norm of the function $x(t)$ is defined in the following manner:

$$\|x\| = \max_{0 \leqslant t \leqslant 1} |x(t)|.$$

The distance between two functions

$$\varrho(x, y) = \max_{0 \leqslant t \leqslant 1} |x(t) - y(t)|$$

---

*) In the spaces L_p, and also in the spaces M and L_M^* (see examples 10 and 14), functions which are equal almost everywhere on $[0, 1]$ are identified.

is the maximal distance between their graphs. The convergence of the sequence $\{x_n\}$ of points of the space $C(0, 1)$ to the point x_0 means the uniform convergence of the sequence of functions $x_n(t)$ to the function $x_0(t)$.

9. The space $C^{(l)}(0, 1)$. The elements of this space are all possible functions defined on the interval $[0, 1]$ and having continuous derivatives on this interval to order l inclusively. The algebraic operations, the operations of addition and multiplication of a function by a number, are defined in the usual manner. The norm of the element $x(t) \in C^{(l)}(0, 1)$ is defined by the formula

$$\|x\| = \sum_{k=0}^{l} \max_{0 \leqslant t \leqslant 1} |x^{(k)}(t)| \qquad (x^{(0)}(t) = x(t)).$$

Convergence in $C^{(l)}(0, 1)$ means the uniform convergence of the sequence of functions and of the sequences of their derivatives of k-th order $(k = 1, 2, ..., l)$.

10. The space $M(0, 1)$. Another example of a normed linear function space is the set $M(0, 1)$ of all measurable and almost everywhere bounded functions $x(t)$ on the interval $[0, 1]$ in which the algebraic operations are defined in the usual way and the norm is defined by the equality

$$\|x\| = \operatorname{vrai\ max}_{0 \leqslant t \leqslant 1} |x(t)| = \inf_{mE=0} \{ \sup_{t \in [0,1] \setminus E} |x(t)| \}.$$

The convergence in $M(0, 1)$ is uniform convergence almost everywhere.

11. The space $V(0, 1)$. Let $x(t)$ be a finite function defined on the interval $[0, 1]$. Let us consider an arbitrary partition τ of the interval $[0, 1]$ $(0 = t_0 < t_1 < \cdots < t_n = 1)$ and form the expression

$$v_\tau = \sum_{k=1}^{n} |x(t_k) - x(t_{k-1})|.$$

If the collection of the sums v_τ corresponding to various subdivisions τ of the interval $[0, 1]$ is bounded, then the function $x(t)$ is called a function of *bounded variation* on the interval $[0, 1]$ and the quantity

$$\overset{1}{\underset{0}{V}}(x) = \sup_\tau v_\tau$$

is called the *total variation* of the function $x(t)$.

Let $V(0, 1)$ be the set of all functions of bounded variation. If the algebraic operations are introduced in the usual manner in the set $V(0, 1)$, then it forms a linear system; if, in addition, we set

$$\|x\| = |x(0)| + \overset{1}{\underset{0}{V}}(x)$$

for $x \in V(0, 1)$, then $V(0, 1)$ becomes a normed linear space.

12. The space $C(-\infty, \infty)$. This is the space whose elements are functions which are continuous and bounded on the entire real line. The norm of $x(t)$ in $C(-\infty, \infty)$ is introduced by the formula

$$\|x\| = \sup_{-\infty < t < \infty} |x(t)|.$$

The convergence of a sequence in $C(-\infty, \infty)$ means uniform convergence on the entire real line.

13. The Hölder space $C_\alpha(0, 1)$. Let $C_\alpha(0, 1)$ be the set of all functions $x(t)$ defined on the interval $[0, 1]$ and satisfying a *Hölder* (or *Lipschitz*) *condition* with index α $(\alpha > 0)$:

$$|x(t_1) - x(t_2)| \leqslant C|t_1 - t_2|^\alpha \quad (t_1, t_2 \in [0, 1]).$$

The norm of $x(t) \in C_\alpha(0, 1)$ is defined by means of the formula

$$\|x\| = |x(0)| + \sup_{t_1, t_2 \in [0, 1]} \frac{|x(t_1) - x(t_2)|}{|t_1 - t_2|^\alpha}.$$

14. The Orlicz spaces $L_M^*(0, 1)$. The Orlicz spaces are generalizations of the spaces $L_p(0, 1)$. The function $M(x)$ $(-\infty < x < \infty)$ is called an *N-function* if it is representable in the form

$$M(x) = \int_0^{|x|} p(\tau) \, d\tau,$$

where $p(\tau)$ is positive for $\tau > 0$, and is a right continuous nondecreasing function satisfying the conditions

$$p(+0) = 0 \quad \text{and} \quad \lim_{\tau \to \infty} p(\tau) = \infty.$$

The *Orlicz class* $L_M(0, 1)$ is the collection of functions $x(t)$ which are measurable on $[0, 1]$ and such that

$$\varrho_M(x) = \int_0^1 M(x(t))\, dt < \infty.$$

All possible linear combinations of functions of the class $L_M(0, 1)$ form the linear system $L_M^*(0, 1)$.

In order to introduce a norm in $L_M^*(0, 1)$, we set

$$q(\sigma) = \sup_{p(\tau) \leqslant \sigma} \tau \quad \text{and} \quad N(y) = \int_0^{|y|} q(\sigma)\, d\sigma.$$

If $q(+0) = 0$, then $N(y)$ is an N-function and is called the *complement* of $M(x)$. We define a norm in $L_M^*(0, 1)$ by the formula

$$\|x\|_M = \sup \left| \int_0^1 x(t)\, y(t)\, dt \right|,$$

where the supremum is taken over all $y \in L_N^*(0, 1)$ for which

$$\varrho_N(y) = \int_0^1 N(y(t))\, dt = 1.$$

The normed linear space thus obtained is called the *Orlicz space* $L_M^*(0, 1)$.

If $M(x) = \dfrac{|x|^p}{p}$, then $L_M^*(0, 1) = L_p(0, 1)$.

In the above, for simplicity, examples were mentioned of function spaces consisting of functions defined on the interval $[0, 1]$. The spaces $L_p(Q)$, $L_M^*(Q)$ and $M(Q)$ of functions defined on a set Q with a countably additive measure; the space $C(Q)$ of bounded continuous functions on a topological space Q; and the spaces $C^1(G)$ and $C_\alpha(G)$ of functions defined on a region G in an n-dimensional Euclidean space are defined analogously.

6. *Completeness of metric spaces. Banach space.* A sequence of points $\{x_n\}$ in a metric space E is called *fundamental* (or *Cauchy*) if, for an

arbitrary $\varepsilon > 0$, there exists a natural number $N = N(\varepsilon)$ such that $\varrho(x_m, x_n) < \varepsilon$ for $m, n \geqslant N$.

Every convergent sequence is fundamental; however, the converse statement, in general, is not valid. Indeed, in the metric space R consisting of the rational numbers with the metric $\varrho(x, y) = |x - y|$, let the sequence $\{x_n\}$ converge to some irrational number. The sequence $\{x_n\}$ is fundamental in the metric of R; however, there is no element in R which will be its limit.

If, in a metric space E, every fundamental sequence converges to some element of the space, then E is called a *complete space*.

A complete normed linear space is called a *Banach space*.

In terms of a norm, the fact that a sequence $\{x_n\}$ is fundamental means that $\|x_n - x_m\| \underset{m,\,n \to \infty}{\longrightarrow} 0$. Therefore the condition of completeness of a normed linear space E appears as follows: if $\|x_m - x_n\| \underset{m,\,n \to \infty}{\longrightarrow} 0$, then there exists an element $x_0 \in E$ such that $\|x_n - x_0\| \underset{n \to \infty}{\longrightarrow} 0$.

All of the normed spaces described in examples 1–14 are complete, and, therefore, are Banach spaces.

An arbitrary finite-dimensional normed linear space is complete, and, hence, is a Banach space. An infinite-dimensional Banach space has dimensionality not less than that of the continuum.

In a Banach space, the principle of nested contracting spheres is valid: *Let there be given a sequence of closed spheres, each of which contains all that follow, whose radii approach zero. Then these spheres have a unique common point.*

This principle plays an important role in the proofs of various existence theorems. In spaces which are not complete, this principle is not valid, and several existence theorems do not hold in such spaces.

Just as the set of rational numbers can be extended to the set of all real numbers, every metric space can be extended to a complete space in an analogous way.

The *completion* $\bar{E}$ of a metric space E is the complete metric space which contains E as an everywhere dense subset. Every metric space has a completion.

We give an example.

15. The space $W_p^l(G)$. Let G be a region in n-dimensional space and $C^{(l)}(G)$ the linear system of all l times continuously differentiable functions $x(t)$ in the region G.

The space $C^{(l)}(G)$ is not complete in the norm which is defined by the equality

$$\|x\| = \left\{ \int\limits_{G} |x(t)|^p \, dt \right\}^{\frac{1}{p}} + \left\{ \int\limits_{G} \left[\sum_{k_1+k_2+\cdots+k_n=l} \left(\frac{\partial^l x}{\partial t_1^{k_1} \partial t_2^{k_2} \ldots \partial t_n^{k_n}} \right)^2 \right]^{\frac{p}{2}} dt \right\}^{\frac{1}{p}}.$$

The completion of this space is called the *space* $W_p^l(G)$. Elements adjoined to $C^l(G)$ for its completion can be identified with functions defined almost everywhere in G and having generalized derivatives of order l whose p-th powers are summable. The spaces $W_p^l(G)$ were introduced by S. L. Sobolev and play an important role in various problems in the theory of partial differential equations.

7. *Compact sets.*   A set T in a metric space E is called *compact**) if from every infinite sequence $\{x_n\} \subset T$ it is possible to extract a subsequence which is convergent to some limit $x \in E$.

Criterion for compactness. In order for a set T in the metric space E to be compact, it is necessary and, in the case of completeness of E, sufficient that for an arbitrary $\varepsilon > 0$ there exists a finite ε-net, that is, a finite set of elements $x_1, \ldots, x_n \in E$ such that for every element $x \in T$ an element x_k can be found situated from x at a distance less than ε:

$$\varrho(x, x_k) < \varepsilon.$$

In a normed linear space, the compactness of a set signifies that it can be approximated arbitrarily closely by a bounded set lying in a finite-dimensional subspace.

The following statements are useful: if a set is compact, then its convex hull is also compact; the closure of a compact set is compact.

Every bounded set in a finite-dimensional Banach space is compact. *A normed linear space in which some sphere is compact is finite dimensional.*

Compactness criteria

1) *Compactness in* $C(0, 1)$. *For compactness of a set $T \subset C(0, 1)$, it is necessary and sufficient that the functions $x(t) \in T$ be uniformly bounded and equicontinuous* (Arzelà's criterion). Uniform boundedness means, in this connection, that a constant K exists such that $|x(t)| \leqslant K$ for all $x(t) \in T$,

---

*) *"Relatively sequentially compact"* is a better name. Compare with §4, no. 4.

and equicontinuity means that for every $\varepsilon > 0$ a number $\delta = \delta(\varepsilon)$ exists such that $|x(t_1) - x(t_2)| < \varepsilon$ for arbitrary $t_1, t_2 \in [0, 1]$ with $|t_1 - t_2| < \delta$ and for arbitrary functions $x(t) \in T$.

2) *Compactness in* $C(-\infty, \infty)$. *For compactness of a set* $T \subset C(-\infty, \infty)$, *it is necessary and sufficient that the functions* $x(t)$ *be uniformly bounded and that for every* $\varepsilon > 0$ *there exists a covering of the axis* $(-\infty, \infty)$ *by a finite number of open sets on each of which the oscillation of an arbitrary function* $x(t) \in T$ *is less than* ε.

3) *Compactness in* $L_p(0, 1)$ $(p \geqslant 1)$. *For compactness of a set* $T \subset L_p(0, 1)$, *it is necessary and sufficient that the set be bounded in* $L_p(0, 1)$ *and that for every* $\varepsilon > 0$ *there exists a number* $\delta > 0$ *such that for* $h < \delta$ *and for an arbitrary function* $x(t) \in T$ *the distance* $\varrho(x, x_h) < \varepsilon$, *where*

$$x_h(t) = \frac{1}{2h} \int\limits_{t-h}^{t+h} x(\tau)\, d\tau$$

(the function $x(t)$ is set equal to zero outside the interval $[0, 1]$) (A. N. Kolmogorov's criterion).

4) Another criterion for compactness in $L_p(0, 1)$ $(p \geqslant 1)$. *For compactness of a set* $T \subset L_p(0, 1)$, *it is necessary and sufficient that the set be bounded in* $L_p(0, 1)$ *and that for every* $\varepsilon > 0$ *there exists a number* $\delta > 0$ *such that*

$$\int\limits_0^1 |x(t+h) - x(t)|^p\, dt < \varepsilon$$

for $|h| < \delta$ (M. Riesz' criterion).

5) *Compactness in* l_p $(p \geqslant 1)$. *For compactness of a set* $T \subset l_p$, *it is necessary and sufficient that* T *be bounded and that for every* $\varepsilon > 0$ *there exists a number* $N = N(\varepsilon)$ *such that the inequality* $\sum\limits_{i=N}^{\infty} |\xi_i|^p < \varepsilon$ *holds for all* $x = \{\xi_1, \xi_2, \ldots, \xi_n, \ldots\} \in T$.

Quantitative characteristics of the "massiveness" of a compact set have been amply studied in the last few years. The minimal number of points in all possible ε-nets of the set T is denoted by $N_\varepsilon(T)$. The number

$$H_\varepsilon(T) = \log_2 N_\varepsilon(T)$$

is called the *ε-entropy* of the set T.

This nomenclature is connected with ideas of information theory. The entropy of a set of information under transmission with a specified precision is the number of binary digits required for the recovery, with this precision, of an arbitrary piece of information.

If we regard a compact set T as a set information, and by the recovery of the point x of T with the necessary precision we understand the indication of some point x' of the ε-neighborhood of the point x, then these two concepts of entropy are analogous.

Another quantitative characterization of the "massiveness" of T is ε-*capacity*. It is defined by the formula

$$C_\varepsilon(T) = \log_2 M_\varepsilon(T),$$

where $M_\varepsilon(T)$ is the maximal number of points with mutual distances greater than ε which can be placed in the compact set T. By means of ε-capacity, ε-entropy can be bounded below, since

$$M_{2\varepsilon}(T) \leqslant N_\varepsilon(T).$$

Asymptotic estimates of ε-entropy for $\varepsilon \to 0$ exist for several important classes of compact sets.

For the cube Q in n-dimensional space E_n,

$$\lim_{\varepsilon \to 0} \frac{H_\varepsilon(Q)}{\log_2 \dfrac{1}{\varepsilon}} = n.$$

The compact set $V_{l,\alpha}(C_1, C_2)$ consisting of all functions having partial derivatives of order l satisfying a Hölder condition such that the derivatives up to order l are bounded by the number C_1, and the Hölder constants of the l-th derivatives are bounded by the number C_2 is considered in the space $C(G)$ where G is a closed bounded n-dimensional region with a smooth boundary. Then

$$\alpha \left(\frac{1}{\varepsilon}\right)^{\frac{n}{l+\alpha}} \leqslant H_\varepsilon\big(V_{l,\alpha}(C_1, C_2)\big) \leqslant \beta \left(\frac{1}{\varepsilon}\right)^{\frac{n}{l+\alpha}},$$

where α and β do not depend on ε.

The sets of analytic functions are considerably less massive in $C(G)$. If the set of all functions of $C(G)$ allowing analytic extension to a fixed region Q of an n-dimensional space and remaining bounded in it by some number C_1 is denoted by $A_Q(C_1)$, then

$$\alpha \log_2^{n+1}\left(\frac{1}{\varepsilon}\right) \leqslant H_\varepsilon(A_Q(C_1)) \leqslant \beta \log_2^{n+1}\left(\frac{1}{\varepsilon}\right).$$

8. *Separable spaces.* A normed linear space is called *separable* if it contains a countable everywhere dense subset.

The spaces R_n, l_p, c, c_0, $L_p(0, 1)$, $C(0, 1)$, $C^{(l)}(0, 1)$ and $W_p^{(l)}(G)$ are separable, while the spaces m, $M(0, 1)$, $V(0, 1)$, $C_\alpha(0, 1)$ are examples of spaces which are not separable.

The space $C(0, 1)$ has the following universality property: *every separable normed linear space is isometric to some subspace of the space $C(0, 1)$.*

§ 3. Linear functionals

1. *Concept of a linear functional. Hyperplane.* Let E be a real (or complex) linear system. If to every element $x \in E$ there is set in correspondence some real (complex) number $f(x)$, then we say that a *functional* $f(x)$ is defined on E.

The functional $f(x)$ is called *linear* if

$$f(\alpha x + \beta y) = \alpha f(x) + \beta f(y) \quad (x, y \in E).$$

A *hyperplane* L in the linear system E is a linear manifold whose complement are one-dimensional.

If $f(x)$ is a linear functional, then the collection of elements for which $f(x) = 0$ forms a hyperplane. Conversely, every hyperplane has an equation of the form $f(x) = 0$ where $f(x)$ is a linear functional.

2. *Continuous linear functionals.* A functional $f(x)$, defined on a normed linear space E, is said to be *continuous* at the point $x_0 \in E$ if $f(x_n) \to f(x_0)$ whenever $x_n \to x_0$.

A functional $f(x)$ is called a *continuous linear functional* if it is simultaneously linear and continuous on E.

The continuity everywhere of a linear functional follows from its continuity at a single point.

We say that a linear functional is *bounded on E* if a non-negative number c exists such that

$$|f(x)| \leqslant c\|x\|$$

for all elements $x \in E$.

The least of the numbers c satisfying this inequality is called the *norm* of the bounded linear functional $f(x)$ and is denoted by $\|f\|$. The formula

$$\|f\| = \sup_{\substack{x \in E \\ x \neq 0}} \frac{|f(x)|}{\|x\|} = \sup_{\|x\|=1} |f(x)|$$

holds true.

In a normed linear space, the concepts of a continuous linear functional and of a bounded linear functional turn out to be equivalent: *in order that the linear functional $f(x)$ be continuous, it is necessary and sufficient that it be bounded.*

In order that the linear functional $f(x)$ be continuous, it is necessary and sufficient that the hyperplane $L_f = \{x : f(x) = 0\}$ defined by it be a (closed!) subspace of the normed linear space E.

For a continuous non-zero linear functional, the formula

$$\varrho(x, L_f) = \frac{|f(x)|}{\|f\|} \quad (x \in E)$$

holds, where $\varrho(x, L_f)$ – the distance from the point x to the subspace L_f – is equal to inf $\|x-y\|$ where y ranges over all of L_f. This formula is analogous to the well-known formula of analytic geometry for the distance from a point to a plane.

If the linear functional $f(x)$, defined on the space E, is not a continuous functional, then the hyperplane L_f is everywhere dense in the space E.

3. *Extension of continuous linear functionals.* Let $f(x)$ be a linear functional defined on a linear manifold M of the linear system E. A linear functional $\tilde{f}(x)$, defined on all of the space E, is called an *extension of the functional $f(x)$* if $\tilde{f}(x) = f(x)$ for all elements $x \in M$.

The following theorem is fundamental.

Hahn-Banach theorem. Let $p(x)$ be a semi-norm on the linear system E, M a linear manifold in E, and f a linear functional on M such that $|f(x)| \leqslant p(x)$ for all elements $x \in M$. Then there exists on extension $\tilde{f}$ of the functional f to all of E such that $|\tilde{f}(x)| \leqslant p(x)$ for all elements $x \in E$.

For normed linear spaces, this theorem has the following corollaries:

1. If a continuous linear functional $f(x)$ is defined on the space M which is a linear manifold in the normed linear space E, then there exists a continuous linear functional $\bar{f}(x)$, which is an extension of $f(x)$ to the entire space E, such that its norm is equal to the norm of the functional $f(x)$:

$$\|\bar{f}\| = \|f\| .$$

2. For any element x_0, different from zero, of the normed linear space E, a continuous linear functional $f(x)$ on E exists such that

$$\|f\| = 1 \quad \text{and} \quad f(x_0) = \|x_0\| .$$

3. Let T be a non-empty convex open set in a normed linear space E and let M be a linear manifold not intersecting T. A closed hyperplane exists containing M and not intersecting T.

4. *Examples of linear functionals.* Mathematical analysis presents many examples of linear functionals. The functional

$$f(x) = \sum_{k=1}^{N} a_k \xi_k$$

is a continuous linear functional on the spaces l_p, m, c and c_0 whose elements are sequences $x = \{\xi_1, \xi_2, ..., \xi_n, ...\}$.

If the sequence $\{a_n\}$ is bounded, then the functional

$$f(x) = \sum_{k=1}^{\infty} a_k \xi_k$$

will be a continuous linear functional on the space l_1. If $\sum_{k=1}^{\infty} |a_k| < \infty$, then this linear functional is continuous on the spaces l_p and m and on the subspaces c and c_0 of the space m.

The functional

$$f(x) = \lim_{k \to \infty} \xi_k$$

is continuous on the space c. Its norm is equal to one.

The integral

$$f(x) = \int_{0}^{1} x(t)\, dt$$

can be used on the function spaces $L_p(0, 1)$, $M(0, 1)$, $C(0, 1)$ and others
as an example of a continuous linear functional. In this connection, the
Lebesgue integral, defined on $L_1(0, 1)$, can be considered as an extension
of the functional generated by the Riemann integral on the space
$C(0, 1) \subset L_1(0, 1)$.

The linear functional

$$P(x) = \int_0^1 x(t)\,\varphi(t)\,dt$$

has a more general form. If $\varphi(t)$ is a bounded measurable function, then
$P(x)$ also is continuous in the spaces $L_p(0, 1)$, $M(0, 1)$ and $C(0, 1)$.

In the space $C(0, 1)$, the value of the function $x(t)$ at a fixed point t_0
will be a continuous linear functional whose norm is equal to one:

$$f(x) = x(t_0), \qquad \|f\| = 1.$$

Analogously, the value of the k-th derivative $x^{(k)}(t_0)$ at the point t_0
will be a continuous linear functional on the space $C^{(l)}(0, 1)$ for $k \leqslant l$.

If the functional $x(t_0)$ is considered on the space $W_p^{(l)}(G)$, then it
can be shown that for $l > n/p$ (n is the dimension of the region G) this

functional is defined on $W_p^l(G)$ and is continuous. If $l < \dfrac{n}{p}$, then this

functional, defined on $C^{(l)}(G)$, will not be continuous in the norm
of $W_p^{(l)}(G)$.

The norms of several functionals which have been considered will be
given in the following section.

§ 4. Conjugate spaces

1. *Duality of linear systems.* The *product* $E \times F$ of two linear systems
E and F is the linear system whose elements are all possible pairs (x, y)
$(x \in E, y \in F)$. The algebraic operations are introduced in $E \times F$ in the
following manner:

1) $(x_1, y_1) + (x_2, y_2) = (x_1 + x_2, y_1 + y_2)$;
2) $\lambda(x, y) = (\lambda x, \lambda y)$.

A functional $B(x, y)$, defined on $E \times F$, is called *bilinear* if it is linear with respect to each of the variables x, y; that is,

$$B(\alpha_1 x_1 + \alpha_2 x_2, y) = \alpha_1 B(x_1, y) + \alpha_2 B(x_2, y),$$
$$B(x, \alpha_1 y_1 + \alpha_2 y_2) = \alpha_1 B(x, y_1) + \alpha_2 B(x, y_2).$$

We say that the *bilinear functional $B(x, y)$ puts the linear systems E and F in duality or that E and F are in duality (with respect to B)* if the following two conditions are satisfied:

1. For every $x \neq 0$ in E, there exists an $y \in F$ such that $B(x,y) \neq 0$;
2. For every $y \neq 0$ in F, there exists an $x \in E$ such that $B(x, y) \neq 0$.

Let E be a linear system over the field of real or complex numbers and let $\tilde{E}$ be the set of all linear functionals $f(x)$ defined on E. The operations of addition of elements f and g and multiplication of an element f by a number λ can be introduced in the following manner in $\tilde{E}$:

1) $h = f + g$ is the functional on E such that

$$h(x) = f(x) + g(x) \qquad (x \in E);$$

2) $f_1 = \lambda f$ (λ is a number, $f \in \tilde{E}$) is the functional on E such that

$$f_1(x) = \lambda f(x).$$

The set $\tilde{E}$ becomes a linear system which is called the *algebraic conjugate space* of the space E.

If we define the bilinear functional $B(x, f) = f(x)$ on $E \times \tilde{E}$, then E and $\tilde{E}$ are placed, in duality, by means of this bilinear functional.

2. *Conjugate space to a normed linear space.* The set E' of all continuous linear functionals defined on the normed linear space E is a linear manifold in the algebraic conjugate space $\tilde{E}$ since the sum of two continuous linear functionals and the product of a continuous linear functional and a number are continuous linear functionals. If we take as the norm of the element $f \in E'$ the norm $\| f \|$ of the functional $f(x)$, then E' becomes a normed linear space which is called the *conjugate space* of the space E.

The space E' is complete, so the conjugate space of a normed linear space is always a Banach space.

The concept of a continuous linear functional for metric linear spaces can be defined in exactly the same way as was done for normed linear spaces. However, there is no assertion of the type of corollary 2 of no.

3, § 3 which guarantees the existence of a non-trivial continuous linear functional. Moreover, examples of metric linear spaces E exist for which the conjugate space E' consists solely of the continuous linear functional which is identically equal to zero on E.

The presence in E of a convex neighborhood of zero different from E is a necessary and sufficient condition for the existence of a non-trivial functional in E.

For concrete normed linear spaces, the problem of the description of their conjugate spaces often arises. The statement of this problem is made more precise. It is a fact that the definition of a conjugate space gives its description. In a narrower sense, by the description of the conjugate space we mean the designation of a concrete Banach space which is isometric to E', the conjugate of the given space E, and a method of computing the values $f(x)$ of a functional $f \in E'$ on the elements $x \in E$. However the posed problem does not have one specific answer since concrete spaces, isometric to the space E', can be constructed in different ways.

For example, let the norm

$$\|x\| = |\xi_1| + \sum_{k=1}^{n-1} |\xi_{k+1} - \xi_k|$$

be defined in E_n.

If the linear functionals on E_n are represented in the form

$$f(x) = a_1 \xi_1 + \sum_{k=1}^{n-1} a_{k+1}(\xi_{k+1} - \xi_k),$$

then the space E_n^1 with the norm

$$\|f\| = \max_{1 \leqslant k \leqslant n} |a_k|$$

will be the conjugate space to E_n.

If the linear functionals are represented in the usual form by

$$f(x) = \sum_{k=1}^{n} b_k \xi_k,$$

then the space E_n^2 with the norm

$$\|f\| = \max_{1 \leqslant k \leqslant n} \left| \sum_{i=k}^{n} b_i \right|$$

will be the conjugate space.

The spaces E_n^1 and E_n^2 are isometric. The correspondence between them is given by the relation $a_k \leftrightarrow \sum\limits_{i=k}^{n} b_i$.

The problem of the description of the conjugate space is well-defined if a method of computing the values of a functional is given analytically in advance, or, as we say, in the form of a functional. Moreover, a linear functional is usually sought, by analogy with linear forms, in the form of a sum of products or in the form of an integral of a product.

Examples

1. The space l_p $(p > 1)$. An arbitrary continuous linear functional defined on the space l_p is representable in the form

$$f(x) = \sum_{i=1}^{\infty} f_i \xi_i,$$

where $\{f_i\} \in l_q, \dfrac{1}{p} + \dfrac{1}{q} = 1$ and $\|f\| = \left(\sum\limits_{i=1}^{\infty} |f_i|^q \right)^{\frac{1}{q}}$.

The space conjugate to the space l_p is isometric to the space l_q, where $\dfrac{1}{p} + \dfrac{1}{q} = 1$.

2. The space l_1. Every continuous linear functional on l_1 is representable in the form

$$f(x) = \sum_{i=1}^{\infty} f_i \xi_i,$$

where $\|f\| = \sup\limits_{1 \leqslant i < \infty} |f_i| < \infty$.

The conjugate space of l_1 is isometric to the space m.

3. The space c_0. A continuous linear functional on c_0 can be given by the equality

$$f(x) = \sum_{i=1}^{\infty} f_i \xi_i, \qquad x = (\xi_i)$$

where $\|f\| = \sum\limits_{i=1}^{\infty} |f_i| < \infty$.

The conjugate space of c_0 is isometric to the space l_1.

4. The space $L_p(0, 1), p > 1$. An arbitrary continuous linear functional

on the space $L_p(0, 1)$ is representable in the form

$$f(x) = \int_0^1 x(t)\alpha(t)\,dt,$$

where $\alpha(t) \in L_q(0, 1), q = \dfrac{p}{p-1}$.

The norm of the functional f is defined by the formula

$$\|f\| = \left\{ \int_0^1 |\alpha(t)|^q\,dt \right\}^{\frac{1}{q}}.$$

The conjugate space of $L_p(0, 1)$ is isometric to the space $L_q(0, 1)$ $\left(\dfrac{1}{p} + \dfrac{1}{q} = 1 \right)$.

5. The space $L_1(0, 1)$. A continuous linear functional on $L_1(0, 1)$ is representable in the form

$$f(x) = \int_0^1 x(t)\alpha(t)\,dt,$$

where $\alpha(t)$ is a function bounded almost everywhere on the interval $[0,1]$ and

$$\|f\| = \operatorname*{vrai\ max}_{0 \leqslant t \leqslant 1} |\alpha(t)|.$$

The conjugate space of the space $L_1(0, 1)$ is isometric to the space $M(0, 1)$.

6. The space $C(0, 1)$. Every continuous linear functional on $C(0, 1)$ is representable in the form of a Stieltjes integral

$$f(x) = \int_0^1 x(t)\,dg(t),$$

where $g(t)$ is a function of bounded variation. The functional $f(x)$ is not changed if an arbitrary constant is added to the function $g(t)$; therefore we set $g(0)=0$. However, even with this condition, different functions $g(t)$ can generate the same functional. These functions can be distinguished by their values at the points of discontinuity lying inside the interval $[0, 1]$. If, for example, only functions $g(t)$ are considered for which

$$g(t) = \frac{g(t+0) + g(t-0)}{2} \quad \text{for} \quad t \in (0, 1),$$

then the correspondence between the functionals $f(x)$ and the functions $g(t)$ becomes one-to-one. In this connection

$$\|f\| = \overset{1}{\underset{0}{V}}(g).$$

The conjugate space of the space $C(0, 1)$ is isometric to the subspace $V_0(0, 1)$ of the space $V(0, 1)$ consisting of all functions $g(t) \in V(0, 1)$ satisfying the conditions: $g(0)=0$ and $g(t)=\frac{1}{2}[g(t+0)+g(t-0)]$ for $t \in (0, 1)$.

3. *Weak and weak* topology.* In the conjugate space E' of a normed linear space E, in addition to the topology induced by the metric of E' (the strong topology), we also define the so-called weak* topology. For every $\alpha > 0$ and every finite number of elements x_i $(i = 1, 2, ..., n)$ of E, we denote by $W(x_1, ..., x_n, \alpha)$ the set of all $f \in E'$ such that $|f(x_i)| \leqslant \alpha$. The topology for which the sets $W(x_1, ..., x_n, \alpha)$ form a fundamental system of neighborhoods of zero is called the *weak* topology* $\sigma(E', E)$ in the conjugate space E', where $x_1, x_2, ..., x_n$ range over all possible finite collections of elements of E and α ranges over the set of all positive numbers.

Analogously, the weak topology is defined in the space E. We denote by $W(f_1, ..., f_n, \alpha)$ $(f_1, ..., f_n \in E', \alpha > 0)$ the set of points x in E such that $|f_i(x)| \leqslant \alpha$. The topology for which the sets $W(f_1, ..., f_n, \alpha)$ form a fundamental system of neighborhoods of zero, where $f_1, ..., f_n$ range over all possible finite collections of elements of E' and α ranges over the set of all positive numbers, is called the *weak topology* $\sigma(E, E')$.

Thus, every set containing a set $W(x_1, ..., x_n, \alpha)$ $(W(f_1, ..., f_n, \alpha))$ is taken for a neighborhood of zero $V_\theta^{(\alpha)}$ in the weak* (weak) topology; every neighborhood V_x^α of the element x is obtained from some neighborhood of zero V_θ^α by means of its "displacement" by the element x:

$$V_x^{(\alpha)} = x + V_\theta^{(\alpha)}.$$

The weak* and weak topologies are locally convex topologies since the sets $W(x_1, ..., x_n, \alpha)$ and $W(f_1, ..., f_n, \alpha)$ are convex.

In the case of an infinite-dimensional space E, the weak topology $\sigma(E, E')$ is an example of a non-metrizable locally convex topology; the weak* topology $\sigma(E', E)$ will also be non-metrizable if the space E is infinite-dimensional and a Banach space.

A sequence of continuous linear functionals $f_1, f_2, ..., f_n, ...$ is called *weakly convergent* to the functional f_0 if it converges to f_0 in the weak* topology $\sigma(E', E)$ of the space E'. In order for the sequence $f_1, f_2, ..., f_n, ...$ to be weakly convergent to the functional f_0, it is necessary and sufficient that $f_n(x) \xrightarrow[n \to \infty]{} f_0(x)$ for every element $x \in E$.

We say that a sequence $x_1, x_2, ..., x_n, ...$ of elements of E is *weakly convergent to the element* x_0 if it converges to x_0 in the weak topology $\sigma(E, E')$ of the space E. The criterion for weak convergence of the sequence $x_1, x_2, ..., x_n, ...$ to the element x_0 is that the equality $\lim_{n \to \infty} f(x_n) = f(x_0)$ be satisfied for every functional $f \in E'$. In contrast to weak convergence of elements (functionals), the convergence of a sequence of elements (functionals) with respect to the norm of the space $E(E')$ is called *strong convergence*. Strong convergence of elements (functionals) always implies weak convergence; however, the converse statement is not true, generally speaking, in the case of an infinite-dimensional space. Strong and weak convergence of elements and functionals are equivalent in a finite-dimensional Banach space. Strong and weak convergence of elements are also equivalent in the space l_1 which is an infinite-dimensional Banach space; however, the weak topology in an infinite-dimensional normed linear space is always weaker than the initial topology (see [5]). Nevertheless, the following assertion is valid: *a convex set T in a normed linear space E has the same closure both in the initial topology of the space E and in the weak topology $\sigma(E, E')$.*

In particular, if the sequence $\{x_n\}$ converges weakly to x_0, then there exists a sequence of linear combinations $\{ \sum_{i=1}^{m} \lambda_i^{(m)} x_i \}$ converging in the norm to x_0.

4. *Properties of a sphere in a conjugate Banach space.* A set S in a topological space E is called *compact* if from every covering of S by open sets of the space E a finite sub-covering can be extracted.*)

---

*) A set which is compact in a separated (Hausdorff) topology is often called *bicompact* in Russian mathematical literature.

For a metric space, this definition of a compact set is equivalent to the definition of a closed compact set given in no. 7, §2.

Every closed sphere is compact in the weak* topology $\sigma(E', E)$ in the space E' conjugate to the normed linear space E.

Every closed sphere in the space E', the conjugate to a separable normed space and equipped with the weak* topology $\sigma(E', E)$, is a compact metrizable space.

Let T be a convex set in the linear system E. We say that the point $x \in T$ is an *extremal point* of the set T if there is no open segment containing x in T.

In the Euclidean space R_n, every boundary point of the unit sphere is extremal; the unit sphere contains only two extremal points $x(t) \equiv 1$ and $x(t) \equiv -1$ in the space $C(0, 1)$, and the unit spheres do not contain extremal points in the spaces c_0, $L_1(0, 1)$.

Every compact convex set in a locally convex linear topological space E is the closed convex hull of the set of its extremal points. (M. G. Krein, D. P. Milman)

It follows from this that the unit sphere S of an infinite-dimensional conjugate Banach space E' contains an infinite set of extremal points; therefore, the spaces $C(0, 1)$, c_0, $L_1(0, 1)$ are Banach spaces which are not isometric to any conjugate Banach spaces.

5. *Factor space and orthogonal complements.* Let M be a (closed!) subspace of the Banach space E. The collection of elements $X = x + M$, where x is a fixed element of E, is called a residue class with respect to the subspace M. The collection of all residue classes will be a linear system if the class $X + Y$, constructed from the element $x + y$ where x and y are elements of the classes X and Y, is understood to be the sum of the classes X and Y. Analogously, the class λX is constructed from the element λx where $x \in X$. With this introduction of the operations of addition and multiplication by numbers, the collection of all residue classes is called the *factor space E/M of the space E with respect to the subspace M.*

The norm

$$\|X\| = \inf_{x \in X} \|x\|$$

is introduced in the factor space E/M. The factor space E/M is a Banach space with respect to this norm.

The collection of all continuous linear functionals on E which are zero on M is denoted by $M^\perp$. The collection $M^\perp$ is a weak* closed (that is, closed in the topology $\sigma(E', E)$) subspace of the conjugate space E' and is called the *orthogonal complement of the subspace M*. Conversely, every weak* closed subspace $M' \subset E'$ is the orthogonal complement of the subspace $M \subset E$ which consists of all elements on which every functional from M' becomes zero.

The relation

$$f(x) = F(X) \qquad (x \in X)$$

establishes a one-to-one correspondence between the functionals $f \in M^\perp$ and the continuous linear functionals F on E/M ($F \in (E/M)'$). In this correspondence, the space conjugate to the factor space E/M is isometric to the space $M^\perp$.

Every functional of E', naturally, generates a continuous functional on M; the functionals of $M^\perp$ generate the null functional on M. Conversely, every continuous functional on M can be extended to all of E without increasing its norm. The space M', conjugate to the space M, is isometric to the space $E'/M^\perp$.

6. *Reflexive Banach spaces.*   Let E be a normed linear space and E' the space conjugate to E. Since E' is a Banach space, it makes sense to speak about the space $E'' = (E')'$ conjugate to E'.

Every element $x_0 \in E$ generates a continuous linear functional $F_{x_0}(f)$ on E' which is defined by the equality $F_{x_0}(f) = f(x_0)$. Thus, a one-to-one correspondence is established between the elements of the space E and some subset $\pi(E)$ of the space E''. This correspondence is an isometry between the spaces E and $\pi(E)$ and is called the *natural mapping* of the space E into the space E''.

A Banach space E is called *reflexive* if it is isometric, with respect to the natural mapping, to its second conjugate space E''.

Since the correspondence which gives the isometry has a special form in this case (the element $F_x(f) \in E''$ corresponds to the element $x \in E$), the existence of an isometry between the spaces E and E'' still does not allow us to conclude that the space E is reflexive. Thus, so-called *quasi-reflexive* Banach spaces E exist for which $E'' = \pi(E) \oplus E_n$ where E_n is n-dimensional. Such spaces E are isomorphic to the spaces E'' but are not reflexive. Examples of non-reflexive quasi-reflexive spaces E exist which are isometric to the spaces E''.

Every subspace M of a reflexive (quasi-reflexive) space E is reflexive (quasi-reflexive); the factor space E/M is reflexive (quasi-reflexive).

Criteria for the reflexivity of a Banach space

1. *In order that the space E be reflexive, it is necessary and sufficient that every continuous linear functional $f(x)$, defined on E, attain its supremum on the unit sphere of the space E; that is, that an element x_f exists such that $\|x_f\|=1$ and $f(x_f)=\|f\|$.*

2. *In order that the Banach space E be reflexive, it is necessary and sufficient that its unit sphere $S(\theta, 1)$ be compact in the weak topology $\sigma(E, E')$.*

3. *In order that the Banach space E be reflexive, it is necessary and sufficient that its unit sphere be a closed set in an arbitrary normed topology comparable (see [5]) with the initial topology of the space.*

A Banach space E is called *uniformly convex* if for every $\varepsilon > 0$ there exists a $\delta > 0$ such that $\|x\|=1$, $\|y\|=1$, and $\left\|\dfrac{x+y}{2}\right\| > 1-\delta$ implies $\|x-y\| < \varepsilon$.

4. *Every uniformly convex Banach space is reflexive.*

The class of uniformly convex Banach spaces does not coincide with the class of all reflexive Banach spaces: an example of a reflexive Banach space which is not uniformly convex can be given.

The spaces l_p, $L_p(0, 1)$ where $p > 1$ are uniformly convex and therefore reflexive. Every finite-dimensional Banach space is reflexive. All the remaining spaces considered in no. 5, § 2 are non-reflexive Banach spaces.

§ 5. Linear operators

1. *Bounded linear operators.* Let E and F be two linear systems. We say that an *operator A with values in F* (an operator *acting* from D to F) is defined on a set $D \subset E$ if to every element $x \in D$ there corresponds an element $y = Ax \in F$. The set D is called the *domain* of the operator and is denoted by $D(A)$. The collection of all elements y of F, representable in the form $y = Ax$ $(x \in D(A))$ is called the *range* of the operator A and is denoted by $R(A)$.

The operator squaring: $Ax(t) = x^2(t)$ is an example of an operator in the space $C(0, 1)$. The entire space $C(0, 1)$ serves as the domain of this operator; the collection of all non-negative functions of $C(0, 1)$ is its range. This same operator, considered on the space $L_2(0, 1)$, will map it onto the collection of non-negative functions of $L_1(0, 1)$.

The operator A is called *linear* if $D(A)$ is a linear manifold in E and

$$A(\alpha_1 x_1 + \alpha_2 x_2) = \alpha_1 A x_1 + \alpha_2 A x_2$$

for arbitrary elements $x_1, x_2 \in D(A)$.

The following serve as examples of linear operators in an arbitrary linear system E: the identity operator I, setting in correspondence to each element of E this same element: $Ix = x$; the operator of similitude: $Ax = \lambda y$ ($x \in E$, λ is a fixed number).

In a finite-dimensional space E_n, the linear transformations of the space serve as examples of linear operators. Such operators can be given by means of a square matrix (a_{ik}); if $x = \{\xi_1, \xi_2, ..., \xi_n\}$ and $y = \{\eta_1, \eta_2, ..., \eta_n\}$, then

$$\eta_i = \sum_{k=1}^{n} a_{ik} \xi_k.$$

The integral operators

$$y(t) = Ax(t) = \int_0^1 K(t, s) x(s) \, ds$$

are the analogues of such operators in function spaces.

If, for example, the kernel $K(t, s)$ is continuous, then this linear operator is defined on the entire space $C(0, 1)$ and maps it onto some part of the space $C(0, 1)$.

The linear operator of differentiation, $Ax(t) = x'(t)$, defined on the continuously differentiable functions, $D(A) = C^{(1)}(0, 1)$, can be considered in the space $C(0, 1)$. The entire space $C(0, 1)$ will be the range of this operator.

If this operator is extended to the collection of absolutely continuous functions, then its range will be the space $L_1(0, 1)$. In the theory of generalized functions (see chapter VIII) the operator of differentiation is extended to the entire space $C(0, 1)$, and, in this connection, it maps the space $C(0, 1)$ onto some space of generalized functions.

Now let E and F be two normed linear spaces. An operator A is said to be *continuous at the point* $x_0 \in D(A)$ if $Ax_n \to Ax_0$ whenever $x_n \to x_0$ ($x_n \in D(A)$). If the operator A is defined and continuous at every point of the space E it is called simply a *continuous operator from E into F*.

A linear operator, defined on E, is called *bounded* if

$$\|Ax\|_F \leqslant C \|x\|_E,$$

where C does not depend on the element $x \in E$.

In order that a linear operator, acting from E into F, be continuous, it is necessary and sufficient that it be bounded.

The smallest of the numbers C in the last inequality is called the *norm of the operator* A and is denoted as follows: $\|A\|_{E \to F}$. If F coincides with E, then it is written simply as $\|A\|$. It follows from the definition that

$$\|A\|_{E \to F} = \sup_{\substack{x \in E \\ x \neq 0}} \frac{\|Ax\|_F}{\|x\|_E} = \sup_{\|x\|_E = 1} \|Ax\|_F.$$

2. *Examples of bounded linear operators. Integral operators. Interpolation theorems.*

1) Operators in finite-dimensional spaces. Every linear operator A given by a matrix (a_{ik}) in a Banach space E_n is bounded. Its norm depends on the norm which is introduced in the space.

If the norm

$$\|x\| = \max_i |\xi_i|$$

is introduced, then

$$\|A\| = \max_{1 \leqslant i \leqslant n} \sum_{k=1}^{n} |a_{ik}|.$$

If

$$\|x\| = \sum_{i=1}^{n} |\xi_i|,$$

then

$$\|A\| = \max_{1 \leqslant k \leqslant n} \sum_{i=1}^{n} |a_{ik}|.$$

If the Euclidean norm

$$\|x\| = \sqrt{\sum_{i=1}^{n} |\xi_i|^2}$$

is introduced, then $\|A\| = \sqrt{\mu_1}$ where μ_1 is the largest eigenvalue of the matrix AA^* here $(A^* = (a_{ki}))$. If the matrix (a_{ik}) is symmetric, then $\sqrt{\mu_1} = \lambda_1$, where λ_1 is the largest eigenvalue of the matrix A.

2) Integral operators. If a linear integral operator with a continuous kernel $K(t, s)$ is considered as an operator from $C(0, 1)$ into $C(0, 1)$, then it is bounded and

$$\|A\|_{C \to C} = \max_{0 \leqslant t \leqslant 1} \int_0^1 |K(t, s)| \, ds.$$

This operator as a bounded operator from $L_1(0, 1)$ into $L_1(0, 1)$ has the norm

$$\|A\|_{L_1 \to L_1} = \max_{0 \leqslant s \leqslant 1} \int_0^1 |K(t, s)| \, dt.$$

If the operator A is considered as an operator from $L_p(0, 1)$ into $L_p(0, 1)$, then the inequality

$$\|A\|_{L_p \to L_p} \leqslant \left\{ \max_{0 \leqslant t \leqslant 1} \int_0^1 |K(t, s)| \, ds \right\}^{1 - \frac{1}{p}} \left\{ \max_{0 \leqslant s \leqslant 1} \int_0^1 |K(t, s)| \, dt \right\}^{\frac{1}{p}}$$

is valid for its norm.

This last assertion follows from a general fact: *if the linear operator A is simultaneously a bounded operator from $C(0, 1)$ into $C(0, 1)$ (or from $M(0, 1)$ into $M(0, 1)$) and from $L_1(0, 1)$ into $L_1(0, 1)$, then it is bounded as an operator from $L_p(0, 1)$ into $L_p(0, 1)$ and*

$$\|A\|_{L_p \to L_p} \leqslant \|A\|_{M \to M}^{1 - \frac{1}{p}} \|A\|_{L_1 \to L_1}^{\frac{1}{p}}.$$

Analogously, *if a linear operator is bounded as an operator from $L_{p_1}(0, 1)$ into $L_{p_1'}(0, 1)$ and as an operator from $L_{p_2}(0, 1)$ into $L_{p_2'}(0, 1)$, then it is bounded as an operator from $L_p(0, 1)$ into $L_{p'}(0, 1)$ where*

$$\frac{1}{p} = \frac{1 - \mu}{p_1} + \frac{\mu}{p_2} \quad \text{and} \quad \frac{1}{p'} = \frac{1 - \mu}{p_1'} + \frac{\mu}{p_2'} \quad \text{and } \mu \text{ is an arbitrary number in} [0, 1].$$

Furthermore

$$\|A\|_{L_p \to L_{p'}} \leqslant \|A\|_{L_{p_1} \to L_{p_1'}}^{1 - \mu} \|A\|_{L_{p_2} \to L_{p_2'}}^{\mu}.$$

(M. Riesz)

The last assertions are called *interpolation theorems* in the theory of operators and allow wide generalizations to other classes of Banach spaces.

3) Operators of potential type. The class of integral operators with discontinuous kernels of the form

$$K(t,s) = \frac{1}{|t-s|^\lambda},$$

where $0<\lambda<1$, are called *operators of potential type*.

If $1-\lambda>\frac{1}{p}$, then the operator with the kernel $K(t,s)$ can be considered as a bounded operator from the space $L_p(0,1)$ into the space $C(0,1)$. If $1-\lambda<\frac{1}{p}$, then the operator will not act from $L_p(0,1)$ into $C(0,1)$, and it can be considered as a bounded operator acting from $L_p(0,1)$ into $L_{p_1}(0,1)$ where $\frac{1}{p_1} = \frac{1}{p}-(1-\lambda)$.

Analogous facts are valid for operators of potential type defined for functions on a bounded n-dimensional region G. In this connection, we understand the quantity $|t-s|$ as the distance between the points t and s; the number λ should be included in the interval $(0,n)$. The operator acts from $L_p(G)$ into $C(G)$ if $1-\frac{\lambda}{n}>\frac{1}{p}$, and into $Lp_1(G)$ if $1-\frac{\lambda}{n}<\frac{1}{p}$ where $\frac{1}{p_1} = \frac{1}{p}-(1-\frac{\lambda}{n})$. It is bounded in both cases.

All these assertions are also valid for operators with kernels of the form $K(t,s)=\frac{A(t,s)}{|t-s|^n}$ where $A(t,s)$ is a continuous kernel.

4) Singular integral operators. In the previous example the kernel of the integral operator had a summable singularity. The kernels of singular integral operators have non-summable singularities of the form $\frac{1}{|t-s|}$. The Hilbert transformation

$$Ax = \int_{-\infty}^{\infty} \frac{x(s)}{t-s}\,ds,$$

where the integral is understood in the sense of a principal value, provides the simplest example of such an operator. It can be shown that this operator is a bounded operator acting from $L_p(-\infty,\infty)$ into $L_p(-\infty,\infty)$ for $1<p<\infty$.

A *many-dimensional (n-dimensional) singular operator* is an integral operator with a kernel of the form

$$K(t, s) = \frac{Q(t, t - s)}{|t - s|^n},$$

where $Q(t, \tau)$, as a function of its second argument τ, is homogeneous of degree zero and has an integral equal to zero on the unit sphere S. *If the integral of $|Q(t, \tau)|^{p'}$ on the sphere S is uniformly bounded with respect to t, then the singular operator is a bounded operator from* $L_p(R_n)$ *into* $L_p(R_n)$ *for* $1 < p < \infty$ $\left(\frac{1}{p} + \frac{1}{p'} = 1 \right)$. (Calderón-Zygmund)

5) Hilbert and Hardy integral operators. The operator

$$Ax = \int_0^\infty \frac{x(s)}{t + s} ds$$

is bounded from $L_p(0, \infty)$ into $L_p(0, \infty)$ for $1 < p < \infty$, and its norm is equal to $\dfrac{\pi}{\sin \frac{\pi}{p}}$. (Hilbert inequality).

The operator

$$Ax = \frac{1}{t} \int_0^t x(s) ds$$

is bounded from $L_p(0, \infty)$ into $L_p(0, \infty)$ for $p > 1$, and its norm is equal to $\dfrac{p}{p-1}$. (Hardy inequality)

6) Differential operators. Linear differential operators, considered as operators taking a space into itself, are not bounded as a rule. In particular, the derivative operator is not bounded in the space $C(0, 1)$; if it is considered as an operator from $C^{(1)}(0, 1)$ into $C(0, 1)$, then it is bounded and its norm is equal to 1.

Analogously, a linear differential operator of order l with continuous coefficients can be considered as a bounded operator from $C^{(l)}(0, 1)$ into $C(0, 1)$.

For the study of linear partial differential operators, generally, either a Hölder space (classical approach) or the spaces W_p^l are used. Thus, an elliptic operator of second order

$$Ax = -\sum_{i,j=1}^{n} a_{ij}(t)\frac{\partial^2 x}{\partial t_i \partial t_j} + \sum_{i=1}^{n} b_i(t)\frac{\partial x}{\partial t_i} + c(t)x,$$

defined in an n-dimensional region G, is considered as a bounded operator from the space $W_2^2(G)$ into the space $L_2(G)$.

3. *Convergence of a sequence of operators.* Let $\{A_n\}$ be a sequence of bounded linear operators acting from a normed linear space E into a normed linear space F.

The sequence $\{A_n\}$ is said to be *convergent in the norm* to the bounded linear operator A_0 from E into F if $\lim_{n\to\infty}\|A_0 - A_n\|_{E\to F} = 0$.

The sequence $\{A_n\}$ is said to be *strongly convergent* to the operator A_0 if $\lim_{n\to\infty}\|A_0 x - A_n x\|_F = 0$ for all $x \in E$.

The sequence $\{A_n\}$ is said to be *weakly convergent* to the operator A_0 if the sequence $\{A_n x\}$ is weakly convergent to $A_0 x$ for all $x \in E$.

Convergence in the norm implies strong convergence, and strong convergence implies weak convergence. The converse assertions are, generally speaking, not true.

If the sequence $\{A_n\}$ converges strongly to A_0 and the sequence of norms of the operators A_n is bounded: $\|A_n\|_{E\to F} \leqslant M$ $(n = 1, 2, \ldots)$, then the operator A_0 is also a bounded linear operator and

$$\|A_0\|_{E\to F} \leqslant \varliminf_{n\to\infty} \|A_n\|_{E\to F}.$$

In the case when E is a Banach space, the last assertion is considerably strengthened: *if a sequence of bounded linear operators A_n acting from a Banach space E into a normed linear space F converges strongly to the operator A_0, then the sequence of norms of the operators is bounded and, hence, the operator A_0 is also bounded.*

In order for a sequence of bounded linear operators, acting from a Banach space E into a Banach space F, to converge strongly to some bounded linear operator, it is necessary and sufficient that: 1) the sequence of norms of the operators A_n be bounded; 2) the sequence $\{A_n x'\}$ be convergent for every element x' of some everywhere dense set $D \subset E$.

The last theorem has many applications to problems connected with convergence and summability of series and integrals, the convergence of interpolation processes, of processes of mechanical quadratures, and so on (see [19]).

The facts mentioned are valid, of course, for the case when the space F coincides with the collection of all real or complex numbers. In this case the word "operator" is replaced by "functional" and "strong convergence of operators" is replaced by "weak convergence of functionals" in all formulations.

4. *Inverse operators.* Let a linear operator A map a linear system E into a linear system F. If the operator A has the property that $Ax=\theta$ only for $x=\theta$, then for every y in the range $R(A)$ of the operator A there corresponds only one element x for which $y=Ax$ (the solution of the equation $y=Ax$ is unique). This correspondence can be considered as an operator B defined on $R(A)$ with values filling E. The operator B is linear. By definition, $BAx=x$; therefore the operator B is called a *left inverse of A.*

If $R(A)=F$, that is, the operator A establishes a one-to-one correspondence between E and F, then the operator B, defined on all of F, is called simply the *inverse operator of A*, and is denoted by A^{-1}. By definition,

$$A^{-1}Ax = x\,(x\in E) \quad\text{and}\quad AA^{-1}y = y\,(y\in F).$$

One of the profound facts of the theory of Banach spaces is the following assertion.

If a bounded linear operator A, mapping a Banach space E onto a Banach space F, has an inverse A^{-1}, then the operator A^{-1} is bounded. (s. Banach).

This theorem ceases to be valid if the completeness of one of the spaces E or F is relinquished. It is generalized to some classes of linear topological spaces, in particular, to complete metric spaces.

The theorem about the inverse operator means, in other words, that from the existence and uniqueness of the solution of the equation

$$Ax = y$$

for every y in F it follows that the solution $x=A^{-1}y$ depends in a continuous manner on the right hand side y.

If a bounded linear operator A from a Banach space E to a Banach space F has an inverse, then bounded linear operators close to it also have inverses: if

$$\|B - A\|_{E \to F} < \frac{1}{\|A^{-1}\|_{F \to E}},$$

then the operator B has an inverse B^{-1}.

5. *Space of operators. Ring of operators.* The operations of addition and multiplication by a number are introduced in a natural manner for linear operators mapping a linear system E into a linear system F. By definition, $A = \alpha_1 A_1 + \alpha_2 A_2$ is the operator for which

$$Ax = \alpha_1 A_1 x + \alpha_2 A_2 x \qquad (x \in E).$$

If E and F are normed, then all the bounded linear operators from E into F form a linear system $L(E, F)$ which can be normed by means of the norm $\|A\|_{E \to F}$. If F is complete, then $L(E, F)$ is a Banach space.

If we consider operators taking a space E into itself, then an operation of multiplication can also be introduced for them: by definition, $A = A_1 A_2$ if

$$Ax = A_1 (A_2 x).$$

Multiplication, generally speaking, is non-commutative: it is possible that $A_1 A_2 \neq A_2 A_1$. If $A_1 A_2 = A_2 A_1$, then we say that the operators A_1 and A_2 *commute*.

If $A_1, A_2 \in L(E, E)$, then $A \in L(E, E)$ where

$$\|A\| \leqslant \|A_1\| \, \|A_2\|.$$

If an operation of multiplication $x \cdot y$ is introduced in a normed linear space so that the space becomes a ring (more precisely, an algebra), and

$$\|x \cdot y\| \leqslant \|x\| \, \|y\|,$$

then it is called a *normed ring (algebra)* (see ch. VI).

The space $L(E, E)$ is a normed ring. The ring has an identity, the identity operator I.

If E is a Banach space, then the collection of operators having inverses forms an open set in this ring.

6. *Resolvent of a bounded linear operator. Spectrum.* Let E be a complex Banach space and $-A$ be a bounded linear operator acting on it.

The complex number λ is called a *regular point* of the operator A if the operator $A - \lambda I$ has an inverse $(A - \lambda I)^{-1}$. In the opposite case, λ is called a *point of the spectrum* of the operator A.

If λ is a regular point of A, then the bounded linear operator $(A - \lambda I)^{-1}$ is called a *resolvent* and is denoted by R_λ.

The regular points form an open set in the complex plane; the spectrum is closed.

All points lying outside the circle of radius $\|A\|$ with center at the origin are regular. All points of the spectrum are in the disk $|\lambda| \leqslant \|A\|$. For $|\lambda| > \|A\|$, the series expansion for the resolvent

$$R_\lambda = -\frac{1}{\lambda}\left(I + \frac{A}{\lambda} + \frac{A^2}{\lambda^2} + \cdots + \frac{A^n}{\lambda^n} + \cdots\right)$$

is valid, where the series converges in the operator norm.

The radius of the smallest disk with center at the origin containing the spectrum of the operator A is called the *spectral radius* r_A of the operator A. The formula (I. M. Gelfand)

$$r_A = \lim_{n \to \infty} \sqrt[n]{\|A^n\|}$$

is valid; moreover, the limit always exists. It follows from the preceding that $r_A \leqslant \|A\|$. Moreover, $r_A \leqslant \sqrt[n]{\|A^n\|}$. On the basis of the Cauchy criterion, the series for the resolvent will converge if $r_A < |\lambda|$ and diverge if $r_A > |\lambda|$. In particular, the series

$$(I - A)^{-1} = -R_1 = I + A + A^2 + \cdots + A^n + \cdots$$

converges if $r_A < 1$ and diverges if $r_A > 1$.

The spectrum of an arbitrary bounded operator is a non-empty set. If

$$r_A = \lim_{n \to \infty} \sqrt[n]{\|A^n\|} = 0,$$

then the spectrum consists of one point, $\lambda = 0$. The Volterra integral operator

$$Ax = \int_0^t K(t, s) x(s)\, ds,$$

where $K(t, s)$ is a function continuous in the triangle $0 \leqslant s \leqslant t \leqslant 1$, is an example of such an operator. If this operator is considered as a bounded operator in the space $C(0, 1)$ or $L_2(0, 1)$, then its spectrum consists of the point $\lambda = 0$. This corresponds to the statement that the equation

$$x(t) - \mu \int_0^t K(t, s) x(s) \, ds = y(t)$$

has a unique solution for an arbitrary right hand side and arbitrary μ.

The Hilbert identity

$$R_\lambda - R_\mu = (\lambda - \mu) R_\lambda R_\mu$$

is valid for any two regular points λ and μ.

The limit in the sense of the operator norm

$$\lim_{\mu \to \lambda} \frac{R_\lambda - R_\mu}{\lambda - \mu} = R_\lambda^2$$

is called, naturally, the *derivative* with respect to λ:

$$\frac{dR_\lambda}{d\lambda} = R_\lambda^2 \, .$$

An operator-valued function of λ is called *analytic* at λ_0 if it can be expanded in a neighborhood of λ_0 in a series of positive integral powers of $(\lambda - \lambda_0)$ which converges with respect to the operator norm.

The resolvent is an operator-valued function of λ which is analytic in the region consisting of regular points of the operator A. For $x \in E$, the function $R_\lambda x$, whose domain consists of regular points of A, is an analytic function with values in E.

Let λ_0 be a pole of the analytic function R_λ. Then any element $R_\lambda x$ has an expansion in a Laurent series

$$R_\lambda x = \frac{e_0}{(\lambda - \lambda_0)^m} + \frac{e_1}{(\lambda - \lambda_0)^{m-1}} + \cdots + \frac{e_{m-1}}{\lambda - \lambda_0} +$$
$$+ f_0 + f_1 (\lambda - \lambda_0) + \cdots + f_n (\lambda - \lambda_0)^n + \cdots$$

The element $e_0 = \lambda e_0(x)$ satisfies the equation

$$Ae_0 = \lambda_0 e_0$$

and is called an *eigenvector* of the operator A corresponding to the *eigenvalue* λ_0. The elements $e_1, ..., e_{m-1}$ satisfy the relations

$$Ae_1 = \lambda_0 e_1 + e_0,$$

$$Ae_2 = \lambda_0 e_2 + e_1, ..., Ae_{m-1} = \lambda_0 e_{m-1} + e_{m-2}$$

and are called *associated vectors* to the eigenvector e_0.

A subspace L of the space E is called *invariant* with respect to the operator A if $x \in L$ implies that $Ax \in L$.

The finite-dimensional subspace which consists of all possible linear combinations of the elements $e_0, e_1, ..., e_{m-1}$ is invariant with respect to the operator A and is called a *radical subspace*. The elements e_k form a basis for it in which the matrix of the operator A has the form of a Jordan cell (see SMB "Higher Algebra" ch. 2, §1, no. 11).

If λ_0 is a simple pole of the resolvent R_λ ($m=1$), then to the eigenvalue λ_0 there correspond only eigenvectors of the operator A. Associated vectors are absent.

By means of the resolvent, the concept of a function of a bounded operator is introduced. If $f(\lambda)$ is a function, analytic in a region containing the spectrum of the operator A, then we define the function $f(A)$ to be the operator

$$f(A) = -\frac{1}{2\pi i} \oint_\Gamma f(\lambda) R_\lambda \, d\lambda,$$

where the contour Γ contains in its interior the spectrum of the operator A. This integral is a generalization of the Cauchy integral. It does not depend on the choice of the contour Γ.

7. *Adjoint operator.* Let E and F be normed linear spaces and let A be a bounded linear operator acting from E into F.

If $g(y)$ is a continuous linear functional on F ($g \in F'$), then the functional

$$f(x) = g(Ax)$$

will be a continuous linear functional on E ($f \in E'$), where

$$\|f\|_{E'} \leqslant \|g\|_{F'} \|A\|_{E \to F}.$$

Thus, to every functional $g \in F'$ there corresponds a functional $f \in E'$; i.e. an operator $A'g = f$ is defined .This operator A' is called the *adjoint* of the operator A.

The adjoint operator is a bounded linear operator; moreover,

$$\|A'\| = \|A\|.$$

If $A, B \in L(E, F)$, then $(\lambda A)' = \lambda A'$ and $(A + B)' = A' + B'$. If $A, B \in L(E, E)$, then $(AB)' = B'A'$.

If an integral operator with continuous kernel $K(t, s)$ is considered, for example, as a bounded operator from $L_p(0, 1)$ into $L_p(0, 1)$, then the integral operator with kernel $K'(t, s) = K(s, t)$, that is, the operator

$$A'x = \int_0^1 K(s, t) x(s) \, ds,$$

considered as an operator from $L_q(0, 1)$ into $L_q(0, 1)$ $\left(\dfrac{1}{p} + \dfrac{1}{q} = 1\right)$, is the adjoint to it.

If E and F are Banach spaces, then the existence of the operator $(A')^{-1}$ where $(A')^{-1} = (A^{-1})'$ is a necessary and sufficient condition for the existence of the inverse operator A^{-1}.

It follows from the last assertion, applied to the operator $A - \lambda I$ and its adjoint $A' - \lambda I$, that the spectra of the operators A and A' coincide.

A more important relation between the properties of an operator and its adjoint is studied in no. 9.

8. *Completely continuous operators.* Let E and F be Banach spaces. A linear operator, acting from E into F, is called *completely continuous* if it maps every bounded set of the space E onto a (relatively) compact set of the space F.

Complete continuity of a linear operator implies continuity. The converse, generally speaking, is not valid. For example, the identity operator I is continuous, but in the case of an infinite-dimensional space E it is not completely continuous.

It suffices for the complete continuity of a linear operator that it map the unit sphere of the space E into a compact set of the space F. The range of a completely continuous operator is separable. A completely continuous operator maps every weakly convergent sequence of elements into a sequence which is convergent with respect to the norm.

The limit with respect to the norm of a sequence of completely continuous operators is again a completely continuous operator. A strong, and hence a weak, limit of a sequence of completely continuous operators need not be a completely continuous operator. Let, for example, E be the Banach space l_1. The projectors P_N, which set in correspondence to every $x = \{\xi_1, \xi_2, ..., \xi_n, ...\}$ the element $P_N x = \{\xi_1, ..., \xi_N, 0, 0, ...\}$, are completely continuous, but their strong limit is the identity operator which is not completely continuous.

A linear combination of completely continuous operators is a completely continuous operator. The product of a completely continuous operator with a bounded operator is a completely continuous operator. The set of all completely continuous operators of $L(E, E)$ forms a closed ideal in the normed ring $L(E, E)$.

The adjoint operator of a completely continuous operator is completely continuous.

A *one-dimensional* linear operator of the form

$$Ax = f_1(x) y_1,$$

where y_1 is a fixed element of F and $f_1(x)$ is a fixed functional from E', is the simplest example of a completely continuous operator. The one-dimensional operator is abbreviated as follows: $A = f_1 \otimes y_1$.

An arbitrary *finite-dimensional* linear operator has the more general form

$$A = \sum_{i=1}^{m} f_i \otimes y_i,$$

where $y_i \in F$ and $f_i \in E'$. By definition,

$$Ax = \sum_{i=1}^{m} f_i(x) y_i.$$

A finite-dimensional operator is completely continuous.

It is not known whether every completely continuous operator is representable as a limit with respect to the norm of finite-dimensional linear operators.

A linear operator is called *atomic* if it is representable in the form

$$Ax = \sum_{i=1}^{\infty} f_i(x) y_i,$$

where $y_i \in F, f_i \in E'$ and $\sum_{i=1}^{\infty} \|f_i\|_{E'} \|y_i\|_F < \infty$.

Atomic operators constitute an important subclass of the class of completely continuous operators.

Let A be a completely continuous operator, defined on and acting into a Banach space E. If it is finite-dimensional, then its spectrum coincides with the finite set of its eigenvalues. In the general case, its spectrum consists of no more than a denumerable number of points. In an infinite-dimensional space, the point $\lambda_0 = 0$ always is a point of the spectrum of a completely continuous operator; moreover, it is the only possible limit point for the set of the other points of the spectrum. Thus, the spectrum of a completely continuous operator consists of a finite number of points or of the point $\lambda_0 = 0$ and a sequence $\lambda_n \to 0$.

All the points λ_n $(n = 1, 2, \ldots)$ are poles of the resolvent R_λ and, hence, eigenvalues of the operator A. To every eigenvalue $\lambda_n \neq 0$ there corresponds only a finite number of linearly independent eigenvectors and associated vectors.

In a finite-dimensional space, the eigenvectors, and associated vectors form a basis. In an infinite-dimensional space, the picture can be considerably more complicated for an arbitrary completely continuous operator. There are completely continuous operators which do not have eigenvalues; their spectrum consists of the single point $\lambda_0 = 0$. The Volterra integral operator (see §4, no. 6) can serve as an example of such an operators. In connection with this, we call completely continuous operators, not having eigenvalues, *Volterric*. In recent years much effort was applied to the discovery of conditions for which the eigenvectors and associated vectors of a completely continuous operator form a complete system in the space E, that is, a system whose closed linear hull coincides with E. Significant results were obtained for the Hilbert space case (see ch. 2, no. 5). This fact is valid in an arbitrary Banach space: although a completely continuous operator may or may not have eigenvectors, it has without fail proper invariant subspaces, that is, invariant subspaces coinciding neither with E nor with $\{\theta\}$.

Numerous examples of completely continuous operators are given by integral operators. If the kernel of an integral operator

$$Ax = \int_0^1 K(t, s) x(s) \, ds$$

is continuous, then it will generate a completely continuous operator from $C(0, 1)$ into $C(0, 1)$. If the kernel satisfies the weaker condition

$$\int_0^1 \int_0^1 |K(t, s)|^q \, dt \, ds < \infty \qquad (q > 1),$$

then the operator will be completely continuous as an operator from $L_p(0, 1)$ into $L_p(0, 1)$ $\left(\dfrac{1}{p} + \dfrac{1}{q} = 1\right)$. The condition mentioned is not necessary. Examples exist of completely continuous integral operators acting from $L_p(0, 1)$ to $L_p(0, 1)$ $(1 < p < \infty)$, whose kernels are not summable, as functions of two variables, to any power > 1.

The equation

$$x - Ax = y$$

is the analogue in the theory of operators of the Fredholm integral equation of the second kind.

The basic facts of the Fredholm theory are valid for this equation in the case when the operator A is completely continuous.

Consider the adjoint equation

$$g - A'g = f.$$

If the initial equation has a solution for an arbitrary right hand side in F, then the adjoint equation has a solution for an arbitrary right hand side in E'; in this case the solutions are unique, that is, the homogeneous equations

$$x - Ax = \theta, \qquad g - A'g = \theta$$

have only the trivial solutions $x = \theta$, $g = \theta$.

The homogeneous equations have the same finite number of linearly independent solutions, and the dimensions of corresponding radical subspaces coincide.

If $g_1, \ldots, g_m$ is a maximal sytem of linearly independent solutions of the homogeneous equation $g - A'g = \theta$, then the initial equation $x - Ax = y$ has a solution only for those right hand sides for which $g_k(y) = 0$ $(k = 1, 2, \ldots, m)$.

The three Fredholm theorems are valid for equations of a more general form:

$$Ux - Ax = y,$$

where A is a completely continuous operator and U is a bounded operator having an inverse U^{-1}. It turns out that these exhaust all linear equations with bounded operators for which the Fredholm theorems are valid.

9. *Operators with an everywhere dense domain of definition. Linear equations.* Integral operators are basic examples of bounded operators. Differential operators in natural norms, as a rule, are unbounded. Therefore, in the application of the theory of bounded operators to differential equations, these equations are reduced to integral equations. Usually this reduction is done by means of the Green's function or other resolvent kernels. The construction of such kernels is not trivial; therefore, recently the theory of unbounded operators has been developed in a form which allows direct application to the theory of differential equations.

Let E be a Banach space and A be a linear operator defined on an everywhere dense set $D(A)$ and acting from $D(A)$ into a Banach space F. For such an operator, the concept of the adjoint operator can be introduced. If $g(y)$ is a bounded linear functional on F, then the functional $g(Ax)$, defined on $D(A)$, can be bounded or unbounded. If $g(Ax)$ is bounded, then it can be extended in an unique manner to a continuous functional $f(x)$ on the entire space E: as $x_n \to x$, $x_n \in D(A)$, we set $g(Ax) = \lim_{n \to \infty} g(Ax_n)$. If these conditions are satisfied, we say that the adjoint operator A' is defined on g:

$$f(x) = A'g(x).$$

It follows from the preceding that $g \in D(A')$ if

$$|g(Ax)| \leqslant C \|x\|$$

for $x \in D(A)$, and for these x

$$A'g(x) = g(Ax).$$

The adjoint operator plays an essential role in the study of the question of the solvability of the equation

$$Ax = y \qquad (y \in F, x \in D(A)). \tag{A}$$

Along with this equation the adjoint equation

$$A'g = f \qquad (f \in E', g \in D(A')) \tag{A'}$$

is considered.

The following assertions are valid for equations (A) and (A'):

1. *In order that equation* (A) *be solvable for an everywhere dense set of right hand sides y from* $F\overline{(R(A)}=F)$, *it is necessary and sufficient that the uniqueness theorem be valid for equation* (A'), *that is,* $A'g=\theta$ *implies* $g=\theta$.

2. *In order that equation* (A') *be solvable for an everywhere dense set of right hand sides f from* E', *it is necessary that the uniqueness theorem be valid for equation* (A).

3. *A necessary and sufficient condition for the solvability of equation* (A') *for every* $f\in E'$ *is that the inequality*

$$\|Ax\|_F \geqslant m\,\|x\|_E \qquad (x\in D(A), \qquad m>0)$$

be satisfied.

4. *A necessary condition for the solvability of equation* (A) *for an arbitrary right hand side* $y\in F$ *is that the inequality*

$$\|A'g\|_{E'} \geqslant m\,\|g\|_{F'} \qquad (g\in D(A'), \qquad m>0)$$

be satisfied.

Properties 1–3 are valid when E and E are normed linear spaces; property 4 is valid when F is a Banach space.

The last inequality, generally speaking, is not sufficient for the solvability of the equation (A), but the following assertion is valid:

5. *If* $D(A')$ *is everywhere dense in* F' *and the last inequality is satisfied, then the equation* (A) *has a "weak solution" for an arbitrary* $y_0\in F$ *in the sense that an element* x_0 *exists satisfying the identity*

$$g(y_0) = A'g(x_0)$$

for arbitrary $g\in D(A')$.

If the weak solution x_0 belongs to the domain of definition of the operator A, then it is a proper solution.

In particular, if the operator A is bounded and defined on the entire space, then the solvability of the equation (A) for arbitrary right hand sides follows from the inequality $\|A'g\|_{E'}\geqslant m\|g_{F'}\|$.

If the space E is reflexive, then uniqueness of the weak solution follows from the uniqueness theorem for equation (A).

10. *Closed unbounded operators.* Let E be a Banach space and A a linear operator defined on an everywhere dense set $D(A)$ acting from $D(A)$ into a Banach space F. An operator A_1 is called an *extension*

of the operator A provided that $D(A_1) \supset D(A)$ and $A_1 x = A x$ for $x \in D(A)$. The *restriction* of the operator A to the set $D \subset D(A)$ is the operator A_2 such that $D(A_2) = D$ and $A_2 x = A x$ $(x \in D)$. If A is bounded, then it can be extended by continuity to a bounded linear operator $\bar{A}$ defined on the entire space E by setting

$$\bar{A}x = \lim_{n \to \infty} A x_n \qquad (x \in E, x_n \in D(A), x_n \to x).$$

An operator A (possibly unbounded) is called *closed* if $x_0 \in D(A)$ and $y_0 = A x_0$ follow from $x_n \to x_0$ $(x_n \in D(A))$ and $A x_n \to y_0$.

A bounded operator defined on the entire space is always closed. The fact that also, conversely, *a closed linear operator defined on an entire Banach space is bounded,* is very important.

If an operator A is not closed, then it allows a closed extension if and only if $y = \theta$ follows from $x_n \to \theta$ $(x_n \in D-A))$ and $A x_n \to y$.

If an operator A has a left inverse operator, bounded on the closed set $R(A)$, then it is closed. In particular, only closed operators can have bounded inverse operators. If we consider the operators as acting from $D(A) \subset E$ into E, then only closed operators can have, for certain λ, resolvents $R_\lambda = (A - \lambda I)^{-1}$. The concepts of regular points and points of the spectrum are introduced for a closed operator as for a bounded operator. The spectrum of a closed operator can fill an arbitrary closed set in the complex plane.

The investigation of closed operators is sometimes reduced to the study of bounded operators by the following scheme: a new norm

$$\|x\|_1 = \|x\|_E + \|Ax\|_F \qquad (x \in D(A))$$

is introduced in the domain $D(A)$ of the closed operator A.

With this norm, $D(A)$ becomes a Banach space E_A. The operator A is bounded as an operator from E_A into F. Moreover, an arbitrary operator B, acting from $D(A)$ into F and allowing closure, will be bounded as an operator from E_A to F:

$$\|Bx\|_F \leqslant C \|x\|_1 = C(\|x\|_E + \|Ax\|_F).$$

The adjoint operator A' of an operator A with everywhere dense domain of definition is always closed. If the domain $D(A')$ is everywhere dense in F', then the initial operator A allows closure.

The following assertions are equivalent for equations (A) and (A') with a closed operator A:

1. *The right hand sides for which equation* (A) *is solvable form a closed subspace of the space F.*

2. *The right hand sides for which equation* (A') *is solvable form a closed subspace of the space E'.*

3. *If we denote the collection of all solutions of the equation* $A'g = 0$ *by N', then equation* (A) *is solvable for those and only those right hand sides y for which* $g(y) = 0$ *for all* $g \in N'$.

4. *If we denote the collection of all solutions of the equation* $Az = 0$ *by N, then the equation* (A') *is solvable for those and only those f for which* $f(z) = 0$ *for all* $z \in N$.

In particular, if we succeed in obtaining the lower estimates

$$\|Ax\|_F \geqslant m \|x\|_E \quad (x \in D(A)) \quad \text{and} \quad \|A'g\|_{E'} \geqslant m \|g\|_{F'} \quad (m > 0)$$

for the closed operator and the operator adjoint to it, then the uniqueness of the solutions of equations (A) and (A'), for arbitrary right hand sides from F and E' respectively, follows.

As remarked, problems from the theory of differential equations are one of the stimuli for the study of unbounded operators. Let Q be linear differential operator of order l with sufficiently smooth coefficients defined in a region G of n-dimensional space. This operator can be considered as an operator acting in $L_p(G)$ whose domain $D(Q) \subset L_p(G)$ consists of all functions with partial derivatives of order l continuous in $\bar{G}$. Let $x(t) \in D(Q)$ and $z(t)$ be a *finitary* function in G; that is, an infinitely differentiable function equal to zero in a neighborhood of the boundary of G. The identity

$$\int_G Qx(t) z(t) \, dt = \int_G x(t) Q'z(t) \, dt$$

is valid where, Q' is the adjoint differential operator. The identity is obtained by integration by parts; the boundary terms vanish because of the finitary nature of the function $z(t)$.

The finitary functions form an everywhere dense set in $L_q \left(\dfrac{1}{p} + \dfrac{1}{q} = 1 \right)$; the operator Q', adjoint to Q, is defined on it; therefore the differential operator Q allows the closure $\bar{Q}$. The domain of definition of the operator $\bar{Q}$ can now contain functions non-differentiable in the classic sense (functions with generalized derivatives, and so on).

The solutions of the equation $\bar{Q}x = y$ belonging to the domain of definition of the closure $\bar{Q}$ of the differential operator Q in various function spaces are called *generalized solutions*.

11. *Remark on complex spaces.* Let E be a complex normed linear space. Sometimes it is more convenient to introduce the operation of multiplication of a linear functional $f(x)$ by a number λ not as indicated in §4, no. 1 but in the following manner: $f_1 = \lambda f$ means that

$$f_1(x) = \bar{\lambda} f(x).$$

The collection of all continuous linear functionals with the operation of multiplication by a number introduced in this way is denoted by E^* and is also called the *conjugate space* of E. All concepts introduced for the space E' are introduced analogously for the space E^*. All facts valid in the space E' are also valid for the space E^*; changes are made only for some formulations:

1. The operator adjoint to the operator A, considered as an operator from F^* to E^*, is denoted by A^*. Then

$$(\lambda A)^* = \bar{\lambda} A^*.$$

2. For an integral operator A with kernel $K(t, s)$, the adjoint operator A^* has kernel $\overline{K(s, t)}$.

3. If the number λ belongs to the spectrum of the operator A, then the number $\bar{\lambda}$ belongs to the spectrum of A^*, and conversely.

§ 6. Spaces with a basis

1. *Completeness and minimality of a system of elements.* A system $\{e_k\}$ of elements $e_1, e_2, \ldots, e_n, \ldots$ is called *complete* in a Banach space E if the linear hull of this system of elements is dense in E. Obviously, a complete system of elements can exist only in a separable space. *In order for the system $\{e_k\}$ to be complete in E, it is necessary and sufficient that there is no linear functional $f \in E'$ different from zero and equal to zero on all elements $e_k (k = 1, 2, \ldots)$ (orthogonal to all e_k).*

The system of elements $\{e_k\}$ is called *minimal* if no element of this system belongs to the closed linear hull of the remaining elements.

In order that the system $\{e_k\}$ be minimal, it is necessary and sufficient that a system of linear functionals exist forming with the given system a biorthogonal system; that is, a system $\{f_k\} \subset E'(k=1, 2, ...)$ such that $f_i(e_i) = \delta_{ij}$). If the system $\{e_k\}$ is complete and minimal, then the system of functionals $\{f_k\}$ is defined in a unique manner.*

In every separable Banach space there is a complete minimal system. Moreover, we can construct a complete minimal system $\{e_k\}$ such that the functionals f_k corresponding to it form a *total set*; that is, $f_k(x)=0$ $(x \in E, k=1, 2, ...)$ implies that $x=0$.

2. *Concept of a basis.* A system of elements $\{e_k\}$ forms a *basis* of the space E if every element $x \in E$ is representable uniquely in the form of a convergent series

$$x = \sum_{k=1}^{\infty} c_k e_k.$$

Every basis is a complete minimal system. However, a complete minimal system may not be a basis in the space. For example, the trigonometric system $e_0(t) = \frac{1}{2}, e_{2n-1}(t) = \sin nt, e_{2n}(t) = \cos nt (n=1,2,...)$ is a complete minimal system in the space $C[-\pi, \pi]$ but it does not form a basis in it.

Examples of bases

1) In the space $L_2[a, b]$ as well as in an arbitrary separable Hilbert space H (see ch. II, §1), every complete orthogonal system of elements forms a basis. Thus the trigonometric system of functions forms a basis in $L_2[-\pi, \pi]$.

We can construct non-orthogonal bases in a Hilbert space. For example, if $\{e_i\}$ is a complete orthonormal system in a Hilbert space H, then the system of elements

$$g_k = \sum_{i=1}^{k} p_i e_i \qquad (k = 1, 2, ...)$$

forms a basis in H if the numbers p_i satisfy the conditions

$$|p_1| > 0, \frac{\sum\limits_{i=1}^{n} p_i^2}{p_{n+1}^2} \leqslant M \qquad (n = 1, 2, ...).$$

---

*) $\delta_{ij} = 0$ if $i \neq j$ and $\delta_{ii} = 1$.

The system of functionals forming a biorthogonal system with $\{g_k\}$ is given by a system of elements of H:

$$f_k = \frac{1}{p_k} e_k - \frac{1}{p_{k+1}} e_{k+1}.$$

2) In the coordinate spaces c_0 and $l_p (p \geqslant 1)$, the system of unit vectors $e_k = \{0, ..., 0, 1, 0, ...\}$ forms a basis. This system does not form a basis in the space c and is not even complete since the element $e_0 = \{1, 1, ...\}$ does not belong to the closure of the linear hull of the elements $e_k (k = 1, 2, ...)$. However, the system $e_0, e_1, e_2, ...$ forms a basis in the space c.

3) We can construct a basis in the space of continuous functions $C[0, 1]$ in the following manner: let $\{r_i\}$ $(i = 0, 1, 2, ...)$ be a sequence of numbers dense in $[0, 1]$, where $r_0 = 0, r_1 = 1, r_i \neq r_j$ for $i \neq j$. We set $e_0(t) \equiv 1$ and $e_1(t) = t$. Then $e_k(t)$ is defined inductively. Let the functions $e_i(t)$ be defined for $i < k$ and the segment $[0, 1]$ be divided by the points $r_2, ..., r_{k-1}$ into $k - 1$ intervals. Let r_k belong to one of these intervals: $r_{s_1} < r_k < r_{s_2}$, $s_1 < k, s_2 < k$. Then we set: $e_k(r_k) = 1, e_k(r_{s_1}) = 0, e_k(r_{s_2}) = 0$, and the function $e_k(t)$ is linearly interpolated on the segments $[0, r_{s_1}]$ $[r_{s_1}, r_k]$ $[r_k, r_{s_2}]$ and $[r_{s_2}, 1]$. The system $\{e_k(t)\}$ $(k = 0, 1, 2, ...)$ forms a basis in $C[0, 1]$.

4) In the spaces $L_p[0, 1]$ $(p \geqslant 1)$, a basis is formed by the system of Haar functions defined as follows:

$$\chi_0^{(0)}(t) \equiv 1; \quad \chi_0^{(1)}(t) = \begin{cases} 1, & 0 \leqslant t < \tfrac{1}{2}, \\ -1, & \tfrac{1}{2} < t \leqslant 1, \\ 0, & t = \tfrac{1}{2}; \end{cases}$$

$$\chi_n^{(k)}(t) = \begin{cases} 2^{\frac{n}{2}}, & \dfrac{2k-2}{2^{n+1}} \leqslant t < \dfrac{2k-1}{2^{n+1}} (k = 1, 2, ..., 2^n), \\ -2^{\frac{n}{2}}, & \dfrac{2k-1}{2^{n+1}} < t \leqslant \dfrac{2k}{2^{n+1}}, \\ 0 & \text{for the remaining values of } t. \end{cases}$$

The functions $\chi_n^{(k)}(t)$ are arranged in a simple sequence $\{e_i(t)\}$ $(i = 1, 2, ...)$ in increasing order of the index n, and for the same n in increasing order of k. The system $\{e_i(t)\}$ is orthogonal and forms a basis in any space $L_p[0, 1]$ $(p \geqslant 1)$. Moreover, it forms a basis in an arbitrary separable Orlicz space on $[0, 1]$.

At present there is no solution of the Schauder problem: does every separable space have a basis?

3. *Criteria for bases.*   Everywhere in this subsection $\{e_i\}$ $(i=1, 2, ...)$ denotes a complete minimal system in a Banach space E, and $\{f_i\}$ the system of functionals, forming a biorthogonal system with $\{e_i\}$.

We define a bounded linear operator on E:

$$S_n x = \sum_{i=1}^{n} f_i(x)\,e_i\,.$$

The operator S_n is a projector: $S_n^2 = S_n$. It projects the entire space onto the n-dimensional space L_n spanned by the elements $e_1, ..., e_n$.

In order for the system $\{e_i\}$ to form a basis, it is necessary and sufficient that the operators S_n be uniformly bounded, that is, the inequality

$$\|S_n x\| = \|\sum_{i=1}^{n} f_i(x)\,e_i\| \leqslant M\,\|x\| \qquad (x \in E)$$

be satisfied, where M is a constant.

If the system $\{e_i\}$ does not form a basis, then an element x can be found on which $\|S_n x\| \leqslant M$ for all $n=1, 2, ...$ but for which the series

$$\sum_{n=1}^{\infty} f_i(x)\,e_i$$

diverges. If the last series converges for an arbitrary $x \in E$, then the system $\{e_i\}$ is a basis. Moreover, if this series converges weakly for arbitrary $x \in E$, then the system $\{e_i\}$ is a basis. The last assertion sometimes is formulated in the following way: every weak basis is a strong basis.

If the closed linear hull of the elements $e_n, e_{n+1}, ...$ is denoted by L^n and the unit sphere in the subspace L_n is denoted by σ_n, then *in order for the system $\{e_i\}$ to be a basis, it is necessary and sufficient that a positive constant α exist such that*

$$\varrho(\sigma_n, L^n) \geqslant \alpha\,,$$

where ϱ is the distance between σ_n and L^n.

If $\{e_i\}$ is a basis, then the system $\{f_i\}$ is a basis in its closed linear hull which may not coincide with E'. If E is reflexive, then this hull does coincide with E', and $\{f_i\}$ is a basis in E'. If $\{f_i\}$ is a basis in the conjugate space E', then $\{e_i\}$ is a basis in the space E.

4. *Unconditional bases.* The system $\{e_i\}$ is called an *unconditional basis* in the space E if it remains a basis for an arbitrary rearrangement of its elements.

The following is an equivalent definition: the basis $\{e_i\}$ is called unconditional if the series

$$\sum_{i=1}^{\infty} f_i(x) f(e_i)$$

is absolutely convergent for arbitrary $x \in E$ and $f \in E'$.

In order for the basis to be unconditional, it is necessary and sufficient that projectors of the form

$$\sum_{i=1}^{k} f_{n_i}(x) e_{n_i}$$

be uniformly bounded for arbitrary finite collections of numbers $(n_1, \ldots, n_k)$ $(n_i \neq n_j$ for $i \neq j)$.

If the unit sphere in the linear hull of the elements of the basis $e_{n_1}, \ldots, e_{n_k}$ is denoted by $S_{n_1, n_2, \ldots, n_k}$ and the closed linear hull of all the remaining elements of the basis is denoted by $L^{n_1, \ldots, n_k}$, then a necessary and sufficient condition for the basis to be unconditional is that a constant $\beta > 0$ exist such that

$$\varrho(S_{n_1, \ldots, n_k}, L^{n_1, \ldots, n_k}) \geqslant \beta$$

for all finite collections $(n_1, \ldots, n_k)$.

Let U be a bounded linear operator acting in the space E and having a bounded inverse. If the system $\{e_i\}$ is a basis, then the system $\{Ue_i\}$ is a basis. If $\{e_i\}$ is an unconditional basis, then $\{Ue_i\}$ is an unconditional basis.

In a Hilbert space H, every orthogonal basis is unconditional. It can be shown that an arbitrary unconditional basis in a Hilbert space is representable in the form $\{Ue_i'\}$ where $\{e_i'\}$ is an orthogonal normalized basis. Such bases are called *Riesz bases*. We can characterize them by the following properties: positive numbers m and M exist such that

$$m \sum_{i=1}^{\infty} |(x, e_i)|^2 \leqslant \|x\|^2 \leqslant M \sum_{i=1}^{\infty} |(x, e_i)|^2$$

for arbitrary $x \in H$.

The system of unit vectors $\{e_k\}$ in the spaces c_0 and l_p $(p \geqslant 1)$ forms an unconditional basis. The system of Haar functions (see no. 2) forms an unconditional basis in all the spaces $L_p[0, 1]$ with $p > 1$. Unconditional bases do not exist in the spaces $C[0, 1]$ and $L[0, 1]$.

The trigonometric system of functions is a basis in the spaces $L_p[-\pi, \pi]$ $(p > 1)$ but is not unconditional.

If the system $\{e_i\}$ is an unconditional basis in E, then the system of functionals $\{f_i\}$ forming a biorthogonal system with $\{e_i\}$ is an unconditional basis in E' provided the space E' is separable.

5. *Stability of a basis.* Let the system $\{e_i\}$ form a basis in the space E and let $\{h_i\}$ be some system of elements of E. The question is: for what conditions will the system $\{e_i + h_i\}$ also be a basis in E? If $\{e_i\}$ is a basis (unconditional basis) and the elements h_i are "sufficiently small" in the sense that

$$\sum_{i=1}^{\infty} \|f_i\| \, \|h_i\| < 1,$$

then the system $\{e_i + h_i\}$ forms a basis (unconditional basis) in E.

An important corollary follows from this last assertion: *if the space E has a basis (unconditional basis) and $\{\varphi_k\}$ is a complete system of elements in E, then a basis (unconditional basis) exists in E of the form*

$$e_i = \sum_{k=1}^{n_i} c_k^{(i)} \varphi_k.$$

A basis of polynomials exists, for example, in the space $C[0, 1]$.

CHAPTER II

LINEAR OPERATORS IN HILBERT SPACE

§ 1. Abstract Hilbert space

1. *Concept of a Hilbert space.* Let H be a linear system, with multiplication by complex numbers, in which to each pair of elements there is assigned a complex number (x, y) having the properties:

a) $(x, y) = \overline{(y, x)}$, in particular (x, x) is real;

b) $(x_1 + x_2, y) = (x_1, y) + (x_2, y)$;

c) $(\lambda x, y) = \lambda(x, y)$ for an arbitrary complex number λ;

d) $(x, x) \geqslant 0$, and $(x, x) = 0$ only for $x = \theta$.

The number (x, y) is called a *scalar product*. If H is a linear system allowing only multiplication by real numbers, then the scalar product is assumed to be real.

The following corollaries follow from axioms a–d:

1) $(x, y_1 + y_2) = (x, y_1) + (x, y_2)$;

2) $(x, \lambda y) = \bar{\lambda}(x, y)$;

3) The Buniakovsky-Schwarz inequality:

$$|(x, y)| \leqslant \sqrt{(x, x)} \sqrt{(y, y)}.$$

In terms of the scalar product in H, a norm

$$\|x\| = \sqrt{(x, x)}$$

can be introduced, after which H becomes a normed linear space. If H is infinite-dimensional and complete with respect to the norm introduced, then it is called a (complex or real) *Hilbert space*. It is obvious from the definition that every Hilbert space is a Banach space.

2. *Examples of Hilbert spaces.* As is known, the scalar product of two vectors $x = \{\xi_1, \xi_2, ..., \xi_n\}$ and $y = \{\eta_1, \eta_2, ..., \eta_n\}$ in n-dimensional Euclid-

ean space is usually found by the formula

$$(x, y) = \sum_{i=1}^{n} \xi_i \eta_i,$$

and, in n-dimensional unitary (complex Euclidean) space, by the formula

$$(x, y) = \sum_{i=1}^{n} \xi_i \bar{\eta}_i.$$

Analogously, a scalar product is introduced in a number of infinite-dimensional spaces after which they become Hilbert spaces.

1. The complex space l_2 becomes a Hilbert space if we set

$$(x, y) = \sum_{i=1}^{\infty} \xi_i \bar{\eta}_i.$$

2. The space $L_2(a, b)$ of complex-valued functions becomes a Hilbert space if we set

$$(x, y) = \int_a^b x(t) \overline{y(t)} \, dt.$$

3. The complex space $L_{2,\varrho}(a, b)$ of functions which are measurable on the segment $[a, b]$ and have on this segment a modulus whose square is summable with weight $\varrho(t)$ ($\varrho(t) > 0$ almost everywhere) will be a Hilbert space if we set

$$(x, y) = \int_a^b x(t) \overline{y(t)} \varrho(t) \, dt.$$

4. The spaces $W_2^l(G)$ of S. L. Sobolev (see ch. 1, § 2, no. 6) are Hilbert spaces with respect to the scalar product

$$(x, y) = \int_G x(t) y(t) \, dt + \int_G \left(\sum_{|\alpha|=l} D^\alpha x(t) D^\alpha y(t) \right) dt.$$

Here

$$D^\alpha x = \frac{\partial^l x}{\partial t_1^{k_1} \dots \partial t_n^{k_n}}, \qquad |\alpha| = k_1 + \dots + k_n.$$

5. The space of functions $x(t)$, defined and measurable on the entire

axis $(-\infty, \infty)$, such that the limit

$$\lim_{T \to \infty} \frac{1}{2T} \int_{-T}^{T} |x(t)|^2 \, dt < \infty$$

exists, will be a Hilbert space if we set

$$(x, y) = \lim_{T \to \infty} \frac{1}{2T} \int_{-T}^{T} x(t)\,\overline{y(t)}\, dt .$$

The spaces given in examples 1–4 are separable; the space given in example 5 is non-separable.

3. *Orthogonality. Projection onto a subspace.* Two elements x and y of a Hilbert space are called *orthogonal*, $x \perp y$, if $(x, y)=0$. An element $x \in H$ is called *orthogonal to the subset* $G \subset H$, $x \perp G$, if $(x, y)=0$ for an arbitrary $y \in G$. Finally, two subsets G and Γ of the space H are called *orthogonal*, $G \perp \Gamma$, if an arbitrary element $x \in G$ is orthogonal to an arbitrary element $y \in \Gamma$.

Let L be a subspace of H. The collection of all elements orthogonal to L forms a subspace M, the so-called *orthogonal complement* to L. The subspaces L and M have only the single element θ in common.

The following is one of the basic properties of a Hilbert space:

If L is a (closed) subspace of the space H, then every $x \in H$ has a unique representation

$$x = y + z ,$$

where $y \in L$ and $z \perp L$.

The element y is called the *projection of x on L*. It has the property that in comparison with other elements of L it has the least distance from x.

Every element of H can be decomposed into a sum of an element of the subspace L and an element from its orthogonal complement M. In other words, H is decomposed into the *orthogonal sum* of L and M: $H = L + M$. In connection with this, we denote $M = H \dot{-} L$.

A highly useful corollary follows from the preceding: *in order for a linear manifold L to be everywhere dense in the space H, it is necessary and sufficient that no element exists which is different from zero and orthogonal to all the elements of the set L.*

4. *Linear functionals.* It follows from the Buniakovsky-Schwarz inequality that the linear functional $f(x)=(x, u)$, for a fixed $u \in H$, is bounded. It can be shown that this exhausts all bounded linear functionals on H; that is, a unique element $u \in H$ can be found such that

$$f(x) = (x, u)$$

for every $f(x) \in H^*$; moreover, $\|f\|_{H^*} = \|u\|_H$.

Thus the conjugate space H^* is isometric to the space H itself. The conjugate space is considered here in the sense studied in chapter I, §5, no. 12. A Hilbert space is *self-conjugate* and, hence, reflexive.

Every linear functional f, defined on $L_2(a, b)$, is representable in the form

$$f(x) = \int_a^b x(t)\bar{u}(t)\,dt,$$

where $u(t) \in L_2(a, b)$ and $\|f\| = \left(\int_a^b |u(t)|^2\,dt \right)^{\frac{1}{2}}.$

Every linear functional f, defined on l_2, is representable in the form

$$f(x) = \sum_{i=1}^\infty \xi_i \bar{c}_i,$$

where $\sum_{i=1}^\infty |c_i|^2 < \infty$ and $\|f\| = \left(\sum_{i=1}^\infty |c_i|^2 \right)^{\frac{1}{2}}.$

Remark. It is sometimes convenient to represent linear functionals on a Hilbert space H not by a scalar product in the space H but by a scalar product in some other Hilbert space. Then the conjugate space H^* to H will be realized by means of elements of another kind. It is convenient, for example, to represent a linear functional $f(x)$ on the space $W_2^l(G)$ by a scalar product in $L_2(G)$, that is, in the form

$$f(x) = \int_G x(t)\bar{u}(t)\,dt.$$

In this connection, $u(t)$ will be a generalized function (see ch. VIII). The collection of these functions forms the space $W_2^{-l} = (W_2^l)^*$.

5. *Weak convergence.* In accordance with the general definition of weak convergence (see ch. I, § 4, no. 3), a sequence of elements $\{x_n\} \subset H$ is called *weakly convergent to the element* x_0 (resp., *weakly fundamental*) if $(x_n, y) \to (x_0, y)$ (resp., $(x_{n+p}, y) - (x_n, y) \to 0$) for an arbitrary element $y \in H$.

The following properties of weak convergence follow from the reflexivity of Hilbert space:

1) If the sequence $\{x_n\}$ converges weakly to x_0 and $\|x_n\| \to \|x_0\|$, then $\|x_n - x_0\| \to 0$, that is, the sequence $\{x_n\}$ converges strongly to x_0.

2) The space H is weakly complete; that is, if the sequence $\{x_n\}$ is weakly fundamental, then it converges weakly to some limit.

3) Bounded sets in the space H are weakly compact; that is, from an arbitrary infinite set of elements of the space H which is bounded in the norm, a weakly convergent sequence can be chosen.

6. *Orthonormal systems.* A system $e_1, e_2 \ldots, e_n, \ldots$ of elements of a Hilbert space H is called *orthonormal* (or *orthonormalized*) if

$$(e_i, e_j) = \delta_{ij},$$

where δ_{ij} is the well-known symbol which is equal to one for $i = j$ and to zero for $i \neq j$. The trigonometric system

$$\frac{1}{\sqrt{2\pi}}, \quad \frac{1}{\sqrt{\pi}} \cos t, \quad \frac{1}{\sqrt{\pi}} \sin t, \quad \frac{1}{\sqrt{\pi}} \cos 2t, \quad \frac{1}{\sqrt{\pi}} \sin 2t, \ldots$$

is an example of such a system in the real space $L_2(-\pi, \pi)$; the system

$$e^{2\pi i n t}, \qquad n = 0, \pm 1, \pm 2, \ldots$$

is an example in the complex space $L_2(0, 1)$.

If an arbitrary system of linearly independent elements $h_1, h_2, \ldots, h_n, \ldots$ is given in H, then an orthonormal system can be easily obtained from it by means of the so-called *Schmidt process of orthonormalization.* Namely, we set $e_1 = \dfrac{h_1}{\|h_1\|}$; then we select c_{21} so that $h_2 - c_{21} e_1$ will be orthogonal to e_1 and set $e_2 = \dfrac{h_2 - c_{21} e_1}{\|h_2 - c_{21} e_1\|}$. We further select c_{32} and c_{31} so that $h_3 - c_{32} e_2 - c_{31} e_1$ will be orthogonal to e_2 and e_1, and set $e_3 = \dfrac{h_3 - c_{32} e_2 - c_{31} e_1}{\|h_3 - c_{32} e_2 - c_{31} e_1\|}$, and so on.

Example. If the system of powers, $1, t, t^2, ..., t^n, ...$, is orthonormalized in the space $L_2(-1, 1)$, then the system of normalized Legendre polynomials is obtained. Orthonormalization of this system in the space $L_{2,\varrho}(-\infty, \infty)$ with weight $\varrho(t) = e^{-t^2}$ gives the system of Čebyšev-Hermite polynomials.

If $\{e_i\}$ is an orthonormal system in H, then the numbers $c_i = (x, e_i)$ are called the *Fourier coefficients of the element* x with respect to this system. The linear combination $\sum\limits_{i=1}^{n} c_i e_i$ gives the best approximation of x in comparison with other combinations of the form $\sum\limits_{i=1}^{n} \alpha_i e_i$; that is,

$$\delta_n = \left\| x - \sum_{i=1}^{n} c_i e_i \right\| \leqslant \left\| x - \sum_{i=1}^{n} \alpha_i e_i \right\|.$$

In other words, $\sum\limits_{i=1}^{n} c_i e_i$ is the projection of the element x onto the subspace L_n, spanned by the elements $e_1, e_2, ..., e_n$.

The formula

$$\delta_n^2 = \|x\|^2 - \sum_{i=1}^{n} |c_i|^2$$

is valid for δ_n.

If the element x belongs to the closed linear hull L of the elements e_i $(i = 1, 2, ...)$, then

$$x = \sum_{i=1}^{\infty} c_i e_i \quad \text{and} \quad \|x\|^2 = \sum_{i=1}^{\infty} |c_i|^2.$$

If $x \notin L$, then the element $x' = \sum c_i e_i$ will be the projection of x onto L, where

$$\|x\|^2 \geqslant \|x'\|^2 = \sum_{i=1}^{\infty} |c_i|^2.$$

The series $\sum\limits_{i=1}^{\infty} c_i e_i$ is called the *Fourier series* of the element x, and the last inequality is called the *Bessel inequality.*

We recall that if L coincides with all of H, that is, the linear combinations of the elements e_i are dense in H, then the system $\{e_i\}$ is called *complete. A necessary and sufficient condition for completeness is the Parseval equality*

$$\|x\|^2 = \sum_{i=1}^{\infty} |c_i|^2$$

for an arbitrary element $x \in H$.

A complete orthonormal system $\{e_i\}$ is a basis for the Hilbert space. An orthonormal basis exists in each separable Hilbert space. All separable Hilbert spaces are isometric to the space l_2.

§ 2. Bounded linear operators in a Hilbert space

1. *Bounded linear operators. Adjoint operators. Bilinear forms.* For a bounded linear operator A, acting in a Hilbert space H, by definition

$$\|A\| = \sup_{\|x\|=1} \|Ax\| = \sup_{(x,x)=1} \sqrt{(Ax, Ax)} = \sup_{\substack{x \in H \\ x \neq O}} \sqrt{\frac{(Ax, Ax)}{(x, x)}}.$$

If y is fixed in the scalar product (Ax, y) then a linear functional of x is obtained:

$$f(x) = (Ax, y),$$

where

$$|f(x)| = |(Ax, y)| \leqslant \|A\| \, \|x\| \, \|y\|.$$

This functional can be represented in the form

$$(Ax, y) = (x, u),$$

where $u \in H$. The correspondence $y \to u$ defines a bounded linear operator $u = A^*y$. By definition

$$(Ax, y) = (x, A^*y).$$

The operator A^* is called the *adjoint operator* to A.

This definition agrees with the definition in chapter I, § 5, no. 12.

A function $l(x, y)$ of two elements x and y of a Hilbert space H is called a *bilinear form* if

$$l(\alpha_1 x_1 + \alpha_2 x_2, \beta_1 y_1 + \beta_2 y_2) = \alpha_1 \bar{\beta}_1 l(x_1, y_1) + \alpha_2 \bar{\beta}_1 l(x_2, y_1) + \\ + \alpha_1 \bar{\beta}_2 l(x_1, y_2) + \alpha_2 \bar{\beta}_2 l(x_2, y_2).$$

The bilinear form is *bounded* if

$$|l(x, y)| \leqslant c \, \|x\| \, \|y\|.$$

The least possible value of c in this inequality, or, what is the same, $\sup_{\|x\|=\|y\|=1} |l(x, y)|$, is called the *norm* of the bilinear form.

If A is a bounded linear operator, then the form

$$l(x, y) = (Ax, y)$$

is a bounded bilinear form. Conversely, to every bounded bilinear form $l(x, y)$, there corresponds a bounded operator A for which the preceding equality is valid. Furthermore, the norm of the bilinear form is equal to the norm of the operator.

The values of the bilinar form $l(x, y)$ are completely defined by the values of the corresponding quadratic form (Ax, x). In fact

$$l(x, y) = [l(x_1, x_1) - l(x_2, x_2)] + i[l(x_3, x_3) - l(x_4, x_4)],$$

where $x_1 = \frac{1}{2}(x + y)$; $x_2 = \frac{1}{2}(x - y)$; $x_3 = \frac{1}{2}(x + iy)$;

$$x_4 = \frac{1}{2}(x - iy).$$

The operator A is uniquely defined by its quadratic form (Ax, x): if $(Ax, x) = (Bx, x)$, then $A = B$.*)

Let the space H be separable and $\{e_i\}$ be an orthonormal basis in H. Then

$$Ae_k = \sum_{i=1}^{\infty} a_{ik} e_i;$$

moreover,

$$a_{ik} = (Ae_k, e_i).$$

If $x = \sum_{j=1}^{\infty} \xi_j e_j$ and $y = \sum_{j=1}^{\infty} \eta_j e_j$, then

$$Ax = \sum_{j=1}^{\infty} \sum_{i=1}^{\infty} \xi_j a_{ij} e_i \quad \text{and} \quad (Ax, y) = \sum_{i, j=1}^{\infty} a_{ij} \xi_j \bar{\eta}_i.$$

The matrix (a_{ik}) is called the *matrix of the operator* A with respect to the basis $\{e_i\}$. The matrix $(a_{ik}^*) = (\bar{a}_{ki})$ will be the matrix of the adjoint operator A^*. It is necessary for the boundedness of the operator A (and this means also for the operator A^*) that

$$\sum_{k=1}^{\infty} |a_{ik}|^2 < \infty \quad \text{and} \quad \sum_{k=1}^{\infty} |a_{ki}|^2 < \infty \qquad (i = 1, 2, \ldots).$$

These conditions are not sufficient for the boundedness of the operator

---

*) The last assertions are valid only in a complex Hilbert space.

given by the matrix. Examples of sufficient conditions are:

1. If

$$\sum_{k=1}^{\infty} |a_{ik}| \leqslant M \quad \text{and} \quad \sum_{k=1}^{\infty} |a_{ki}| \leqslant M \quad (i = 1, 2, \ldots),$$

where M does not depend on i, then the operator A is bounded.

2. If

$$\sum_{i,k=1}^{\infty} |a_{ik}|^2 < \infty,$$

then the operator A is bounded. The number $\{ \sum_{i,k=1}^{\infty} |a_{ik}|^2 \}^{\frac{1}{2}}$ is sometimes called the *abolute norm* of the operator A. This norm does not depend on the choice of the orthogonal basis $\{e_i\}$ in H.

Effectively verifiable necessary and sufficient conditions for the boundedness of an operator given in matrix form are not known.

In a function space, the most prevalent class of linear operators is the class of integral operators of the form

$$Ax = \int_a^b K(t,s)x(s)\,ds.$$

It suffices for the boundedness of the integral operator in the Hilbert space $L_2(a, b)$ that a number M exist such that

$$\int_a^b |K(x, y)|\,dy \leqslant M \quad \text{and} \quad \int_a^b |K(x, y)|\,dx \leqslant M.$$

Also, the summability of the square of the kernel $K(t, s)$ with respect to both variables:

$$\int_a^b \int_a^b |K(t,s)|^2\,dt\,ds < \infty$$

is a sufficient condition for boundedness.

2. *Unitary operators.* A linear operator U mapping a Hilbert space H onto all of H with preservation of the norm:

$$\|Ux\| = \|x\|,$$

is called *unitary*.

In the coordinate Hilbert space l_2, the operator mapping the element x onto the element y by means of a fixed permutation of the coordinates of the element x can serve as an example of a unitary operator.

In the complex space $L_2(a, b)$, the operator of multiplication by the function e^{ict}, where c is a real number, is a unitary operator.

The shift (or displacement) operator

$$A_s x = x(t + s)$$

is a unitary operator in the space $L_2(-\infty, \infty)$.

Indeed

$$\int_{-\infty}^{\infty} |x(t)|^2 \, dt = \int_{-\infty}^{\infty} |x(t + s)|^2 \, dt .$$

Analogous unitary operators arise by the consideration of shift operators on functions defined on groups with invariant measures or in dynamical systems.

Unitary operators have the properties:

1) $(Ux, Uy) = (x, y)$ $(x, y \in H)$.

2) The operator U^{-1}, the inverse of the unitary operator, exists, and

$$U^{-1} = U^*$$

(this property can serve as the definition of a unitary operator).

3) The product of unitary operators is again a unitary operator. Unitary operators form a group.

If λ is an *eigenvalue* of the unitary operator U, that is, an element $e \neq 0$ exists such that

$$Ue = \lambda e,$$

then $|\lambda| = 1$.

Analytic descriptions of all unitary operators can be given for the space $L_2(a, b)$ (see [39]).

A number of transformations used in analysis generate unitary operators. The Fourier-Plancherel transformation, given by the formula

$$g(t) = \frac{1}{\sqrt{2\pi}} \frac{d}{dt} \int_{-\infty}^{\infty} \frac{e^{-its} - 1}{-is} f(s) \, ds = Uf(x)$$

or by the simpler formula

$$g(t) = \frac{1}{\sqrt{2\pi}} \int\limits_{-\infty}^{\infty} e^{-its} f(s)\, ds,$$

in which the integral must be understood as the limit in the mean (with respect to t) of the integral from $-N$ to N as $N \to \infty$, is a particularly important example of these transformations. The operator U is unitary in $L_2(-\infty, \infty)$. The inverse operator is given by the formula

$$U^{-1}g(t) = f(t) = \frac{1}{\sqrt{2\pi}} \int\limits_{-\infty}^{\infty} e^{its} g(s)\, ds.$$

An *isometric* operator is a generalization of a unitary operator. Such a linear operator maps a subspace H_1 of a Hilbert space H onto a subspace H_2 of the same or another Hilbert space with preservation of the scalar product and, hence, of the norm. In the case where $H_1 = H_2 = H$, the isometric operator becomes a unitary operator.

3. *Self-adjoint operators.* A bounded linear operator coinciding with its adjoint, $A = A^*$, is called *self-adjoint*. For a self-adjoint operator,

$$(Ax, y) = (x, Ay) = \overline{(Ay, x)}.$$

A bilinear form having the property that

$$l(x, y) = \overline{l(y, x)}$$

is called *Hermitian*. To every bounded hermitian form there corresponds a bounded self-adjoint operator. The quadratic form (Ax, x) corresponding to a self-adjoint operator is real. The numbers

$$m = \inf_{\|x\|=1} (Ax, x) \quad \text{and} \quad M = \sup_{\|x\|=1} (Ax, x)$$

respectively are called the *lower* and *upper* bounds of the self-adjoint operator.

The norm of the operator A is equal to the largest of the numbers $|m|$ and $|M|$:

$$\|A\| = \max(|m|, |M|) = \sup_{\|x\|=1} |(Ax, x)|.$$

If the lower bound is non-negative, that is,

$$(Ax, x) \geqslant 0$$

for arbitrary $x \in H$ and $A \neq 0$, then the operator is called *positive*.

If a bounded operator is given by a matrix (a_{ik}), then it will be self-adjoint if and only if the matrix corresponding to it is hermitian, that is,

$$a_{ik} = \overline{a_{k_i}}.$$

A bounded integral operator in $L_2(a, b)$ with kernel $K(t, s)$ will be self-adjoint if $K(t, s) = \overline{K(s, t)}$. Every bounded self-adjoint operator in $L_2(a, b)$ is representable in the form of an integral operator, but by the kernel $K(t, s)$ we must now understand not a usual function but a generalized function (see ch. VIII.)

The eigenvalues of a self-adjoint operator are real; the eigenvectors corresponding to distinct eigenvalues are mutually orthogonal.

Let A be an arbitrary bounded operator. It is representable in the form

$$A = \frac{A + A^*}{2} + i\frac{A - A^*}{2i} = A_1 + iA_2,$$

where the operators A_1 and A_2 are self-adjoint. The operators Re A $= \frac{A + A^*}{2}$ and Im $A = \frac{A - A^*}{2i}$ are called the *real* and *imaginary parts* of the operator A.

If A is an arbitrary bounded operator, then the operators AA^* and A^*A are self-adjoint and positive.

If Re $A = \frac{A + A^*}{2}$ is a negative operator, then the operator A is called *dissipative*.

4. *Self-adjoint completely continuous operators.* If the self-adjoint operator A is completely continuous, then the space H can be decomposed into the orthogonal sum of two subspaces: $H = H_0 + H'$ where $Ax_0 = 0$ for arbitrary $x_0 \in H_0$ and where an orthonormal basis $\{x_i\}$ exists in the space H' consisting of eigenvectors of the operator A corresponding to non-zero eigenvalues λ_i. Thus, for an arbitrary element $x \in H$

$$x = x_0 + \sum_i c_i x_i = x_0 + \sum_i (x, x_i) x_i$$

and

$$Ax = \sum_i \lambda_i c_i x_i = \sum_i \lambda_i (x, x_i) x_i.$$

In particular, it follows that a self-adjoint completely continuous operator, not vanishing on the entire space, has at least one eigenvalue different from zero. The space H_0 consists of eigenvectors of the operator A corresponding to the eigenvalue $\lambda = 0$. Selecting in this space an arbitrary orthonormal basis $\{e_i'\}$, we obtain an orthonormal basis $\{e_i'\} + \{e_i\}$ for the space H consisting of eigenvectors of the operator A.

The eigenvalues and eigenvectors of a self-adjoint completely continuous operator can be obtained by the following process: the form $A(x, x)$ on the unit sphere of the space H attains its largest absolute value at some element x_1. It can be shown that $Ax_1 = \lambda_1 x_1$, where $\lambda_1 = (Ax_1, x_1) = \pm \max_{\|x\|=1} |A(x, x)| = \pm \|A\|$ (the sign $+$ or $-$ coincides with the sign of (Ax_1, x_1)).

Let H_1 be the orthogonal complement of x_1 in H. The subspace H_1 is invariant with respect to the operator A. If the operator A annihilates every element of H_1, then the process comes to a stop; if $Ax \neq 0$, then the form $(Ax, x) \neq 0$, and on the unit sphere of the space H_1 it attains its largest absolute value at some element x_2. In this connection, $Ax_2 = \lambda_2 x_2$ where $\lambda_2 = \pm \max_{\|x\|=1, x \in H_1} |(Ax, x)|$ and $x_2 \in H_1$.

It follows from the construction that $|\lambda_2| \leqslant |\lambda_1|$.

The continuation of this process gives a finite or denumerable complete system of eigenvalues and eigenvectors of the operator A in H'.

Let $\lambda_1^+, \lambda_2^+, \ldots$ be the positive eigenvalues of the operator A arranged in decreasing order, and $\lambda_1^-, \lambda_2^-, \ldots$ be the negative eigenvalues arranged in increasing order (multiple eigenvalues are repeated as many times as their multiplicity). The eigenvalues have the following *minimaximal property*: let $z_1, \ldots, z_n$ be arbitrary elements of H and $M(z_1, z_2, \ldots, z_n)$ be the maximum of the form (Ax, x) on all elements x satisfying the conditions

$$\|x\| = 1 \quad \text{and} \quad (x, z_1) = (x, z_2) = \cdots = (x, z_n) = 0.$$

Then the smallest value of the function $M(z_1, z_2, \ldots, z_n)$ for all possible systems $(z_1, z_2, \ldots, z_n)$ of elements of H will be equal to λ_n^+. Analogously, $\lambda_n^- = \max_{z_i \in H} m(z_1, z_2, \ldots, z_n)$, where $m(z_1, z_2, \ldots, z_n)$ is the minimum of the form (Ax, x) on the elements x satisfying the preceding conditions.

A self-adjoint completely continuous operator will be positive if and only if all its eigenvalues are non-negative.

The properties of self-adjoint completely continuous operators are generalizations of properties of integral operators with symmetric kernels, considered in the theory of integral equations. The equation

$$x - \mu A x = y$$

is a generalization of the integral equation.

If A is a self-adjoint completely continuous operator and the number $1/\mu$ does not coincide with any of its eigenvalues, then the formula

$$x = y + \mu \sum_i \frac{\lambda_i}{1 - \mu \lambda_i} (y, x_i) x_i .$$

gives the solution of the preceeding equation.

If $1/\mu$ coincides with one of the eigenvalues of the operator A, then a solution exists only under the condition that the element y is orthogonal to all the eigenvectors corresponding to the eigenvalue $1/\mu$. In this case, one of the solutions can be obtained by the same formula if terms in it containing $\lambda_i = \frac{1}{\mu}$ are discarded.

5. *Completely continuous operators.* Besides the basic definition of a completely continuous operator, according to which an operator is called completely continuous if it maps every bounded set into a (relatively) compact set, equivalent definitions exist in a Hilbert space.

1. A linear operator A is completely continuous if it maps every weakly convergent sequence into a strongly convergent sequence, that is, if $x_n \overset{w}{\to} x_0$ implies that $A x_n \to A x_0$ in the norm.

2. A linear operator A is completely continuous if the equality

$$\lim_{n \to \infty} (A x_n, y_n) = (A x_0, y_0)$$

is valid for arbitrary sequences $\{x_n\}$ and $\{y_n\}$ which are weakly convergent to x_0 and y_0; that is, the form (Ax, y) is a weakly continuous function of x and y.

If A is completely continuous, then A^* is completely continuous.

This assertion is useful: if AA^* is completely continuous, then A is completely continuous.

A finite-dimensional operator in a Hilbert space H is representable in the form

$$Ax = \sum_{k=1}^{n} (x, x_k) y_k = \sum_{k=1}^{n} x_k \otimes y_k,$$

where x_k and y_k $(k = 1, 2, ..., n)$ are fixed elements of H. A representation is possible for completely continuous operators which is analogous to the above representation of a finite-dimensional operator. The numbers $\mu > 0$, for which non-zero solutions of the system

$$\begin{cases} Ax = \mu y, \\ A^* y = \mu x \end{cases}$$

exist, are called *singular values* of the operator A, and the corresponding solutions x, y are called *associated fundamental Schmidt elements*. The numbers μ^2 are eigenvalues of the positive self-adjoint completely continuous operators AA^* and A^*A. Therefore, there exist only a denumerable number of singular values μ_i; moreover $\mu_i \to 0$ for $i \to \infty$.*)

The representation

$$A = \sum_{i=1}^{\infty} \mu_i x_i \otimes y_i$$

holds, where x_i, y_i are associated fundamental elements corresponding to the singular values μ_i. The series converges in the operator norm. Explicitly, the preceding equality has the form

$$Ax = \sum_{i=1}^{\infty} \mu_i (x, x_i) y_i.$$

An analogous representation

$$A^* x = \sum_{i=1}^{\infty} \mu_i y_i \otimes x_i$$

holds for the adjoint operator.

If A is self-adjoint, then $x_i = y_i$, and the previously examined representation is obtained.

If the operator A is given by the matrix (a_{ik}), then it suffices for its complete continuity that

$$\sum_{i, k=1}^{\infty} |a_{ik}|^2 < \infty.$$

---

*) There may be only a finite number of singular values μ_j. (Editor)

Analogously, it suffices for the complete continuity of an integral operator in $L_2(a, b)$ that

$$\int_a^b \int_a^b |K(t, s)|^2 \, dt \, ds < \infty.$$

Neither of these conditions is necessary.

An integral operator with a symmetric kernel satisfying the last condition is called a *Hilbert-Schmidt operator*.

The question of the completeness of a system of eigenvectors and associated vectors of a completely continuous operator is important in the theory of completely continuous operators (see ch. I, §5, no. VIII.) A basis of eigenvectors exists for a self-adjoint completely continuous operator. However, if a finite-dimensional operator is added to a self-adjoint completely continuous operator, then the operator obtained may not always have a complete system of eigenvectors and associated vectors. If, for example, the one-dimensional operator

$$t \int_0^1 (1 - s) x(s) \, ds$$

is added to the integral operator with symmetric kernel

$$K(t, s) = \begin{cases} (t - 1)s & \text{for} \quad s \leqslant t \\ (s - 1)t & \text{for} \quad s \geqslant t, \end{cases} \quad (0 \leqslant s, t \leqslant 1)$$

then a Volterra operator

$$\int_0^t (t - s) x(s) \, ds$$

is obtained which does not have eigenfunctions.

From the available list of criteria for completeness we deduce the following: let A be a completely continuous operator such that the values of the form (Ax, x) for arbitrary $x \in H$ are contained in the sector of the complex plane:

$$|\arg \xi| \leqslant \frac{\pi}{2\varrho} \quad (\varrho \geqslant 1).$$

The system of eigenvectors and associated vectors of the operator A is complete in the space H if its singular values μ_n *arranged in decreasing order have the property*

$$\lim_{n \to \infty} n^{1/\varrho}\mu_n = 0,$$

in particular, if the series

$$\sum \mu_n^{1/\varrho}$$

converges.

These conditions are not very suitable in that the singular values of the operator A appear in them. For $\varrho > 1$ in these conditions, the singular values can be replaced by the eigenvalues of the real or imaginary part of the operator A $\left(\text{the operator } \dfrac{A+A^*}{2} \text{ or } \dfrac{A-A^*}{2i}\right).$

The system of eigenvectors and associated vectors is complete when $\varrho = 1$, that is, for a dissipative operator, if the operator has finite *trace*

$$\sum \mu_n < \infty.$$

However, this assertion becomes invalid if the singular values are replaced by the eigenvalues of the real or imaginary part of the operator A. If both the real and the imaginary parts of the dissipative operator A have a finite trace, then the system of eigenvectors and associated vectors is complete in the space H.

6. *Projective operators.* The self-adjoint operators having the simplest structure are projective operators. Let L be a (closed) subspace of the space H. The operator which sets in correspondence to every element x its projection y on the subspace L:

$$y = P_L x$$

is called the projector onto the subspace L, or, more precisely, the *projective operator* P_L. By definition, $P_L x = x$ for an element $x \in L$.

A projective operator is self-adjoint; its square is equal to itself and, hence, it is positive. Conversely, if a bounded linear operator P has the properties $P^* = P$ and $P^2 = P$, then it is the projector of the space H onto its range of values.

The norm of a projective operator is equal to 1.

If L is finite-dimensional, then P_L is finite dimensional and, consequently, completely continuous. If L is infinite-dimensional, then P_L is not completely continuous.

If L_1 and L_2 are orthogonal subspaces, then $P_{L_1} P_{L_2} = 0$, and conversely. In this case the operators P_{L_1} and P_{L_2} are called *orthogonal*.

Properties of projective operators

1) In order for the sum of two projective operators P_{L_1} and P_{L_2} to be a projective operator, it is necessary and sufficient that these operators be orthogonal. If this condition is satisfied, then

$$P_{L_1} + P_{L_2} = P_{L_1 + L_2}.$$

2) In order for the product of two projective operators P_{L_1} and P_{L_2} to be a projective operator, it is necessary and sufficient that the operators P_{L_1} and P_{L_2} commute. If this condition is satisfied, then

$$P_{L_1} P_{L_2} = P_{L_1 \cap L_2}.$$

The projective operator P_1 is called a *part* of the projective operator P_2 if

$$P_1 P_2 = P_2 P_1 = P_1.$$

3) The projective operator P_1 is a part of the projective operator P_2 if and only if the subspace L_1 is a part of the subspace L_2.

4) In order for the projective operator P_{L_2} to be a part of the projective operator P_{L_1}, it is necessary and sufficient that the inequality

$$\|P_{L_2} x\| \leqslant \|P_{L_1} x\|$$

be satisfied for all $x \in H$.

5) The difference $P_{L_1} - P_{L_2}$ of two projective operators is a projective operator if and only if P_{L_2} is a part of P_{L_1}.

If this condition is satisfied, then $P_{L_1} - P_{L_2}$ is the projector onto $L_1 \dot{-} L_2$ (the orthogonal complement of L_2 with respect to L_1).

6) A series of mutually orthogonal projective operators

$$\sum_{n=1}^{\infty} P_n$$

is always strongly convergent, and its sum is a projective operator P. The subspace L, onto which this operator projects, is called the *orthogonal sum of the subspaces L_n* onto which the operators P_n project:

$$L = \sum_{n=1}^{\infty} \dot{+} L_n.$$

§ 3. Spectral expansion of self-adjoint operators

1. *Operations on self-adjoint operators.* The sum of two bounded self-adjoint operators is again a self-adjoint operator. Moreover, an arbitrary linear combination of self-adjoint operators with real coefficients is a self-adjoint operator. The sum of positive operators is also a positive operator.

A product of bounded self-adjoint operators will be self-adjoint if and only if these operators commute. If, in this connection, the factors are positive, then the product is positive.

The set of self-adjoint operators is closed with respect to weak convergence; that is, the limit of a weakly convergent (see ch. I, § 4, no. 3) sequence of self-adjoint operators is a self-adjoint operator.

In the set of self-adjoint operators, an order relation can be introduced by setting $A \geqslant B$ if $A - B$ is a positive operator. In this connection, inequalities between the operators have the basic properties of ordinary inequalities between real numbers. However, for two different self-adjoint operators, it is impossible to talk about one always being greater than the other, since it is possible that the form $((A-B)x, x)$ will be greater than zero for one x and less than zero for another x. In this case, the operators A and B are called *non-comparable*. Since there exist comparable and non-comparable self-adjoint operators, we say that a *partial ordering* or a *semi-ordering* exists in the set of all self-adjoint operators. The existence of a partial ordering allows the introduction, in the set of self-adjoint operators, of several concepts such as, for example, sets of operators bounded above or below, lower and upper bounds for the bounded set of operators, monotone increase and monotone decrease of a sequence of operators, and others.

The following is an important property of bounded sequences of self-adjoint operators:

If $\{A_n\}$ is a monotone increasing sequence of mutually commutative self-adjoint operators, bounded above by a self-adjoint operator B which commutes with all the A_n, then the sequence $\{A_n\}$ converges strongly to a self-adjoint operator $A \leqslant B$, and

$$A = \sup_n A_n.$$

To every self-adjoint operator A there corresponds the partially ordered ring K_A of all bounded self-adjoint operators commuting with A. The ring K_A, generally speaking, is non-commutative. This ring contains the operator A and an arbitrary polynomial

$$P(A) = a_0 + a_1 A + a_2 A^2 + \cdots + a_n A^n$$

in A with real coefficients. The correspondence between operator polynomials and polynomials in a real variable is linear and multiplicative, that is, if

$$P(t) = \alpha Q(t) + \beta R(t),$$

then

$$P(A) = \alpha Q(A) + \beta R(A),$$

and if

$$P(t) = Q(t) R(t),$$

then

$$P(A) = Q(A) R(A).$$

A more profound fact is the *positiveness* of this correspondence in the sense that if $P(t) \geqslant 0$ on $[m, M]$, where m and M are lower and upper bounds for the operator A, then $P(A) \geqslant 0$. It follows from the positiveness of the correspondence that if a monotone increasing sequence of polynomials $\{P_n(t)\}$, uniformly bounded on the segment $[m, M]$ by the number K, converges to a function $F(t)$, then the sequence of polynomials $\{P_n(A)\}$ is also monotone increasing and bounded by the operator KI and, hence, has a limit $B = \lim_n P_n(A)$. This operator is, naturally, denoted by $F(A)$ and called a *function of the operator A*. The operator $F(A)$ belongs to K_A and, moreover, commutes with an arbitrary operator from K_A.

In particular, we can introduce the function $B = \sqrt{A}$ for a positive operator A. The operator B is positive and $B^2 = A$. It is defined uniquely by these properties. $\sqrt{A}$ can be defined as the limit of the sequence of polynomials B_n defined by the recurrence relation

$$B_0 = 0,$$
$$B_{n+1} = B_n + \tfrac{1}{2}(A - B_n^2).$$

The operator A^2 is positive for an arbitrary self-adjoint operator A; therefore, naturally we denote $\sqrt{A^2} = |A|$.

2. *Resolution of the identity. The spectral function.* Functions corresponding to characteristic functions of intervals of the real axis are an important class of functions of operators. Since the square of a characteristic function is equal to itself, the square of the corresponding self-adjoint operator will be equal to itself; that is, the operator will be projective. We denote, in particular, by E_λ the operator corresponding to the characteristic function of the half-axis $(-\infty, \lambda)$ (i.e. to the function equal to zero for $t \geqslant \lambda$ and unity for $t < \lambda$).

A certain set of projective operators $E_\lambda (-\infty < \lambda < \infty)$ is called a *resolution of the identity generated by the operator A* and has these properties:

1) $E_\lambda \leqslant E_\mu$ or, what is the same, $E_\lambda E_\mu = E_\lambda$ for $\lambda < \mu$;

2) E_λ is continuous from the left with respect to λ, that is $E_{\lambda - 0} = \lim\limits_{\mu \to \lambda - 0} E_\mu = E_\lambda$;

3) $E_\lambda = 0$ for $\lambda \in (-\infty, m)$ and $E_\lambda = I$ for $\lambda \in (M, \infty)$, where m and M are the lower and upper bounds of the operator A;

4) the operator E_λ commutes with an arbitrary operator from K_A.

The operator function E_λ is called the *spectral function* of the operator A; the operator $E_\Delta = E_\beta - E_\alpha$ is called the *spectral measure of the interval* $\Delta = [\alpha, \beta]$. This measure has the property of orthogonality: if $\Delta_1 \cap \Delta_2 = 0$, then $E_{\Delta_1} E_{\Delta_2} = 0$.

The operator A can be reconstructed from its spectral function or measure. It can be shown that

$$A = \int_m^{M+0} \lambda \, dE_\lambda,$$

where the integral on the right is an abstract Stieltjes integral.

The *abstract Stieltjes integral*

$$\int_a^b f(\lambda) \, dE_\lambda$$

with respect to spectral measure is understood to be the limit with respect to the operator norm of integral sums

$$\sum_{k=1}^n f(v_k) E_{\Delta_k},$$

where the $\varDelta_k$ are the subintervals into which the interval $[a, b]$ is partitioned and v_k is an arbitrary point inside $\varDelta_k$.

From the spectral representation of the operator follow the formulas

$$Ax = \int\limits_m^{M+0} \lambda \, dE_\lambda x,$$

$$(Ax, x) = \int\limits_m^{M+0} \lambda \, d(E_\lambda x, x),$$

$$\|Ax\|^2 = \int\limits_m^{M+0} \lambda^2 \, d(E_\lambda x, x).$$

3. *Functions of a self-adjoint operator.* The spectral representation of an operator allows the introduction of a broader class of functions of the operator, which includes the functions defined previously. We set

$$f(A) = \int\limits_m^{M+0} f(\lambda) \, dE_\lambda,$$

if the last integral exists. In particular, it exists for an arbitrary continuous function. The correspondence between functions of a real variable and functions of the operator has the following properties:

1) If

$$f(\lambda) = af_1(\lambda) + bf_2(\lambda),$$

then

$$f(A) = af_1(A) + bf_2(A).$$

2) If

$$f(\lambda) = f_1(\lambda) f_2(\lambda),$$

then

$$f(A) = f_1(A) f_2(A).$$

3) $\bar{f}(A) = [f(A)]^*$,

where the $\overline{}$ above the function denotes transition to the complex conjugate function.

4) $\|f(A)\| \leqslant \max\limits_{m \leqslant \lambda \leqslant M} |f(\lambda)|$.

5) It follows from $AB=BA$ that $f(A)B=Bf(A)$ for an arbitrary bounded linear operator B.

6) If $f(\lambda)\leqslant\varphi(\lambda)$ everywhere on $[m, M]$, then $f(A)\leqslant\varphi(A)$.

4. *Unbounded self-adjoint operators.* If A is an unbounded linear operator with an everywhere dense domain $D(A)$ in H, then its adjoint operator A^* will be defined for those elements y for which the functional (Ax, y) is bounded (see ch. I, §5, no. 10). In a Hilbert space, this means that

$$(Ax, y) = (x, y^*),$$

where $y^*\in H$ and $A^*y=y^*$.

An unbounded operator is called *self-adjoint* if $A=A^*$. In distinction from the case of bounded operators, this means not only the presence of the identity

$$(Ax, y) = (x, Ay) \qquad (x, y\in D(A)),$$

but also the coincidence of the domains $D(A)$ and $D(A^*)$ of the operators A and A^*. Thus in order to test the self-adjointness of an operator, it is necessary to show, for every element y for which the functional (Ax, y) is bounded, that $y\in D(A)$ and then test the validity of the preceding identity.

An unbounded self-adjoint operator is always closed.

With some modifications, the basic results of spectral theory stated above for bounded self-adjoint operators remain true for unbounded self-adjoint operators; in particular, the spectral theorem is true. Precisely: let A be an unbounded self-adjoint operator with domain $D(A)$. Then the operator generates a set of projective operators E_λ, $-\infty<\lambda<+\infty$, having the properties:

1) $E_\lambda\leqslant E_\mu$ for $\lambda<\mu$;

2) E_λ is continuous from the left;

3) $E_{-\infty} = \lim\limits_{\lambda\to-\infty} E_\lambda = 0$, $E_{+\infty} = \lim\limits_{\lambda\to+\infty} E_\lambda = I$;

4) $BE_\lambda=E_\lambda B$ if B is an arbitrary bounded operator which commutes with A. In this connection, a bounded operator B is called *permutable* (or *commutative*) with the unbounded operator A if $x\in D(A)$ implies $Bx\in D(A)$ and $ABx=BAx$.

The element x belongs to $D(A)$ if and only if

$$\int_{-\infty}^{\infty} \lambda^2 \, d\,(E_\lambda x, x) < \infty .$$

For these elements x,

$$Ax = \int_{-\infty}^{\infty} \lambda \, dE_\lambda x \quad \text{and} \quad \|Ax\|^2 = \int_{-\infty}^{\infty} \lambda^2 \, d\,(E_\lambda x, x).$$

The integral $\displaystyle\int_{\infty}^{\infty} \lambda dE_\lambda x$ is understood to be the limit of the proper

integral $\displaystyle\int_{a}^{b} \lambda \, dE_\lambda x$ in the sense of strong convergence as $a \to -\infty$, $b \to \infty$.

A self-adjoint operator A is called *semi-bounded below* if $(Ax, x) \geqslant a(x, x)$ for all $x \in D(A)$. In this case

$$Ax = \int_{a}^{\infty} \lambda \, dE_\lambda x .$$

An operator which is *semi-bounded above* is defined analogously.

If the function $f(\lambda)$ is finite and measurable with respect to all measures generated by the functions $\sigma(\lambda) = (E_\lambda z, z)$ $(z \in H)$, then an operator $f(A)$ can be defined. This operator, generally speaking, is not bounded. Its domain $D(f(A))$ is the collection of elements x for which

$$\int_{-\infty}^{\infty} f^2(\lambda) \, d\,(E_\lambda x, x) < \infty .$$

The set $D(f(A))$ is dense in H; the operator $f(A)$ is given by the formula

$$(f(A)x, y) = \int_{-\infty}^{\infty} f(\lambda) \, d\,(E_\lambda x, y)$$

$$\left(x \in D(f(A)), y \in H\right)$$

and is self-adjoint (for a real function $f(\lambda)$).

If the function $f(\lambda)$ is bounded $(-\infty < \lambda < \infty)$, then the operator $f(A)$ will also be bounded.

The resolvent is an important example of a bounded function of an operator. If λ_0 does not belong to the spectrum of the operator A, then the spectral representation

$$R_{\lambda_0}x = \int_{-\infty}^{\infty} \frac{1}{\lambda - \lambda_0} dE_\lambda x$$

is valid for the resolvent R_{λ_0}. Hence, in particular, we have the following bound for the resolvent:

$$\|R_{\lambda_0}x\| \leqslant \frac{1}{d}\|x\|,$$

where d is the distance from the point λ_0 to the spectrum of the operator A. In this connection $d \geqslant |\mathrm{Im}\,\lambda_0|$ and, hence, $\|R_{\lambda_0}\| \leqslant \dfrac{1}{|\mathrm{Im}\,\lambda_0|}$.

If the operator A is semi-bounded below, then the function e^{-A}:

$$e^{-A}x = \int_{a}^{\infty} e^{-\lambda} dE_\lambda x$$

will be a bounded operator.

The collection of all functions of a self-adjoint operator A allows an "extrinsic" description. If the collection of all bounded linear operators commuting with A is denoted by $R(A)$, then the set of functions of A coincides with the collection of all closed operators commuting with an arbitrary operator from $R(A)$.

5. *Spectrum of a self-adjoint operator.*   The spectrum of a self-adjoint operator A is a closed set on the real axis, consisting of all points of increase of the function E_λ. Jumps of the function E_λ correspond to eigenvalues of the operator A; the operator $E_{\lambda+0} - E_{\lambda-0}$ is the projector onto the eigenspace corresponding to the eigenvalue λ. The eigenvalues form the *discrete* or *point* spectrum of the operator A.

If the eigenvectors of the operator A form a complete system in the space H, then we say that the operator has a *pure point* spectrum. In this case, the spectrum of the operator consists of the set of eigenvalues and limit points of this set.

In the general case, the space H can be decomposed into the orthogonal sum of subspaces H_1 and H_2 invariant with respect to A where the operator A has a pure point spectrum in H_1 and does not have eigenvectors in H_2. The spectrum of the operator A in the subspace H_2 is called the *continuous spectrum*. The continuous spectrum and the point spectrum can intersect.

The points of the continuous spectrum, the limit points of the set of eigenvalues, and the eigenvalues with an infinite multiplicity form the *limit spectrum* of the operator A.

The limit spectrum of a self-adjoint operator consists of the single point 0 only in the case when the operator is completely continuous.

Example. In the space $L_2[0, 1]$, the integral operator with symmetric kernel $K(t, \tau)$, having the property that

$$\int_0^1 |K(t, \tau)|^2 \, d\tau$$

exists for almost all $t \in [0, 1]$, generates a self-adjoint operator, the so-called *Carleman operator*. This operator can be unbounded. The point 0 always is a point of the limit spectrum of the Carleman operator.

In order for the point λ_0 to be a point of the spectrum of the operator A, it is necessary and sufficient that a sequence of elements $x_n \in D(A)$ with $\|x_n\| = 1$ exist such that $\|Ax_n - \lambda_0 x_n\| \to 0$. In order for λ_0 to be a point of the limit spectrum, it is necessary and sufficient that a sequence of elements x_n exist, weakly converging to zero and having the preceding properties.

The addition to a self-adjoint operator of a completely continuous operator does not change the limit spectrum of the operator. On the other hand, a completely continuous operator with arbitrarily small norm can be annexed to an arbitrary self-adjoint operator such that its spectrum becomes purely point.

6. *Theory of perturbations.* The last assertion of the preceding subsection can be applied to the theory of perturbations which studies the change of spectral properties for small changes of the operators.

Let a set of self-adjoint operators $A(\varepsilon)$ depending on the parameter ε be given, and let D be the set of x for which the limit

$$\lim_{\varepsilon \to 0} A(\varepsilon) x = A_0 x$$

exists. (It is assumed that $x \in D(A(\varepsilon))$ for $0 < \varepsilon < \varepsilon_0(x)$.) If the self-adjoint operator A is the closure of the operator A_0, then the relation

$$E_\lambda = \lim_{\varepsilon \to 0} E_\lambda(\varepsilon)$$

is valid for the spectral functions $E_\lambda(\varepsilon)$ and E_λ of the operators $A(\varepsilon)$ and A and for an arbitrary λ not belonging to the point spectrum of the operator A. The limit is understood in the strong sense. Uniform convergence of $E_\lambda(\varepsilon)$ to E_λ (with respect to the operator norm) under the indicated conditions, generally speaking, does not hold. Moreover, it may not hold even if it is required that the operators $A(\varepsilon)$ be bounded and uniformly convergent to the operator A.

If the new norm $\|x\|_1 = \|x\| + \|Ax\|$ is introduced in the domain $D(A)$ of the self-adjoint operator A, then $D(A)$, with this norm, will be a Banach space H_1 (see ch. I, §5, no. 10). If all the operators $A(\varepsilon)$ are defined on $D(A)$ and converge to A uniformly with respect to the norm $\|x\|_1$:

$$\|A(\varepsilon)x - Ax\| \leqslant C_\varepsilon \|x\|_1 \qquad (x \in D(A)),$$

where $C_\varepsilon \to 0$ as $\varepsilon \to 0$, then the spectral function $E_\lambda(\varepsilon)$ converges uniformly to the function E_λ at an arbitrary point λ not belonging to the spectrum of the operator A; that is

$$\lim_{\varepsilon \to 0} \|E_\lambda(\varepsilon) - E_\lambda\| = 0.$$

If λ_0 is an isolated point of the spectrum which is an eigenvalue of finite multiplicity m and Δ is an interval separating it from the remaining part of the spectrum, then under the preceding conditions for a sufficiently small ε the spectrum of the operator $A(\varepsilon)$ in the interval Δ consists of m eigenvalues (taking into account their multiplicity). These eigenvalues $\lambda_k(\varepsilon)$ $(k = 1, 2, ..., m)$ tend to the point λ_0 as $\varepsilon \to 0$. However, we must keep in mind that although $E_\Delta(\varepsilon)$ converges uniformly to E_Δ, the eigenvectors $e_k(\varepsilon)$, corresponding to the eigenvalues $\lambda_k(\varepsilon)$, may not have a limit as $\varepsilon \to 0$. If λ_0 is a simple eigenvalue, then the eigenvectors $e(\varepsilon)$ of the operators $A(\varepsilon)$ approach an eigenvector e of the operator A.

The operator $A(\varepsilon)$ is called an *analytic function of ε* if

$$A(\varepsilon) = A + \varepsilon A_1 + \varepsilon^2 A_2 + \cdots,$$

where the operators A_i and A act from H_1 to H, $D(A_i) = D(A) = H_1$, and the series converges with respect to the operator norm. Then $E_\lambda(\varepsilon)$ is also an analytic function of ε in a neighborhood of $\varepsilon = 0$ for every λ not belonging to the spectrum of the operator A.

For the above considered case of an isolated eigenvalue λ_0 of multiplicity m,

$$\lambda_k(\varepsilon) = \lambda_0 + \varepsilon \lambda_k^{(1)} + \varepsilon^2 \lambda_k^{(2)} + \cdots$$

and

$$e_k(\varepsilon) = e_k + \varepsilon e_k^{(1)} + \varepsilon^2 e_k^{(2)} + \cdots.$$

Let the operator A have a complete system of eigenvectors $\{e_n\}$ with corresponding eigenvalues λ_n.

If λ_n is an isolated simple eigenvalue of the operator A, then formulas for the determination of the coefficients of the expansion in powers of ε of the eigenvalue $\lambda_n(\varepsilon)$ of the operator $A(\varepsilon) = A + A_1 \varepsilon$ can be obtained. Here, only formulas or first and second approximations are mentioned:

$$\lambda_n(\varepsilon) = \lambda_n + \varepsilon \lambda_n^{(1)} + \varepsilon^2 \lambda_n^{(2)} + \cdots,$$

where

$$\lambda_n^{(1)} = (A_1 e_n, e_n) \quad \text{and} \quad \lambda_n^{(2)} = {\sum_m}' \frac{|(A_1 e_n, e_m)|^2}{\lambda_n - \lambda_m}$$

(the prime on the summation sign means that the term for $m = n$ is omitted).

The expansion

$$e_n(\varepsilon) = e_n + \varepsilon e_n^{(1)} + \varepsilon^2 e_n^{(2)} + \cdots,$$

where $e_n^{(1)} = {\sum_m}' \dfrac{(A_1 e_n, e_m)}{\lambda_n - \lambda_m} e_m,$

$$e_n^{(2)} = {\sum_m}' {\sum_k}' \frac{(A_1 e_m, e_k)(A_1 e_k, e_n)}{(\lambda_n - \lambda_k)(\lambda_n - \lambda_m)} e_m -$$
$$- {\sum_m}' \frac{(A_1 e_n, e_n)(A_1 e_m, e_n)}{(\lambda_n - \lambda_m)^2} e_m - \tfrac{1}{2} e_n \sum \frac{|(A_1 e_n, e_m)|^2}{(\lambda_n - \lambda_m)^2}$$

is valid for the eigenvector $e_n(\varepsilon)$.

In physics, these formulas are called the *formulas of perturbation theory*.

If the operator A has parts of continuous spectrum, then analogous formulas occur where, besides the sums, integrals occur.

In the case of an eigenvalue of multiplicity m, in order to obtain the coefficients of the powers of ε, one must find eigenvalues and eigenfunctions of m-dimensional operators.

Problems of perturbation theory are particular cases of the more general problem of the study of the behavior of the function $f(A(\varepsilon))$ as ε varies, where $f(\lambda)$ is a given function. The function $E_\lambda(\varepsilon)$ is precisely a function of such type (see no. 2). For the expansion in powers of ε of such functions it is natural to apply Taylor's formula, assuming the functions $f(\lambda)$ and $A(\varepsilon)$ are sufficiently smooth. Then

$$f(A(\varepsilon)) = f(A) + \varepsilon \frac{df(A(\varepsilon))}{d\varepsilon}\bigg|_{\varepsilon=0} + \varepsilon^2 \frac{d^2f(A(\varepsilon))}{d\varepsilon^2}\bigg|_{\varepsilon=0} + \cdots$$

There are special formulas for the derivatives of functions of operators with respect to a parameter. Here, only the formula for the first derivative is mentioned. It is valid under the assumption that the operator A has a finite absolute norm (see ch. I, § 2, no. 1).

If $x = \Sigma_k c_k e_k$, then

$$\left(\frac{df(A(\varepsilon))}{d\varepsilon}\bigg|_{\varepsilon=0}\right)x = \sum_k \sum_m c_k \frac{f(\lambda_m) - f(\lambda_k)}{\lambda_m - \lambda_k}\left(\left(\frac{dA}{d\varepsilon}\right)_{\varepsilon=0} e_k, e_m\right) e_m,$$

where it is assumed that

$$\frac{f(\lambda_m) - f(\lambda_k)}{\lambda_m - \lambda_k} = f'(\lambda_k) \quad \text{for} \quad m = k.$$

7. *Multiplicity of the spectrum of a self-adjoint operator.* The spectrum of a self-adjoint operator A is called *simple* if an element $u \in H$ exists such that the closed linear hull of all elements of the form $E_\Delta u$, where Δ is an arbitrary interval of the real axis, coincides with H. In this case, u is called a *generating* element.

The formulas

$$x = \int_{-\infty}^{\infty} f(\lambda) dE_\lambda u, \qquad y = \int_{-\infty}^{\infty} g(\lambda) dE_\lambda \mu$$

and

$$(x, y) = \int_{-\infty}^{\infty} f(\lambda)\overline{g(\lambda)}\, d\sigma(\lambda),$$

are valid for arbitrary x and $y \in H$, where

$$\sigma(\lambda) = (E_\lambda u, u)$$

and $f(\lambda)$ and $g(\lambda)$ are certain functions with square-integrable moduli with respect to the measure $\sigma(\lambda)$. The function $\sigma(x)$ is a non-decreasing function of bounded variation on $(-\infty, \infty)$ and is called the *spectral function of the operator A.*

It can be shown that, to an arbitrary function $f(\lambda)$ $(-\infty<\lambda<\infty)$ with a square-integrable modulus with respect to the measure $\sigma(\lambda)$, there corresponds some element x for which

$$x = \int\limits_{-\infty}^{\infty} f(\lambda)\, dE_\lambda u .$$

Thus, the last formula establishes an isometric correspondence between the space H and the space $L_{2,\sigma}$ of all functions $f(\lambda)$ on the axis $(-\infty, \infty)$ for which

$$\|f\|_{L_{2,\sigma}} = \int\limits_{-\infty}^{\infty} |f(\lambda)|^2\, d\sigma(\lambda) < \infty .$$

The formula

$$Ax = \int\limits_{-\infty}^{\infty} \lambda f(\lambda)\, dE_\lambda u$$

is valid for the operator A and, hence, under the isometric correspondence it maps into the operator Λ of multiplication by the independent variable λ:

$$\Lambda f(\lambda) = \lambda f(\lambda),$$

defined for all the functions $f(x) \in L_{2,\sigma}$, for which $\lambda f(\lambda) \in L_{2,\sigma}$.

A collection of elements $u_1, u_2, ..., u_n$ is called a *generating basis* for the operator A if the closed linear hull of the set of all elements $E_A u_k$ $(k=1, 2, ..., n)$ coincides with H. The spectrum of the operator A is called *n-multiple* if the minimal number of elements in a generating basis for the operator A is equal to n. The corresponding basis is called a *minimal* generating basis.

Numerous examples of self-adjoint operators with a finite-multiple spectrum are given by ordinary differential operators (see § 5).

If $u_1, ..., u_n$ is a minimal generating basis for the operator A, then the formulas

$$x = \sum_{k=1}^{n} \int_{-\infty}^{\infty} f_k(\lambda)\, dE_\lambda u_k, \qquad y = \sum_{k=1}^{n} \int_{-\infty}^{\infty} g_k(\lambda)\, dE_\lambda u_k$$

and

$$(x, y) = \sum_{i, j=1}^{n} \int_{-\infty}^{\infty} f_i(\lambda)\, \overline{g_j(\lambda)}\, d\sigma_{ij}(\lambda),$$

where

$$\sigma_{ij}(\lambda) = (E_\lambda u_i, u_j),$$

are valid. The matrix $\sigma(\lambda) = (\sigma_{ij}(\lambda))$ is Hermitian for every $\lambda\,(-\infty < \lambda < \infty)$ and is continuous from the left, and the difference $\sigma(\mu) - \sigma(\lambda)$ for $\mu > \lambda$ is a non-negative definite matrix. The space H is isometric to the Hilbert space $L_{2,\sigma}$ of the vector functions $f(\lambda) = \{f_1(\lambda), \ldots, f_m(\lambda)\}\ (-\infty < \lambda < \infty)$, for which

$$\|f\|_{L_{2,\sigma}} = \sum_{i, j=1}^{n} \int_{-\infty}^{\infty} f_i(\lambda)\, \overline{f_j(\lambda)}\, d\sigma_{ij}(\lambda) < \infty,$$

and the scalar product is introduced by the formula

$$(f, g) = \sum_{i, j=1}^{n} \int_{-\infty}^{\infty} f_i(\lambda)\, \overline{g_j(\lambda)}\, d\sigma_{ij}(\lambda),$$

where the integrals are understood in a particular sense (see [37]).

Under the isometric correspondence, the operator A transforms again into the operator of multiplication of all components of the vector function $f(\lambda)$ by the independent variable λ:

$$Ax = \sum_{k=1}^{n} \int_{-\infty}^{\infty} \lambda f_k(\lambda)\, dE_\lambda u_k.$$

In the general case, for a self-adjoint operator acting in a separable Hilbert space H, the space can be represented in the form of an orthogonal sum of subspaces $H_k\,(k = 1, 2, \ldots)$ such that each subspace H_k is invariant with respect to the operator A and the operator A has simple spectrum in it.

In conclusion, it remains to remark that sometimes it is convenient to use non-eigenvectors as generating elements (see [37]). In this case all the formulas are unchanged, but the functions $\sigma(\lambda)$ need not be of bounded variation on $(-\infty, \infty)$.

8. *Generalized eigenvectors.* In § 2, no. 5 it was remarked that if A is a completely continuous self-adjoint operator, then its eigenvectors $e_k (k=1, 2, ...)$ form a basis in the space H; that is,

$$x = \sum_{k=1}^{\infty} c_k e_k$$

for every $x \in H$.

The formula

$$x = \int_{-\infty}^{\infty} dE_\lambda x$$

is a generalization of the preceding for the case of an arbitrary self-adjoint operator. A more natural generalization would be the formula

$$x = \int_{-\infty}^{\infty} e_\lambda d\varrho(\lambda),$$

where e_λ is an element of H satisfying the equation $Ae_\lambda = \lambda e_\lambda$ (an eigenvector or θ), and the weight $d\varrho(\lambda)$ plays the role of the coefficients c_k in the series expansion. However, the simplest non-completely continuous self-adjoint operators in a Hilbert space, such as the operator of multiplication by x in $L_2(a, b)$ or the operator of differentiation in $L_2(-\infty, \infty)$, do not have eigenvectors in these spaces. In fact, if the relation $xy(x) = \lambda y(x)$ is satisfied for some function $y(x) \in L_2(a, b)$, then the function $y(x)$ must be equal to zero for $x \neq \lambda$ and can be different from zero only for $x = \lambda$. But in the space $L_2(a, b)$ there is no non-zero element having this property. Nevertheless, the operator of multiplication by x has eigenfunctions, namely delta functions $\delta(x-\lambda)$ which are generalized functions (see ch. VIII, §1) and do not belong to $L_2(a, b)$.

The examples mentioned suggest seeking expansions in eigenvectors not belonging to the space H. The difficulty, consisting in having available only concepts connected with the space H to construct elements not belonging to it, is overcome in the following manner: a more restricted linear topological or Banach or Hilbert space Φ is constructed in the initial Hilbert space H. A topology is introduced in Φ so that the functionals (φ, h) $(h \in H)$ are continuous on Φ; then the space H is embedded in the larger space Φ^* in which eigenvectors of the operator A are sought. Eigenvectors of the operator A belonging to Φ^* and not belonging to H are called *generalized eigenvectors.*[*])

It turns out that a space Φ^* can be constructed with respect to the space H so that every self-adjoint operator in H has a complete system of eigenvectors in Φ^*. In the case of a self-adjoint operator A with a simple spectrum and the generating element u, the expansion with respect to the generalized eigenvectors has the form

$$\varphi = \int_{-\infty}^{\infty} g(\lambda) e_\lambda \, d\sigma(\lambda)$$

for arbitrary $\varphi \in \Phi$, where $\sigma(\lambda) = (E_\lambda u, u)$ and the function $g(\lambda)$ is defined by the equality

$$g(\lambda) = (\varphi, e_\lambda).$$

The last formulas are analogous to the inversion formulas in the theory of the Fourier transform where, the role of e_λ is played by the function $e^{i\lambda x}$ and $\sigma(\lambda) = \lambda$ (see § 2, no. 2). The equation

$$\|\varphi\|^2 = (\varphi, \varphi) = \int_{-\infty}^{\infty} |g(\lambda)|^2 \, d\sigma_\lambda = \int_{-\infty}^{\infty} |(\varphi, e_\lambda)|^2 \, d\sigma_\lambda$$

is a valid analogue of the Parseval equality. For a n operator with an arbitrary spectrum, the formulas acquire a more complicated form:

$$\varphi = \sum_{i=1}^{\infty} \int_{-\infty}^{\infty} (\varphi, e_\lambda^{(i)}) e_\lambda^{(i)} \, d\sigma_\lambda^{(i)}$$

and

$$\|\varphi\|^2 = \sum_{i=1}^{\infty} \int_{-\infty}^{\infty} |(\varphi, e_\lambda^{(i)})|^2 \, d\sigma_\lambda^{(i)}.$$

The spectrum of an operator A is called a *Lebesgue* spectrum if the func-

---

[*]) The triple of spaces Φ, H and Φ^* ($\Phi \subset H \subset \Phi^*$) is called *equipped Hilbert space*.

tions $\sigma(\lambda)$ and λ are equivalent, that is $\sigma(\lambda)$ and λ are absolutely continuous with respect to one another. In this case $d\sigma(\lambda)$ can be replaced in all formulas by $\varrho(\lambda)d\lambda$ where $\varrho(\lambda)$ is a function which is summable on an arbitrary finite interval.

§ 4. Symmetric operators

1. *Concept of a symmetric operator, deficiency indices.* A linear operator A with an everywhere dense domain $D(A)$ is called *symmetric* if

$$(Ax, y) = (x, Ay)$$

for arbitrary $x, y \in D(A)$.

Every self-adjoint operator is symmetric, but the converse does not hold. The domain of definition of the operator A^*, adjoint to the symmetric operator A, can be larger than the domain of definition of the operator A. On $D(A)$, obviously, $Ax = A^*x$; therefore, the operator A^* is an extension of the operator A.

A symmetric operator always allows closure, and its closure is again a symmetric operator. *If a symmetric operator is defined on the entire space, then it is bounded.* If the range of a symmetric operator coincides with the entire space, then it is self-adjoint.

The point λ_0 is called a *point of regular type* for the operator A if

$$\|Ax - \lambda_0 x\| \geqslant k \|x\|, \qquad k > 0,$$

for all $x \in D(A)$. In other words, this means that the operator $A - \lambda_0 I$ has a bounded left inverse. If, besides this, the operator A is closed, then the range $\mathfrak{M}_{\lambda_0}$ of the operator $A - \lambda_0 I$ will be a closed set. If $\mathfrak{M}_{\lambda_0}$ coincides with the entire space, then the point λ_0 will be a regular point of the operator A. The orthogonal complement $\mathfrak{N}_{\lambda_0}$ of the subspace $\mathfrak{M}_{\lambda_0}$ is called the *deficiency subspace*. The dimension n_{λ_0} of the deficiency subspace $\mathfrak{N}_{\lambda_0}$ is called the *deficiency index* of the operator A at the point λ_0.

If a connected set of points of regular type is given for a symmetric operator, then the deficiency index is the same at all the points of this set. For a symmetric operator, all non-real numbers are points of regular type; therefore, the deficiency index n_+ will be the same for all the points of the upper half-plane. Analogously, the deficiency index n_- is the same for all points of the lower half-plane. If there is at least one point of regular type on the real axis, then $n_+ = n_-$.

A closed symmetric operator will be self-adjoint if and only if its deficiency indices are equal to zero. The pair of (finite or infinite) numbers (n_+, n_-) shows the degree of deviation of the symmetric operator from a self-adjoint operator.

One must note that the deficiency subspaces $\mathfrak{N}_{\lambda_0}$ consist of all the solutions of the equation

$$A^* y = \bar{\lambda}_0 y.$$

Thus the deficiency index n_{λ_0} coincides with the number of linearly independent solutions of this equation.

2. *Self-adjoint extensions of symmetric operators.* The question arises whether every symmetric operator can be extended to a self-adjoint operator. The answer follows: *in order for a symmetric operator to be extendible to a self-adjoint operator, it is necessary and sufficient that the deficiency indices n_+ and n_- of the operator be identical.*

As was indicated above, this will hold, for example, if there are points of regular type on the real axis.

One must note that it is a question here of the extension of the operator in an initial Hilbert space H. If the deficiency indices of the operator A are not identical, then the space H can be extended to a larger Hilbert space H_1 such that the deficiency indices become equal in the larger space. In this larger space, self-adjoint extensions of the operator A will exist.

Let A_0 be a closed symmetric operator. Every symmetric and, in particular, self-adjoint extension of the operator A_0 is a restriction of the operator A_0^*. Therefore, in the construction of such an extension, the question does not arise of how we define it on new elements, but only in what its domain of definition is; that is, on what elements of $D(A_0^*)$ it is defined. In order to find self-adjoint extensions of the operator A_0, it is necessary to find linear subsets in $D(A_0^*)$, containing $D(A)$, on which the operator A_0^* generates a self-adjoint operator.

It turns out that the set $D(A_0^*)$ has the following structure:

$$D(A^*) = D(A) \oplus \mathfrak{N}_\lambda \oplus \mathfrak{N}_{\bar{\lambda}},$$

where λ is some non-real number. The sum on the right is direct, that is, an arbitrary element $y \in D(A^*)$ is representable uniquely in the form

$$y = x + z_\lambda + z_{\bar\lambda},$$

where $x \in D(A)$, $z_\lambda \in \mathfrak{N}_\lambda$ and $z_{\bar\lambda} \in \mathfrak{N}_{\bar\lambda}$.

If the deficiency indices n_+ and n_- are equal, then an arbitrary self-adjoint extension $\tilde A$ of the operator A_0 can be constructed in the following manner: select some linear operator V which isometrically maps the space $\mathfrak{N}_\lambda$ onto $\mathfrak{N}_{\bar\lambda}$. Then the domain $D(\tilde A)$ of the operator $\tilde A$ will consist of all elements of the form

$$y = x + z_\lambda + V z_\lambda,$$

where $x \in D(A)$ and $z_\lambda \in \mathfrak{N}_\lambda$.

As already remarked above, the values of the operator $\tilde A$ will coincide with the values of the operator A^*, that is,

$$\tilde A y = A_0 x + \bar\lambda z_\lambda + \lambda V z_\lambda.$$

If the deficiency indices are not equal, for example, $n_+ < n_-$, then the construction mentioned gives all the maximal symmetric extensions of the operator A_0, that is, those symmetric extensions which cannot be extended further with retention of symmetry.

The method of constructing self-adjoint extensions described here is due to J. von Neumann. Practically, it is not very effective, since it requires finding solutions of the equation $A^* y = \bar\lambda y$ and constructing of an isometric operator V.

In the following subsection other methods for the construction of self-adjoint extensions are considered.

3. *Self-adjoint extensions of semi-bounded operators.* A symmetric operator A_0 is called *semi-bounded below* if for an arbitrary $x \in D(A_0)$

$$(A_0 x, x) \geqslant a(x, x).$$

All real numbers not exceeding the number a will be points of regular type; therefore, the deficiency indices of the operator A_0 are identical.

For the construction of self-adjoint extensions of the operator A_0, without loss of generality, we can assume it to be *positive definite*, that is, $a > 0$. Otherwise, we can consider the operator $A_0 + kI$ with sufficiently large positive k. If $A_0 + kI$ is a self-adjoint extension of this operator, then $A_0 + kI - kI$ is a self-adjoint extension of the operator A_0.

Let the operator A_0 be positive definite. In its domain $D(A_0)$, a new scalar product can be introduced by the formula

$$[x, y] = (A_0 x, y).$$

The completion of $D(A_0)$ with respect to the norm generated by this scalar product will be a Hilbert space H_0. It turns out that the elements of the completion are identified naturally with some elements from H, and therefore H_0 can be considered as a linear subset of the space H. On the intersection of this subset H_0 with the domain of the adjoint operator $D(A_0^*)$, the operator A_0^* is self-adjoint. Thus a self-adjoint extension A_μ of the operator A_0 is obtained which is the restriction of the adjoint operator A_0^* to $D(A_\mu) = H_0 \cap D(A_0^*)$. The operator A_μ is called the *Friedrichs* or *strict extension* of the operator A_0. The operator A_μ is positive definite and has the same lower bound as the operator A_0:

$$(A_\mu x, x) \geqslant a(x, x).$$

The strict extension A_μ is the simplest. For its construction nothing needs to be known besides the form $(A_0 x, x)$ generated by the operator A_0. In connection with this, the method of Friedrichs for the construction of self-adjoint extensions is one of basic importance in the theory of partial differential equations. For a more detailed description of the domain of definition of the strict extension, it is necessary to study in more detail the nature of convergence in the norm $\sqrt{(A_0 x, x)}$ on $D(A_0)$ and the structure of the domain of definition of the operator A_0^* (see § 5 and § 6).

The set H_0 is the domain of definition of the square root of the operator A_μ:

$$H_0 = D(A_\mu^{\frac{1}{2}}).$$

For an arbitrary positive self-adjoint extension $\tilde{A}$ of the operator A_0, the domain $D(\tilde{A}^{\frac{1}{2}})$ contains the set H_0. The strict extension has the following extremal property: for an arbitrary self-adjoint positive definite extension $\tilde{A}$ of the operator A_0

$$(A_\mu^{-1} x, x) \leqslant (\tilde{A}^{-1} x, x).$$

The operator $B = \tilde{A}^{-1} - A_\mu^{-1}$, which by virtue of the preceding is a bounded positive self-adjoint operator, plays an essential role. On $R(A_0)$ the operator B is equal to zero and, hence, we can consider it as an operator acting in the orthogonal complement U of $R(A_0)$:

$$H = R(A_0) \oplus U \qquad \text{and} \qquad BU \subset U.$$

The subspace U is the deficiency subspace $\mathfrak{N}_0$, consists of all solutions of the equation $A^*u = 0$ in H, and is called the *null space* of the operator A_0^*.

The following assertions are valid:

1) *The domain of definition of the adjoint operator A_0^* decomposes into a direct sum*:

$$D(A_0^*) = D(A_0) \oplus A_\mu^{-1} U \oplus U.$$

2) *The domain of definition of an arbitrary positive definite self-adjoint extension A of the operator A_0 decomposes into a direct sum*:

$$D(A) = D(A_0) \oplus (A_\mu^{-1} + B) U,$$

where B *is some bounded self-adjoint positive operator acting in the subspace U.*

3) *For an arbitrary operator B having the properties described above, the restriction of the operator A_0^* to the set $D(A_0) \oplus (A_\mu^{-1} + B) U$ is a self-adjoint positive definite operator.*

Thus, knowledge of the strict extension A_μ allows the reduction of the description of an arbitrary positive definite self-adjoint extension to the description of the operator B. The operator B acts in a smaller space than H, the space U. In the theory of boundary value problems in partial differential equations, the subspace U is mapped one-to-one onto some space of functions defined on the boundary of the region, and the operator B is connected with the boundary conditions.

By means of the operator B, we can describe the structure of the domain of definition of the square root of an arbitrary self-adjoint positive definite extension A of the operator A_0:

$$D(A^{\frac{1}{2}}) = D(A_\mu^{\frac{1}{2}}) \oplus R(B^{\frac{1}{2}}) = H_0 \oplus R(B^{\frac{1}{2}}).$$

The importance of the theory of semi-bounded symmetric operators is illustrated by the following example. Let a self-adjoint differential expression of order $2m$ be given in an n-dimensional region G of Euclidean space with sufficiently smooth boundary:

$$Lu = (-1)^m \sum_{|\alpha|=|\beta|=m} D^\alpha (a_{\alpha\beta} D^\beta u),$$

where

$$\alpha = (\alpha_1, ..., \alpha_n), \quad |\alpha| = \alpha_1 + \alpha_2 + \cdots + \alpha_n, D^\alpha = \frac{\partial^{|\alpha|}}{\partial x_1^{\alpha_1} ... \partial x_n^{\alpha_n}}$$

and β, $|\beta|$ and D^β are defined analogously. The coefficients $a_{\alpha\beta}$ are assumed to be sufficiently smooth functions of $x=(x_1 ..., x_n)$, and $a_{\alpha\beta}=a_{\beta\alpha}$. The expression L is called *elliptic* if the inequality

$$\sum_{|\alpha|=|\beta|=m} a_{\alpha\beta} \xi_1^{\alpha_1} ... \xi_n^{\alpha_n} \xi_1^{\beta_1} ... \xi_n^{\beta_n} \geqslant \lambda \sum_{k=1}^n \xi_k^{2m}$$

is valid for arbitrary real $\xi_1, ..., \xi_n$ where $\lambda > 0$ and is independent of $x \in G$.

The operator L_0, defined by the equality $L_0 u = Lu$ on the set $D(L_0)$ of all *finitary* functions $u(x)$ (that is, infinitely differentiable functions equal to zero near the boundary of the region G) is symmetric and semi-bounded in the space $H = L_2(G)$. Moreover, constants $c > 0$ and k exist such that

$$(A_0 u, u) = \int_G (Lu \cdot u + ku^2) \, dx \geqslant c \left[\int_G \sum_{|\alpha|=m} |D^\alpha u|^2 \, dx + \int_G |u|^2 \, dx \right]$$

for the operator $A_0 = L_0 + kI$. The metric introduced by means of the form $(A_0 u, u)$ on $D(L_0)$ turns out to be equivalent to the metric of the Sobolev space W_2^m. The space H_0 is a subspace $\mathring{W}_2^m$ of the space W_2^m. A solution of the equation

$$A_\mu u = f,$$

where A_μ is the strict extension of the operator A_0 and $f \in L_2(G)$, is called a *generalized solution* of the first boundary value problem for the equation $Lu + ku = f$.

The theory of extension is illustrated in more detail by the example of the elliptic expression of second order in § 6.

4. *Dissipative extensions.* A linear operator B with an everywhere dense domain $D(B)$ is called *dissipative* if

$$\mathrm{Re} \, (Bx, x) \leqslant 0$$

for arbitrary $x \in D(B)$.

If, in the preceding relations, the equality sign holds for all $x \in D(B)$, then the operator is called *conservative*.

If A_0 is a symmetric operator, then for the operator $B_0 = iA_0$ we have that

$$\text{Re}(B_0 x, x) = \text{Re}(iA_0 x, x) = 0,$$

and, hence, the operator B_0 is conservative. In a number of problems, the question of the construction of dissipative extensions of the operator B_0 arises; the most interesting are the *maximal dissipative extensions*, that is, those which cannot be extended further with retention of dissipativeness.

The domain of definition of the adjoint operator B_0^* can be decomposed into the direct sum

$$D(B_0^*) = D(B_0) \oplus V_+ \oplus V_-,$$

where $V_{\pm}$ is the collection of all solutions of the equation

$$B_0^* v = \pm v.$$

The decomposition is orthogonal with respect to the scalar product

$$[x, y] = (A_0^* x, A_0^* y) + (x, y) \quad (x, y \in D(A_0^*) = D(B_0^*)).$$

We can indicate the general form of maximal dissipative extensions of the operator B_0 which are restrictions of the operator B_0^*. To this end it is sufficient to describe their domain of definition.

For an arbitrary maximal dissipative extension $B \subset B_0^*$ of the operator B_0, the domain of definition has the form

$$D(B) = D(B_0) \oplus (I + C) V_-,$$

where C is a contractive operator (i.e. $\|C\| \leqslant 1$) acting from V_- to V_+. For any such operator C, the operator $Bx = B_0^* x$ on the set indicated is a maximal dissipative extension of the operator B_0.

§ 5. Ordinary differential operators

1. *Self-adjoint differential expressions.*   An expression of the form

$$l(y) = q_0(x) y^{(n)} + q_1(x) y^{(n-1)} + \cdots + q_n(x) y$$

with real coefficients $q_i(x)$ $(i = 0, 1, \ldots, n)$ is called an *ordinary linear differential expression* of n-th order.

The expression

$$l^*(y) = (-1)^n (q_0 y)^{(n)} + (-1)^{n-1} (q_1 y)^{(n-1)} + \cdots + q_n y$$

is called the *adjoint* differential expression.

The expression $l(y)$ is called *self-adjoint* if $l(y) \equiv l^*(y)$.

Every self-adjoint differential expression with coefficients differentiable a sufficient number of times is representable in the form

$$l(y) = (-1)^n (p_0 y^{(n)})^{(n)} + (-1)^{n-1} (p_1 y^{(n-1)})^{(n-1)} + \cdots + p_n y.$$

In the sequel it is assumed that the coefficients $p_i(x)$ $(i = 0, 1, ..., n)$ are defined on a finite or infinite interval $[a, b]$ and have continuous derivatives of orders $n - i$ on $[a, b]$. Besides this, it is assumed that the function $\dfrac{1}{p_0(x)}$ is summable on every finite segment $[\alpha, \beta] \subset (a, b)$.

It is convenient to call the expressions defined by the formulas

$$y^{[0]} = y,$$

$$y^{[k]} = \frac{d^k y}{dx^k} \qquad \text{for} \quad k = 1, 2, ..., n-1,$$

$$y^{[n]} = p_0 \frac{d^n y}{dx^n},$$

$$y^{[n+k]} = p_k \frac{d^{n-k} y}{dx^{n-k}} - \frac{d}{dx}(y^{[n+k-1]}) \quad \text{for} \quad k = 1, 2, ..., n$$

quasi-derivatives corresponding to $l(y)$ *of the function* y.

It follows from the definition that

$$l(y) = y^{[2n]}.$$

On every segment $[\alpha, \beta] \subset (a, b)$ the *Lagrange identity*

$$\int_\alpha^\beta l(y) \bar{z} \, dx - \int_\alpha^\beta y l(\bar{z}) \, dx = [y, z] \Big|_\alpha^\beta$$

is valid, where

$$[y, z] = \sum_{k=1}^n \left\{ y^{[k-1]} \bar{z}^{[2n-k]} - y^{[2n-k]} \bar{z}^{[k-1]} \right\}.$$

In the Hilbert space $L_2[a, b]$ we consider the linear everywhere dense set D_0' consisting of finitary functions, that is, infinitely differentiable functions equal to zero outside some segment $[\alpha, \beta]$ (depending on the function) contained entirely in the interval $[a, b]$. An operator L_0' is defined on D_0' by the equality $L_0'y = l(y)$. It follows from the Lagrange formula that the operator L_0' will be a symmetric operator.

The closure A_0 of the operator L_0' will also be a symmetric operator. Self-adjoint extensions of the operator A_0 are studied in the theory of differential operators.

2. *Regular case.* The self-adjoint expression $l(y)$ is called *regular* if the interval (a, b) is finite and the function $\dfrac{1}{p_0(x)}$ is summable on the entire interval (a, b).

If $l(y)$ is regular, then the domain $D(A_0)$ consists of all the functions having absolutely continuous quasi-derivatives on $[a, b]$ up to the $(2n-1)$-st order inclusively and a quasi-derivative of order $2n$ belonging to $L_2[a, b]$ and satisfying the boundary conditions

$$y^{[k]}(a) = y^{[k]}(b) = 0 \quad \text{for} \quad k = 0, 1, \ldots, 2n - 1.$$

The deficiency indices of the operator A_0 are equal to $2n$. The adjoint operator is given by the equality $A_0^* y = l(y)$, and it is defined on the set $D(A_0^*)$ of all functions having absolutely continuous quasi-derivatives up to the $(2n-1)$-st order inclusively and the quasi-derivative $y^{[2n]} \in L_2[a, b]$ on $[a, b]$.

Every self-adjoint extension A of the operator A_0 satisfies the equality $Ay = l(y)$ on the functions of $D(A_0^*)$ satisfying the system of boundary conditions

$$\Gamma_j y = \sum_{k=1}^{2n} [\alpha_{jk} y^{[k-1]}(a) + \beta_{jk} y^{[k-1]}(b)] = 0 \quad (j = 1, 2, \ldots, 2n),$$

where

$$\sum_{v=1}^{n} [\alpha_{jv} \bar{\alpha}_{k, 2n-v+1} - \alpha_{j, 2n-v+1} \bar{\alpha}_{kv}] =$$

$$= \sum_{v=1}^{n} [\beta_{jv} \bar{\beta}_{k, 2n-v+1} - \beta_{j, 2n-v+1} \bar{\beta}_{kv}] \quad (j, k = 1, 2, \ldots, 2n).$$

Conversely, *for arbitrary α_{jk} and β_{jk} satisfying the last condition, the operator $Ay = l(y)$ generates a self-adjoint operator on the set of all functions of $D(A_0^*)$ satisfying the system of boundary conditions $\{\Gamma_j y = 0\}$.*

If $p_0(x) > 0$, then the operator A_0 is semi-bounded below.

The strict extension of the operator A_0 corresponds to the system of boundary conditions

$$y^{[k]}(a) = 0 \quad \text{and} \quad y^{[k]}(b) = 0 \quad \text{for} \quad k = 0, 1, ..., n - 1.$$

The resolvent of an arbitrary self-adjoint extension of the operator A_0 is an integral operator of Hilbert-Schmidt type (see § 2, no. 5). Consequently, the resolvent of an arbitrary self-adjoint extension A is a completely continuous operator, the spectrum of the operator A is discrete, and the operator A has a complete system of eigenvectors.

3. *Singular case.* If the interval (a, b) is infinite or the function $\dfrac{1}{p_0(x)}$ is not summable on (a, b), then the expression $l(y)$ is called *singular*. In this case the picture obtained is considerably more complicated.

The domain $D(A_0^*)$ of the operator A_0^* is obtained the same as in the regular case. The domain of definition of the operator A_0 itself does not always allow description by means of the boundary conditions. It follows directly from the Lagrange identity that $D(A_0)$ consists of all the functions y of $D(A_0^*)$ for which

$$[y, z]\,|_a^b = 0$$

for all $z \in D(A_0^*)$.

The deficiency indices n_+ and n_- of the operator A_0 are always identical (a consequence of the fact that the coefficients of the expression $l(y)$ are real) and can be equal to an arbitrary integer m between 0 and $2n$: $0 \leqslant m \leqslant 2n$. Recall that the deficiency index is equal to the number m of linearly independent solutions in $L_2[a, b]$ of the equation

$$l(y) = \lambda y$$

for non-real λ.

A description of all self-adjoint extensions of the operator cannot always be given in terms of systems of boundary conditions. The conditions distinguishing the domain of definition of a self-adjoint extension from the set $D(A_0^*)$ are given in an implicit form (see [37]).

The resolvent of every self-adjoint extention is an integral operator with a Carleman kernel (see § 3. no. 6).

If the deficiency index of the operator A_0 is equal to $2n$, then the kernel is a Hilbert-Schmidt kernel. In this case the spectrum of an arbitrary self-adjoint extension is discrete. In the general case, the spectrum consists of discrete and continuous parts. The continuous part of the spectrum is the same for all self-adjoint extensions.

More can be said in the case when the expression $l(y)$ is regular at one of the ends of the interval (a, b). Let a be finite and $\dfrac{1}{p_0(x)}$ be summable on every interval (a, c) where $a < c < b$. Then the domain of definition of the operator A_0 consists of all the functions of $D(A_0^*)$ for which

$$y^{[k]}(a) = 0, \quad k = 0, 1, ..., 2n - 1,$$

and $[y, z]^b = 0$ for all $z \in D(A_0^*)$.

The deficiency index can be an arbitrary integer between n and $2n$: $n \leqslant m \leqslant 2n$.

If the deficiency index is equal to n, then the second condition $[y, z]^b = 0$ is satisfied for all $y, z \in D(A_0^*)$; therefore, the domain $D(A_0^*)$ is described by only the first condition $y^{[k]}(a) = 0$ $(k = 0, 1, ..., 2n - 1)$. In this case, an arbitrary self-adjoint extension is also described by means of boundary conditions at the regular end: the domain of the extension consists of all functions from $D(A_0^*)$ satisfying the conditions

$$\Gamma_j y = \sum_{k=1}^{n} \alpha_{jk} y^{[k-1]}(a) = 0 \quad (j = 1, 2, ..., n),$$

where

$$\sum \left[\alpha_{jv} \bar{\alpha}_{k, 2n-v+1} - \alpha_{j, 2n-v+1} \bar{\alpha}_{kv} \right] = 0 \quad (j, k = 1, 2, ..., n).$$

Conversely, the above system of boundary conditions selects the domain of definition of a self-adjoint extension of the operator A_0 from $D(A_0^*)$ if the α_{jk} satisfy the last system of equalities.

If the expression $l(y)$ is singular at both ends of (a, b), then the interval (a, b) can be decomposed by an interior point c into two intervals: (a, c) and (c, b) in each of which $l(y)$ will be regular at the end c. If we denote by A_0^+ and A_0^- the operators generated by $l(y)$ on the intervals (c, b) and (a, c) and by m^+ and m^- the deficiency indices of the operators A_0^+ and A_0^-, then the important formula

$$m = m^+ + m^- - 2n$$

is valid for the deficiency index of the operator A_0 on the entire interval (a, b). In particular, if the deficiency indices of the operators A_0^+ and A_0^- are equal to n, then the operator A_0 will be self-adjoint on (a, b). Conversely, if A_0 is self-adjoint on (a, b), then, since $m^+ \geqslant n$ and $m^- \geqslant n$, the operators A_0^+ and A_0^- have deficiency index n.

4. *Criteria for self-adjointness of the operator A_0 on* $(-\infty, \infty)$. In this subsection several simple criteria are given, in terms of the coefficients of the expression $l(y)$, allowing us to establish the self-adjointness of the operator A_0 generated by $l(y)$ on the entire axis $-\infty < x < \infty$. As was indicated above, these criteria are simultaneously criteria that on the half-axes $[0, \infty)$ and $(-\infty, 0]$ the corresponding operators A_0^+ and A_0^- have deficiency index equal to n.

If the coefficients of the expression $l(y)$ are constant:

$$p_0(x) = a_0 \neq 0, \; p_1(x) = a_1, \ldots, p_n(x) = a_n,$$

then this expression assumes the form

$$l(y) = a_0 \frac{d^{2n}y}{dx^{2n}} + a_1 \frac{d^{2n-2}y}{dx^{2n-2}} + \cdots + a_n y.$$

The operator A_0, generated on $(-\infty, \infty)$ by the expression $l(y)$ with constant coefficients, is self-adjoint.

Several criteria establish that the operator A_0 is self-adjoint if its coefficients are, in a well-known sense, close to constant coefficients. The operator A_0 is self-adjoint on $(-\infty, \infty)$ in each of the cases described below:

1) the limits $\lim\limits_{x \to \infty} p_0 = a_0 \neq 0, \; \lim\limits_{x \to \infty} p_1 = a_1, \ldots, \lim\limits_{x \to \infty} p_n = a_n$ exist;

2) the functions $\dfrac{1}{p_0}, p_1, \ldots, p_n$ differ from some numbers $\dfrac{1}{a_0}, a_1, \ldots, a_n$, by functions which are summable on $(-\infty, \infty)$;

3) the functions $\left(\dfrac{1}{p_0}\right)', p_1, p_2, \ldots, p_n$ are summable on $(-\infty, \infty)$ and $\lim\limits_{x \to \infty} p_0(x) > 0$.

All these criteria can be generalized if the following property is used: the deficiency index of the operator A_0 is not changed by the addition of a function bounded on $(-\infty, \infty)$ to the coefficient $p_n(x)$. Thus, in particular, the self-adjointness of the operator A_0, generated on $(-\infty, \infty)$ by the expression

$$l(y) = (-1)^n \frac{d^{2n}y}{dx^{2n}} + q(x)y,$$

where $q(x)$ is a function bounded on $(-\infty, \infty)$, is implied.

Stronger assertions are valid for $n=2$, that is, for the expression

$$l(y) = -y'' + q(x)y.$$

The operator A_0, generated by this expression on $(-\infty, \infty)$, will be self-adjoint if the function $q(x)$ is only bounded below or, more generally, if for sufficiently large $|x|$

$$q(x) \geqslant -kx^2 \quad (k > 0).$$

The operator A_0 is also self-adjoint if $q(x) \in L_2(-\infty, \infty)$.

Other criteria for self-adjointness and non-self-adjointness of the operator A_0 generated by the self-adjoint differential expression $l(y)$ are given in [37].

5. *Nature of the spectrum of self-adjoint extensions.* As was indicated, the spectrum of self-adjoint extensions can be both discrete and continuous in the singular case. If we consider the expression $l(y)$ on $(0, \infty)$, then for the satisfaction of condition 3) of the preceding subsection, the continuous part of the spectrum of every self-adjoint extension of the operator A_0 on $[0, \infty)$ coincides with the entire positive half-axis $\lambda \geqslant 0$. Points of the discrete spectrum can be found on the negative part as well as on the positive part of the axis.

If $p_0(x) > 0$ and the conditions 1) of the preceding subsection are satisfied where $a_1, a_2, \ldots, a_{n-1}$ are positive, then only the discrete part of the spectrum can be found in the interval $(-\infty, a_n)$. Only the point $\lambda = a_n$ can be a point of condensation of the discrete spectrum on $(-\infty, a_n)$.

The question of the nature of the spectrum is one of the most important in the theory of differential operators. It is of special value in problems of quantum mechanics. It is discussed in chapter VII for the differential operators of quantum mechanics.

6. *Expansion in terms of eigenfunctions.* In the regular case, a complete orthonormal system of eigenfunctions exists for a self-adjoint extension A in terms of which an arbitrary function from $L_2(a, b)$ can be expanded in a Fourier series. If the function belongs to the domain of definition of the self-adjoint extension, that is, is sufficiently smooth and satisfies the corresponding boundary conditions, then its Fourier series is uniformly convergent.

In the singular case, for a self-adjoint extension, continuous spectrum can appear, and instead of expansions in series there appear expansions in integrals which are also called *expansions in terms of eigenfunctions* of the differential operator $l(y)$.

Let $u_1(x, \lambda), u_2(x, \lambda), \ldots, u_{2n}(x_1\lambda)$ be a system of solutions of the equation $l(y) = \lambda y$ satisfying the initial conditions

$$u_j^{[k-1]}(x_0) = \begin{cases} 1 & \text{if } j = k, \\ 0 & \text{if } j \neq k, \end{cases}$$

where x_0 is a fixed point of (a, b).

For every self-adjoint extension A of the operator A_0 generated by the expression $l(y)$, there exists a matrix function

$$\sigma(\lambda) = (\sigma_{jk}(\lambda)) \quad (j, k = 1, 2, \ldots, 2n)$$

such that for an arbitrary function $f(x) \in L_2(a, b)$ the formula

$$f(x) = \int_{-\infty}^{\infty} \sum_{j, k=1}^{2n} \varphi_j(\lambda) u_k(x, \lambda) \, d\sigma_{jk}(\lambda)$$

is valid, where the integral converges in the mean square sense. The vector function $(\varphi_1(\lambda), \ldots, \varphi_{2n}(\lambda))$ belongs to $L_{2,\sigma}$. The inversion formulas

$$\varphi_j(\lambda) = \int_a^b f(x) u_j(x, \lambda) \, dx \quad (j = 1, 2, \ldots, 2n),$$

are valid for it where the integral converges in $L_{2,\sigma}$.

The analogue of the Parseval equality holds:

$$\int_a^b |f(x)|^2 \, dx = \int_{-\infty}^{\infty} \sum_{j, k=1}^{2n} \varphi_j(\lambda) \overline{\varphi_k(\lambda)} \, d\sigma_{jk}(\lambda).$$

The multiplicity of the spectrum of the operator A does not exceed $2n$. The kernel of the resolvent of the operator A is given by the formula

$$K(x, s, \mu) = \int_{-\infty}^{\infty} \sum_{j, k=1}^{2n} \frac{u_k(x, \lambda) u_j(s, \lambda)}{\lambda - \mu} \, d\sigma_{jk}(\lambda),$$

where the integral converges in $L_2(a, b)$ in each of the variables x and s for a fixed value of the other.

In the case when the expression $l(y)$ is regular at one of the endpoints of the interval (a, b), for example at the end a, and the corresponding operator A_0 has deficiency index n, the preceding expansions are simplified. Every self-adjoint extension in this case is described by means of a system of n boundary conditions at the end a. In the expansion, not every solution $u(x, \lambda)$ of the equation $l(y) = \lambda y$ can be taken, but only those solutions which satisfy the corresponding boundary conditions at the end a. Of them, n solutions will be linearly independent. The matrix $\sigma(\lambda)$ will be of order n. Thus for the expression of second order

$$l(y) = -\frac{d}{dx}\left(p\frac{dy}{dx}\right) + qy$$

on the interval (a, ∞) $(a > -\infty)$ the expansions assume the form

$$f(x) = \int_{-\infty}^{\infty} \varphi(\lambda) u(x, \lambda) \, d\sigma(\lambda)$$

and

$$\varphi(\lambda) = \int_{a}^{\infty} f(x) u(x, \lambda) \, dx,$$

where $\sigma(\lambda)$ is a numerical non-decreasing function and $u(x, \lambda)$ is a solution of the equation $l(y) = \lambda y$ satisfying the boundary condition

$$(pu' - \theta u)_{x=a} = 0.$$

The real coefficient θ corresponds to the given self-adjoint extension.

The function $\sigma(\lambda)$ can be found in the following manner: let $u_1(x, \lambda)$ and $u_2(x, \lambda)$ be two solutions of the equation $l(y) = \lambda y$ such that

$$u_1(a, \lambda) = 1, \quad p(a)u_1'(a, \lambda) = 0,$$
$$u_2(a, \lambda) = 0, \quad p(a)u_2'(a, \lambda) = -1.$$

Since the deficiency index is equal to unity, then for every non-real λ only one combination of the form $u_1(x, \lambda) + M(\lambda) u_2(x, \lambda)$ belongs to $L_2(a, b)$. The function $\sigma(\lambda)$ is found from the function $M(\lambda)$:

$$\sigma(\lambda) = \lim_{\delta \to +0} \lim_{\varepsilon \to +0} \frac{1}{\pi} \int_{\delta}^{\delta+\lambda} \operatorname{Im}[M(\lambda + i\varepsilon)] \, d\lambda.$$

7. Examples

1. Let the differential expression

$$l(y) = -y''$$

be considered on the interval $(0, \infty)$. Then

$$y_1 = e^{\frac{1-i}{\sqrt{2}}x} \quad \text{and} \quad y_2 = e^{-\frac{1-i}{\sqrt{2}}x}$$

will be linearly independent solutions of the equation $-y'' = iy$. Of these only $y_2 \in L_2(0, \infty)$. The deficiency index of the corresponding operator A_0 is equal to 1. The self-adjoint extensions are defined by the boundary conditions

$$y'(0) = \theta y(0),$$

where θ is a real number. Here

$$u(x, \lambda) = \cos\sqrt{\lambda}x + \frac{\theta}{\sqrt{\lambda}} \sin\sqrt{\lambda}x$$

will be a solution of the equation $l(y) = \lambda y$ satisfying this condition. Calculation shows that

$$M(\mu) = \frac{1}{\theta - i\sqrt{\mu}}.$$

If $\theta \geqslant 0$, then $\sigma(\lambda) = 0$ for $\lambda < 0$ and $\sigma'(\lambda) = \frac{\sqrt{\lambda}}{\pi(\lambda + \theta^2)}$. Hence, self-adjoint extensions for $\theta \geqslant 0$ have a simple continuous Lebesgue spectrum on $(0, \infty)$.

The expansions in terms of eigenfunctions have the form

$$f(x) = \frac{1}{\pi} \int_0^\infty \Phi(\lambda) \left\{ \cos\sqrt{\lambda}x + \frac{\theta}{\sqrt{\lambda}} \sin\sqrt{\lambda}x \right\} \frac{\sqrt{\lambda}}{\lambda + \theta^2} d\lambda,$$

where

$$\Phi(\lambda) = \int_0^\infty f(x) \left\{ \cos\sqrt{\lambda}x + \frac{\theta}{\sqrt{\lambda}} \sin\sqrt{\lambda}x \right\} dx.$$

Formulas for Fourier cosine transformations are obtained when $\theta = 0$.

2. A *Bessel differential expression* has the form

$$l(y) = -y'' + \frac{v^2 - \frac{1}{4}}{x^2} y \quad (v \geqslant 0).$$

If this expression is considered on the interval $(0, \infty)$, then it will be singular on both ends. The corresponding operator A_0 will be self-adjoint for $v \geqslant 1$; it will have deficiency index $(1, 1)$ for $0 \leqslant v < 1$. Since the general solution of the equation $l(y) = \lambda y$ has the form

$$y = A \sqrt{x} J_v(x \sqrt{\lambda}) + B \sqrt{x} Y_v(x \sqrt{\lambda}),$$

these facts are easily established by the asymptotic behavior of the Bessel functions as $z \to 0$ and $z \to \infty$.

All self-adjoint extensions of the operator A_0 have discrete spectrum on the interval $(0, 1)$. The deficiency index of A_0 is equal to 1 on the interval $(1, \infty)$; the continuous spectrum of self-adjoint extensions fills out the positive half-axis. If we consider the self-adjoint estension corresponding to the boundary condition $y(1) = 0$, then the expansion in terms of eigenfunctions has the form

$$f(x) = \int_0^\infty \sqrt{x} \, \frac{J_v(x\sqrt{\lambda}) Y_v(\sqrt{\lambda}) - Y_v(x\sqrt{\lambda}) J_v(\sqrt{\lambda})}{2(J_v^2(\sqrt{\lambda}) + Y_v^2(\sqrt{\lambda}))} \, \Phi(\lambda) \, d\lambda,$$

where

$$\Phi(\lambda) = \int_1^\infty \sqrt{x} \, \{J_v(x\sqrt{\lambda}) Y_v(\sqrt{\lambda}) - Y_v(x\sqrt{\lambda}) J_v(\sqrt{\lambda})\} f(x) \, dx.$$

These formulas are called *The Weber inversion formulas*.

The expansions have the form

$$f(x) = \int_0^\infty \sqrt{x} J_v(x\sqrt{\lambda}) \Phi(\lambda) \, d\lambda,$$

where

$$\Phi(\lambda) = \int_0^\infty \sqrt{x} J_v(x\sqrt{\lambda}) f(x) \, dx,$$

on the interval $(0, \infty)$ for $v \geqslant 1$.

These formulas yield the *Hankel transformations*.

8. *Inverse Sturm-Liouville problem.* Problems on the restoration of a differential expression of given order by the spectral characteristics of a self-adjoint operator generated by this expression are called *inverse problems* in the spectral theory of differential operators. Here one variant of the statement of the inverse problem is considered.

Let the spectral function $\sigma(\lambda)$, corresponding to some self-adjoint extension A of the operator A_0 generated by the expression $l(y)$ on $(0, \infty)$, be known for some differential expression of second order

$$l(y) = -y'' + q(x)y.$$

We are required to find the coefficient $q(x)$ in the expression $l(y)$ and the form of the boundary condition of the corresponding operator A.

There are a number of methods of solution for this problem, one of which is set forth here.

We set

$$\tau(\lambda) = \begin{cases} \sigma(\lambda) - \dfrac{2}{\pi}\sqrt{\lambda} & \text{for} \quad \lambda > 0 \\[2mm] \sigma(\lambda) & \text{for} \quad \lambda < 0 \end{cases}$$

and find the functions

$$F(x, y) = \int\limits_{-\infty}^{\infty} \frac{\sin\sqrt{\lambda}x \, \sin\sqrt{\lambda}y}{\lambda} d\tau(\lambda)$$

and

$$f(x, y) = \frac{\partial^2 F(x, y)}{\partial x \, \partial y}.$$

The integral equation

$$f(x, y) + \int\limits_{0}^{x} f(y, s) K(x, s) ds + K(x, y) = 0$$

has a unique solution $K(x, y)$ for every fixed x. The function $q(x)$ is defined by the formula

$$q(x) = \frac{1}{2} \frac{dK(x, x)}{dx}.$$

The boundary condition, defining the given self-adjoint extension, has the form

$$y'(0) - \theta y(0) = 0, \quad \text{where} \quad \theta = K(0, 0).$$

Eigenfunctions $u(x, \lambda)$ of the equation $l(y) = \lambda y$ satisfying the boundary conditions $u(0, \lambda) = 1$, $u'(0, \lambda) = \theta$ can be found by the formula

$$\varphi(x, \lambda) = \cos \sqrt{\lambda} x + \int_0^x K(x, t) \cos \sqrt{\lambda} t \, dt.$$

The construction described here can be justified under these conditions:

1) the integral

$$\int_{-\infty}^0 e^{\sqrt{|\lambda|} x} \, d\sigma(\lambda)$$

exist for an arbitrary real x,

2) the function

$$a(x) = \int_1^\infty \frac{\cos \sqrt{\lambda} x}{\lambda} \, d\tau(\lambda)$$

has continuous derivatives to the 4-th order inclusively.

If the function $q(x)$ has a continuous derivative, then the indicated conditions are necessary for the solvability of the inverse problem.

§ 6. Elliptic differential operators of second order

1. *Self-adjoint elliptic differential expressions.* Let a differential expression of second order be given in self-adjoint form:

$$lu = - \sum_{i,k=1}^n \frac{\partial}{\partial x_i} \left(a_{ik}(x) \frac{\partial u(x)}{\partial x_k} \right) + c(x) u(x),$$

where $x = (x_1, \ldots, x_n)$ is a point of n-dimensional space. It is assumed that the coefficients $a_{ik}(x)$ and $c(x)$ and, hence, the expression lu are defined in some region G and on its boundary. The matrix $(a_{ik}(x))$ is symmetric.

The expression lu is called *elliptic* if all the eigenvalues of the matrices

$(a_{ik}(x))$ are bounded below, uniformly on $G+\Gamma$, by a positive constant. This definition agrees with the usual definition (see §4, no. 3).

In the sequel it is assumed that the region G is bounded, the boundary Γ is sufficiently smooth, the coefficients $a_{ik}(x)$ have partial derivatives of first order which are continuous in the closed region $G+\Gamma$, the function $c(x)$ is continuous in $G+\Gamma$ and $c(x)\geqslant 0$.

Green's formula for functions of the class $C^{(2)}(G+\Gamma)$ (see ch. I, § 1, no. 5)

$$\int\limits_G luv\,dx = \int\limits_G ulv\,dx - \int\limits_\Gamma \left(\frac{\partial u}{\partial v}v - u\frac{\partial v}{\partial v}\right)ds$$

is obtained by integration by parts. Here $\dfrac{\partial}{\partial v}$ denotes differentiation in the direction of the *co-normal* at a point of the boundary Γ:

$$\frac{\partial}{\partial v} = \sum_{i,k=1}^{n} a_{ik}(x)\cos(\boldsymbol{n},x_k)\frac{\partial}{\partial x_i}$$

where we denote by $\boldsymbol{n}$ the unit outer normal vector to the surface Γ.

2. *Minimal and maximal operators. L-harmonic functions.* In the Hilbert space $L_2(G)$ we consider the set D_0' consisting of finitary functions on G. The equality $L_0'u=lu$ on D_0' defines a linear operator which, by virtue of Green's formula, will be symmetric. The operator L_0' is positive definite. The closure L_0 of the operator L_0' will also be a symmetric operator which is called the *minimal operator* generated by the expression lu.

The domain of definition of the minimal operator consists of all those functions of the space $W_2^2(G)$ (see ch. I, § 2, no. 6), for which $u|_\Gamma=0$ and $\dfrac{\partial u}{\partial v}\bigg|_\Gamma=0$.

The operator $L=L_0^*$, the adjoint to L_0 in the space $L_2(G)$, is called the *maximal operator*. The operator L can be obtained as the closure of the operator $L'v$ defined by the formula $L'v=lv$ on the set of all infinitely differentiable functions $v(x)$ on $G+\Gamma$.

The deficiency index of the operator L_0 is equal to ∞. The deficiency subspace U, orthogonal to the range of the operator L_0, consists of all solutions of the equation $Lu=0$. These solutions are called *L-harmonic functions.* A smooth *L*-harmonic function *u* defines some function φ on the boundary Γ, the values of which coincide with the boundary values of the function *u*:

$$\varphi(s) = u(s)\,(s \in \Gamma).$$

The operator γ, realizing the correspondence $u \to \varphi$, can be extended by continuity to all the space *U*. Thus a "generalized" boundary value γu is set in one-to-one correspondence to each *L*-harmonic function. For a measure of the extensiveness of the deficiency space *U*, it can be shown that the collection of boundary values of all *L*-harmonic functions contains the space $L_2(\Gamma)$ and further $L_p(\Gamma)$ for $p \geqslant \dfrac{2(n-1)}{n}$ when $n > 2$ and for $p > 1$ when $n = 2$. The collection of boundary values can be characterized completely in terms of generalized functions.

In connection with the complicated nature of boundary values of *L*-harmonic functions, the domains of definition of self-adjoint extensions of the operator L_0, generally speaking, do not allow description by means of boundary conditions in classical form. These domains are described by means of boundary operators not allowing, in the general case, representation within the scope of mathematical analysis. Here the description of self-adjoint extensions corresponding to only three basic boundary value problems of mathematical physics is given.

3. *Self-adjoint extensions corresponding to basic boundary value problems.* The strict extension L_μ of the operator L_0 is defined by the formula $L_\mu u = lu$ on the set $\overset{0}{W}{}_2^2(G)$ of all functions of $W_2^2(G)$, which vanish on the boundary Γ.

The set $\overset{0}{W}{}_2^1(G)$ consisting of all functions of $W_2^1(G)$ vanishing on Γ is the domain of definition of the square root $L_\mu^{\frac{1}{2}}$ of the strict extension L_μ.

Thus, $D(L_\mu)$ and $D(L_\mu^{\frac{1}{2}})$ are characterized by the same boundary condition but by different conditions of smoothness.

The solutions of the equation $L_\mu u = f$ are called solutions of the *first homogeneous boundary value problem* for the equation $lu = f$. The important inequality

$$C_1 \|u\|_{W_2^2} \leqslant \|L_\mu u\|_{L_2} \leqslant C_2 \|u\|_{W_2^2},$$

is valid for these solutions.

The representation

$$D(L) = D(L_0) \oplus L_\mu^{-1} U \oplus U$$

is valid for the domain of definition of the operator L adjoint to L_0 (see §4, no. 3). The functions of $D(L_0)$ and $L_\mu^{-1} U$ belong to $W_2^2(G)$, and consequently the "non-smoothness" of the functions of the domain $D(L)$ is due only to L-harmonic terms from U.

The operator $L^0 u = lu$, defined on all the functions of $W_2^2(G)$ satisfying the boundary condition $\dfrac{\partial u}{\partial v} = 0$, defines another self-adjoint extension of the operator L_0. If $c(x) \not\equiv 0$, then the operator L^0 is positive definite on $L_2(G)$; if $c(x) \equiv 0$, then it is positive definite on the subspace of $L_2(G)$ orthogonal to the function identically equal to 1.

The domain of definition of the square root of the operator L^0 coincides with the space $W_2^1(G)$. Thus, here the boundary condition is removed in the passage from $D(L^0)$ to $D((L^0)^{1/2})$.

The solutions of the equation $L^0 u = f$ are called *solutions of the second homogeneous boundary value problem* for the equation $lu = f$.

The inequality

$$C_1 \|u\|_{W_2^2(G)} \leqslant \|L^0 u\|_{L_2(G)} \leqslant C_2 \|u\|_{W_2^2(G)}$$

is valid for these solutions.

If in $W_2^2(G)$ the collection of all functions satisfying the boundary condition $\dfrac{\partial u}{\partial v} + \sigma(s)u\Big|_{s \in \Gamma} = 0$, where the function $\sigma(s) \geqslant 0$ is sufficiently smooth, is considered, then the operator L^σ, defined by the equality $L^\sigma u = lu$ on this collection, will be self-adjoint and will have the same properties as L^0. If the function $\sigma(s)$ is only continuous, or more generally measurable, then the picture becomes considerably more complicated. The domain of definition of the corresponding self-adjoint extension already can contain functions not belonging to $W_2^2(G)$ for which the concept of a co-normal derivative can be undefined not only in the classical sense but also in the sense of Sobolev. In connection with this, the operator of differentiation in the co-normal direction is extended. It is defined on functions of $W_2^2(G)$ with the help of the usual generalized derivatives. It is additionally defined on some set of (not smooth) L-harmonic functions. Let $\varphi(s)$ be the boundary value of the L-harmonic function $u(x)$ of $W_2^2(G)$:

$$\varphi = \gamma u, \quad u = \gamma^{-1}\varphi.$$

Then the operator

$$P'(\gamma u) = P'\varphi = \frac{\partial(\gamma^{-1}\varphi)}{\partial v}\Bigg|_\Gamma = \frac{\partial u}{\partial v}\Bigg|_\Gamma$$

is defined.

The operator P', defined on such functions φ, is symmetric and positive in $L_2(\Gamma)$. Its deficiency index is equal to zero. The closure P of the operator P' is a positive self-adjoint operator. An L-harmonic function $\gamma^{-1}\varphi$ corresponds to each function $\varphi \in D(P)$. The collection $U_P(G)$ of all these L-harmonic functions already contains all the functions which are needed to define the operator of differentiation in the co-normal direction. If $w = v + u$ where $v \in W_2^2(G)$ and $u \in U_P(G)$ then we set

$$D_v w = \frac{\partial v}{\partial v} + P(\gamma u).$$

It turns out that the operator L, considered on all those functions of the domain of definition of the operator D_v for which

$$D_v w(s) + \sigma(s) w(s) = 0 \quad (s \in \Gamma),$$

generates the self-adjoint operator L^σ.

The solutions of the equation $L^\sigma w = f$ are called *solutions of the third boundary value problem* for the equation $lu = f$. These solutions may not belong to $W_2^2(G)$ but obviously belong to $W_2^1(G)$.

The solutions $w(x)$ have a "strong co-normal derivative" in the sense that they can be approximated by smooth functions in $W_2^1(G)$ such that

$$\left\| \frac{\partial w_n}{\partial v} - D_v w \right\|_{L_2(\Gamma)} \to 0.$$

The construction mentioned in the theory of the third boundary value problem goes through not only for bounded measurable functions $\sigma(s)$ but also for $\sigma \in L_{2n-2}(\Gamma)$.

The domain of definition of the square root of the operator L^σ coincides with the space $W_2^1(G)$.

The described extension of the operator of co-normal differentiation to the operator D_v is also justified by the fact that Green's formula

$$\int_G (Lw)v\,dx = \int_G \left[\sum_{i,k} a_{ik}(x)\frac{\partial w}{\partial x_i}\frac{\partial v}{\partial x_k} + cwv\right]dx - \int_\Gamma (D_v w)v\,ds$$

remains valid for an arbitrary function w in the domain of definition of the operator D_v and $v \in W_2^1$.

All the self-adjoint extensions of the operator L_0 considered have completely continuous resolvents; therefore, the spectrum of every one of these extensions is discrete and the eigenfunctions form an orthogonal basis in $L_2(G)$. The resolvent will be a Hilbert-Schmidt operator if the number of independent variables $n \leqslant 3$. For $n > 3$ only some power of the resolvent will be a Hilbert-Schmidt operator.

§ 7. Hilbert scale of spaces

1. *Hilbert scale and its properties.* As already remarked in preceding paragraphs, for several problems the limitations of one Hilbert space become restricting; and for the investigation of different aspects of the problem, different Hilbert spaces are introduced (see § 3, no. 8; § 6). In connection with this, the concept of a Hilbert scale of spaces was recently introduced.

Let H_0 be a Hilbert space and J an unbounded self-adjoint positive definite operator in H_0 such that

$$(Jx, x) \geqslant (x, x) \quad (x \in D(J))$$

We denote by H_α for $\alpha \geqslant 0$ the domain of definition of the power J^α of the operator J:

$$H_\alpha = D(J^\alpha)$$

(for the definition of J^α, see § 3, no. 4). The space H_α is a Hilbert space with respect to the scalar product

$$(x, y)_\alpha = (J^\alpha x, J^\alpha y).$$

A scalar product is introduced in the space H_0 for $\alpha < 0$ by this same formula, and the space obtained by the completion of H_0 with respect to the corresponding norm is denoted by $H_\alpha (\alpha < 0)$.

The set of Hilbert spaces $\{H_\alpha\}\,(-\infty<\alpha<\infty)$ which is obtained is called a *Hilbert scale of spaces.*

A Hilbert scale of spaces has the following properties:

1) If $\alpha<\beta$, then $H_\beta\subset H_\alpha$; the space H_β is everywhere dense in the space H_α and

$$\|x\|_\alpha \leqslant \|x\|_\beta .$$

2) If $\alpha<\beta<\gamma$, then the inequality

$$\|x\|_\beta \leqslant \|x\|_\alpha^{\frac{\gamma-\beta}{\gamma-\alpha}} \|x\|_\gamma^{\frac{\beta-\alpha}{\gamma-\alpha}}$$

is valid for $x \in H_\gamma$.

3) *The spaces H_α and $H_{-\alpha}$ are mutually conjugate with respect to the scalar product in H_0. In particular,*

$$|(x, y)_0| \leqslant \|x\|_\alpha \|y\|_{-\alpha} \quad (x \in H_\alpha, y \in H_{-\alpha}).$$

(We understand by the functional $(x, y)_0$ the scalar product in H_0 if $x, y \in H_0$ and its extension by continuity if $x \in H_\alpha$ and $y \in H_{-\alpha}$).

Let H_0 and H_1 be two Hilbert spaces with scalar products $(x, y)_0$ and $(x, y)_1$ and norms $\|x\|_0$ and $\|x\|_1$ respectively. It is assumed that $H_1 \subset H_0$, H_1 is everywhere dense in H_0, and

$$\|x\|_0 \leqslant \|x\|_1 \quad (x \in H_1).$$

It turns out that there exists an unbounded self-adjoint positive definite operator J on H_0 whose domain of definition is the space H_1 and such that

$$\|x\|_1 = \|Jx\|_0 .$$

The operator J is called a *generating* operator for the pair (H_0, H_1). A Hilbert scale of spaces can be constructed with respect to the operator J which will include the spaces H_0 and H_1.

The operator J, originally defined on the space H_1 and mapping it onto the space H_0, can be extended to the spaces H_α. Thus, the operator J can be assumed to be extended to an operator $\tilde{J}$ defined on all the spaces $H_\alpha(-\infty<\alpha<\infty)$ and mapping H_α one-to-one onto $H_{\alpha-1}$. An arbitrary operator $J^l(l>0)$ generates the same Hilbert scale of spaces and can also be extended to an operator defined on the entire scale and mapping H_α onto $H_{\alpha-l}$.

2. *Example of a Hilbert scale. The spaces W_2^α.* As the space H_0, we

take the space $L_2(R_n)$ where R_n is n-dimensional space. We denote by $\hat{u}(\xi)$ the Fourier transformation of the function $u(x) \in L_2(R_n)$:

$$\hat{u}(\xi) = \int e^{i(x,\xi)} u(x)\, dx.$$

Let J be the operator setting in correspondence to the function $u(x)$ the function $v(x)$ whose Fourier transformation has the form

$$\widehat{Ju}(\xi) = \hat{v}(\xi) = \sqrt{1 + |\xi|^2}\, \hat{u}(\xi).$$

The Hilbert scale constructed with respect to the operator J can be described as follows: the spaces H_α for $\alpha \geqslant 0$ consist of all functions for which

$$\|u\|_\alpha^2 = \int (1 + |\xi|^2)^\alpha |\hat{u}(\xi)|^2\, d\xi < \infty.$$

For $\alpha < 0$ the spaces H_α are obtained by the completion of $L_2(R_n)$ with respect to the above norm.

The scale of spaces obtained is denoted by $\{W_2^\alpha(R_n)\}$. Apparently, it is basic for many problems of analysis and the theory of partial differential equations.

For positive integers α, the spaces $W_2^\alpha(R_n)$ coincide with the Sobolev spaces (see § 1, no. 2).

The question arises whether a Hilbert scale of spaces exists containing the Sobolev spaces $W_2^l(G)$ defined for a region G of n-dimensional space. The answer to this question is not known. However, for every N we can construct a Hilbert scale of spaces $H_\alpha^{(N)}$ containing all the spaces $W_2^l(G)$ for $0 \leqslant l \leqslant N$. The construction of such a scale can be carried out by means of an extension to the whole space R_n of functions defined in the region G with preservation of smoothness.

For $0 \leqslant \alpha \leqslant N$, the norms in the spaces $H_\alpha^{(N)}$ will be equivalent to the norms in the spaces $W_2^\alpha(G)$: for $\alpha = m$ where m is an integer,

$$\|u\|_{W_2^m} = \int\limits_G \left\{ |u|^2 + \sum_{|\beta|=1}^m |D^\beta u|^2 \right\} dx;$$

for $\alpha = m + \lambda$, where m is an integer and $0 < \lambda < 1$,

$$\|u\|_{W_2^{m+\nu}} = \int\limits_G \left\{ |u|^2 + \sum_{|\beta|=1}^m |D^\beta u|^2 \right\} dx +$$
$$+ \sum_{|\beta|=m} \int\limits_G \int\limits_G \frac{|D^\beta u(x) - D^\beta u(y)|^2}{|x-y|^{n+2\lambda}}\, dx\, dy.$$

There is no effective description of the norm for negative indices, but it can be defined as a norm in the conjugate space

$$\|u\|_{W_2^{-\alpha}} = \sup_{\|v\|_{W_2^\alpha}=1} \left| \int_G u(x)v(x)\,dx \right| \quad (\alpha > 0).$$

Thus, for an arbitrary set of bounded indices α, the spaces $W_2^\alpha(G)$ have the properties of the spaces of a Hilbert scale. In particular:

1. The space $W_2^\beta(G)$ is contained and is everywhere dense in the space $W_2^\alpha(G)$ for $\alpha < \beta$ and

$$\|u\|_{W_2^\alpha} \leqslant C \|u\|_{W_2^\beta} \quad (u \in W_2^\beta).$$

2. The inequality

$$\|u\|_{W_2^\beta} \leqslant C \|u\|_{W_2^\alpha}^{\frac{\gamma-\beta}{\gamma-\alpha}} \|u\|_{W_2^\gamma}^{\frac{\beta-\alpha}{\gamma-\alpha}} \quad (u \in W_2^\gamma(G))$$

holds for $\alpha < \beta < \gamma$.

The constants C in the last inequalities depend on the form of the region G and on the maximal absolute value of the index of the norms of the spaces appearing in these inequalities. It is often convenient to write the last inequality in an equivalent form:

$$\|u\|_{W_2^\beta} \leqslant C(\varepsilon^{-\frac{\beta-\alpha}{\gamma-\alpha}} \|u\|_{W_2^\alpha} + \varepsilon^{\frac{\gamma-\beta}{\gamma-\alpha}} \|u\|_{W_2^\gamma}),$$

where ε is an arbitrary positive number.

3. *Operators in a Hilbert scale.* The following interpolation theorem is valid for Hilbert scales: *let there be given two Hilbert scales $\{H_\alpha\}$ and $\{F_\alpha\}$ and a linear operator A which is a bounded operator, acting from the space H_{α_0} to the space F_{α_1} and from the space H_{β_0} to the space F_{β_1}. Then the operator A is a bounded operator acting from an arbitrary space $H_{\alpha_0(\mu)}$ to the space $F_{\alpha_1(\mu)}$ where $\alpha_0(\mu) = (1-\mu)\alpha_0 + \mu\beta_0$ and $\alpha_1(\mu) = (1-\mu)\alpha_1 + \mu\beta_1$.*

In connection with the indicated situation, the majority of operators arising in applications are naturally considered not as operators acting from one space to another but as operators acting from a series of spaces of one scale to a corresponding series of spaces of another scale. Thus, for example, the operator γ, setting in correspondence to an L-harmonic function its boundary value (see § 6), is a bounded operator effecting a one-to-one mapping of an arbitrary space of the scale $W_2^\alpha(G)$ with $\alpha \geqslant 0$ to the space $W_2^{\alpha-\frac{1}{2}}(\Gamma)$.

The self-adjoint operator A, generated by an elliptic differential operator of order $l=2m$ and a system of boundary conditions (with specific conditions on the coefficients and region), generates a homeomorphism between the subspace $W_2^l(\text{bd.})$ of the space $W_2^2(G)$ consisting of all functions of W_2^2 satisfying the boundary conditions and the space $L_2(G)$:

$$AW_2^l(\text{bd.}) = L_2.$$

Ordinarily it turns out that in this connection the restriction of the operator A to $W_2^l(\text{bd.}) \cap W_2^{l+s}(G) = W_2^{l+s}(\text{bd.})$ $(s \geqslant 0)$ establishes a homeomorphism between the spaces $W_2^{l+s}(\text{bd.})$ and $W_2^s(G)$. Further, the operator A allows extension to the closure $W_2^{l-s}(\text{bd.})$ $(0 \leqslant s \leqslant l)$ of the set $W_2^l(\text{bd.})$ in the metric of the space $W_2^{l-s}(G)$ and this effects a homeomorphism between the space $W_2^{l-s}(\text{bd.})$ and the space $W_2^{-s}(\text{bd.})$ conjugate to $W_2^s(\text{bd.})$. Finally, the operator A can be extended to all spaces $W_2^{-s}(G)$ so that it establishes a homeomorphism between the spaces $W_2^{-s}(G)$ and $W_2^{-l-s}(\text{bd.})$.

A non-self-adjoint elliptic operator generates an analogous system of homeomorphisms for specified boundary conditions.

4. *Theorems about traces.* Let the spaces $\{H_\alpha\}$ form a Hilbert scale. We consider functions $x(t)$ $(-\infty < t < \infty)$ with values in the Hilbert space H_1 and having continuous derivatives of l-th order in the space H_0 (in the sense of the norm of the space H_0 (see ch. III, § 1, no. 1)). We introduce the norm

$$\|x\|_{\mathfrak{A}}^2 = \int_{-\infty}^{\infty} \left\{ \|x(t)\|_{H_1}^2 + \left\| \frac{d^l x}{dt^l} \right\|_{H_0}^2 \right\} dt$$

in the set $\mathfrak{A}$ of all such functions.

The space $\mathfrak{A}$ is completed with respect to this norm, and we are concerned about what can be said regarding the values of the functions obtained and their derivatives of order less than l at an arbitrary point of the real axis, for example at the point $t=0$.

This question can be formulated, in another way, as follows: if the sequence of functions $x_n(t) \in \mathfrak{A}$ is fundamental in the norm of this space, then what can be said about the convergence of the values of these functions and their derivatives at the point $t=0$?

It turns out that the operators setting the elements $x^k(0)$ $(k=0, 1, \ldots, l-1)$ in correspondence to the functions $x(t) \in \mathfrak{A}$ are continuous operators from the space $\mathfrak{A}$ to the spaces H_{α_k} where $\alpha_k = 1 - \dfrac{2k+1}{2l}$. Thus, for an arbitrary function from the completion $\overline{\mathfrak{A}}$ of the space $\mathfrak{A}$, we can talk about the values at a point of the function itself and of its derivatives of order less than l. These values belong to the spaces H_{α_k} respectively.

Conversely, if a set of elements $x_0, x_1, \ldots, x_{l-1}$ is given such that $x_k \in H_{\alpha_k}$, then a function $x(t) \in \overline{\mathfrak{A}}$ can be constructed for which $x^k(0) = x_k$ $(k = 0, 1, \ldots, l-1)$.

Assertions of the type mentioned are important in the consideration of non-homogeneous boundary value problems and in other problems where it is required to be able to extend functions on manifolds of smaller dimension to a manifold of greater dimension with the guarantee of specified differentiability properties. We come here to the corresponding result for the scale of spaces W_2^α (see no. 2).

Let a set of functions $u_0(s), u_1(s), \ldots, u_{l-1}(s)$ be given on the boundary Γ of the region G. In order for this set to be able to serve as values on the boundary for a function from $W_2^l(G)$ and its normal derivatives, it is necessary and sufficient that the functions $u_k(s)$ belong to the corresponding spaces $W_2^{l-(k+\frac{1}{2})}(\Gamma)$.

Propositions of the type indicated are called *theorems about traces*.

CHAPTER III

LINEAR DIFFERENTIAL EQUATIONS IN
A BANACH SPACE

§ 1. Linear equations with a bounded operator

1. *Linear equations of first order. Cauchy problem.*   A linear differential equation of first order has the form

$$\frac{dx}{dt} = A(t)x + f(t),$$

where $f(t)$ is a given function with values in a Banach space E, $x = x(t)$ is an unknown function with values in E and $A(t)$ is, for each fixed t, a linear operator acting in the space E. The derivative $\frac{dx}{dt}$ is understood to be the limit, in the norm of the space E, of the difference quotient $\frac{x(t + \Delta t) - x(t)}{\Delta t}$ as $\Delta t \to 0$.

In this section we consider the case where $A(t)$ is a bounded operator for every t. Under this condition, the properties of the solutions of this linear equation are analogous to the properties of the solutions of a system of linear differential equations, which can be considered as a linear equation in a finite-dimensional Banach space.

The problem of finding a solution of the equation for $0 \leqslant t < \infty$, satisfying a given initial condition

$$x(0) = x_0,$$

is called the *Cauchy problem* for the equation considered.

A linear equation is called *homogeneous* if $f(t) \equiv 0$.

2. *Homogeneous equations with a constant operator.*   A unique solution

of the Cauchy problem exists for the homogeneous equation

$$\frac{dx}{dt} = Ax$$

with a constant bounded operator A and can be written in the form

$$x(t) = e^{At}x_0.$$

The operator e^{At} is defined by the series

$$e^{At} = I + tA + \frac{t^2 A^2}{2!} + \cdots + \frac{t^n A^n}{n!} + \cdots,$$

which converges in the operator norm. The bound of the norm of each term of the series gives the inequality

$$\|e^{At}\| \leqslant e^{t\|A\|}.$$

The operators e^{At}, $-\infty < t < \infty$, form a one-parameter group of bounded operators:

$$e^{At}e^{A\tau} = e^{A(t+\tau)}.$$

The estimate of the norm of the operator e^{At} mentioned above is rough cince it takes into account only the norm of the operator A and does not sonsider the distribution of its spectrum. A more precise estimate is contained in the following assertion: *if the real parts of all the points of the spectrum of the operator A are less than the number σ, then*

$$\|e^{At}\| \leqslant N e^{\sigma t}.$$

Conversely, if this inequality is satisfied, it follows that the real parts of the points of the spectrum of the operator A do not exceed $\sigma (\mathrm{Re}\lambda \leqslant \sigma)$.

In particular, it is necessary for the boundedness of all solutions of the equation on the half-axis $0 \leqslant t < \infty$ that the spectrum of the operator A eli in the closed left-hand half-plane, and sufficient that it lie in the open left-hand half-plane.

It is necessary for the boundedness of all solutions on the entire axis $-\infty < t < \infty$ that the spectrum of the operator A lie on the imaginary axis. The fact that this condition is not sufficient is verified by the example of a finite-dimensional operator with multiple elementary divisors.

Let λ_0 be an eigenvalue of the operator A to which the eigenvector e_0 and the associated vectors $e_1, e_2, \ldots, e_{m-1}$ correspond (see ch. I, §5, no. 6).

Then the equation has particular solutions of the form

$$x_0(t) = e^{\lambda_0 t} e_0, \; x_1(t) = e^{\lambda_0 t}(e_1 + te_0), \; \ldots, \; x_{m-1}(t) =$$
$$= e^{\lambda_0 t}\left(e_{m-1} + te_{m-2} + \cdots + \frac{t^{m-1}}{(m-1)!}e_0\right).$$

If the eigenvectors and associated vectors of the operator A form a basis in the space E (see ch. I, §6), then an arbitrary solution is representable in the form of a series of particular solutions of the form indicated. In particular, if the eigenvectors $\{e_n\}$ of the operator A form a basis in E, then the general solution of the equation has the form

$$x(t) = \sum c_n e^{\lambda_n t} e_n.$$

3. *Case of a Hilbert space.* Let a homogeneous equation with a constant operator be considered in a Hilbert space H.

If the operator A is self-adjoint, then the operator e^{tA} is also self-adjoint and positive definite. If $A = iB$, where B is a self-adjoint operator, then the operator e^{iBt} is unitary.

In order for all solutions of the homogeneous equation in a Hilbert space to be bounded on the entire axis it is necessary and sufficient that the operator A be *similar* to an operator iB where B is a self-adjoint operator, that is,

$$A = Q(iB)Q^{-1},$$

where the operators Q and Q^{-1} are bounded.

As in the general case, it is sufficient for the boundedness of all solutions on the half-axis $0 \leqslant t < \infty$ that the spectrum of A lie in the open left-hand half-plane.

In a Hilbert space, a criterion can be given generalizing the well-known Lyapunov theorem: *in order for the spectrum of the operator A to lie in the open left-hand half-plane it is necessary and sufficient that a bounded self-adjoint positive definite operator W exist such that the operator $WA + A^*W$ is negative definite.*

In other words, it is necessary and sufficient that there exists a positive definite form (Wx, x) for which

$$\frac{d(Wx, x)}{dt} \leqslant -\beta(x, x) \quad (\beta > 0)$$

for an arbitrary solution $x(t)$ of the differential equation.

4. *Equations of second order.* For the equation of second order

$$\frac{d^2x}{dt^2} + Bx = 0$$

with a bounded linear operator B in the Banach space E, the *Cauchy problem* consists of finding a solution satisfying the initial conditions

$$x(0) = x_0 \quad \text{and} \quad x'(0) = x_1 .$$

The solution of this problem is given by the formula

$$x(t) = \cos t \sqrt{B}\, x_0 + \frac{\sin t \sqrt{B}}{\sqrt{B}}\, x_1 ,$$

where the bounded operators $\cos t \sqrt{B}$ and $\dfrac{\sin t \sqrt{B}}{\sqrt{B}}$ are defined by the series

$$\cos t \sqrt{B} = I - \frac{t^2 B}{2!} + \frac{t^4 B^2}{4!} - \frac{t^6 B^3}{6!} + \cdots ,$$

$$\frac{\sin t \sqrt{B}}{\sqrt{B}} = tI - \frac{t^3 B}{3!} + \frac{t^5 B^2}{5!} + \frac{t^7 B^3}{7!} + \cdots$$

The series converge in the operator norm.

In order for all solutions of the equation of second order to be bounded on the whole axis $-\infty < t < \infty$, it is necessary and sufficient that the operator $\dfrac{\sin t \sqrt{B}}{\sqrt{B}}$ be uniformly bounded with respect to t.

In a Hilbert space, it is necessary and sufficient for the boundedness of all solutions on the axis $-\infty < t < \infty$ that the operator B be similar to a positive definite operator.

5. *Homogeneous equation with a variable operator.* Now in the equation

$$\frac{dx}{dt} = A(t)x ,$$

let the bounded operator $A(t)$ on the Banach space E depend continuously on t. The solution of the Cauchy problem for this equation exists and is unique. It can be obtained by the method of successive approximations applied to the integral equation

$$x(t) = x_0 + \int_0^t A(\tau) x(\tau) \, d\tau.$$

Finally, the solution can be written in the form

$$x(t) = U(t) x_0,$$

where the operator $U(t)$ is the sum of the series

$$U(t) = I + \int_0^t A(\tau) \, d\tau + \int_0^t A(\tau) \int_0^\tau A(\tau_1) \, d\tau_1 \, d\tau + \cdots$$

which converges in the operator norm.

The rough estimate

$$\|U(t)\| \leqslant e^{t \max_{0 \leqslant \tau \leqslant t} \|A(\tau)\|}$$

is valid for the bounded operator $U(t)$.

The operator $U(t)$ can be considered as the solution of the Cauchy problem

$$\frac{dU}{dt} = A(t) U, \quad U(0) = I$$

for a differential equation in the space of bounded operators acting on E.

A bounded inverse operator $V(t) = U^{-1}(t)$ exists for every t. This operator is the solution of the Cauchy problem for the operator differential equation

$$\frac{dV}{dt} = -VA(t), \quad V(0) = I,$$

which is called the adjoint to the preceding.

If we consider a more general Cauchy problem for the original equation in which the initial condition is given not at the time $t=0$ but at an arbitrary time t_0:

$$x(t_0) = x_0,$$

then its solution can be written in the form

$$x(t) = U(t) U^{-1}(t_0) x_0 = U(t, t_0) x_0.$$

The operator $U(t, \tau) = U(t) U^{-1}(\tau)$ is called a *resolving* operator. It has the properties

$$U(t, s) U(s, \tau) = U(t, \tau) \quad \text{and} \quad U(t, t) = I.$$

In the case where $A(t)$ is constant: $A(t) \equiv A$, the resolving operator $U(t, \tau) = e^{A(t-\tau)}$.

It is assumed in this section that the operator $A(t)\,(0 \leqslant t < \infty)$ is uniformly bounded: $\|A(t)\| \leqslant M$. Then the estimate

$$\|x(t)\| \leqslant e^{Mt} \|x(0)\|$$

is valid for the solutions of the original equation.

The number

$$\sigma = \varlimsup_{t \to \infty} \frac{\ln \|x(t)\|}{t}$$

is called the *index of exponential growth* of the solution.

Always $\sigma \leqslant M$. We call the least upper bound of the numbers σ, for all solutions of the equation, the *leading index* σ_s. The formula

$$\sigma_s = \varlimsup_{t \to \infty} \frac{\ln \|U(t)\|}{t}$$

is valid for it.

The *special index*, defined by the formula

$$\sigma^* = \lim_{\tau, t - \tau \to \infty} \frac{\ln \|U(t, \tau)\|}{t - \tau}$$

is an important characteristic of the equation.

The inequality

$$\|x(t)\| \leqslant N_\varepsilon e^{(\sigma^* + \varepsilon)(t - t_0)} \|x(t_0)\| \quad (t \geqslant t_0),$$

where N_ε depends only on ε, is valid for every solution for arbitrary $\varepsilon > 0$.

The relation $\sigma_s \leqslant \sigma^*$ holds between the leading and special indices.

If the operator $A(t)$ is constant, then the leading and special indices coincide. In the general case they do not coincide. For example, the leading index is equal to 1 and the special index is equal to $\sqrt{2}$ for the ordinary equation $\dfrac{dx}{dt} = (\sin \ln t + \cos \ln t) x$.

The leading and special indices are not changed by a translation of the argument; that is, by the passage to the equation $\dfrac{dx}{dt} = A(t+a)x$. If, in the equation, we make the substitution for the unknown function $x = Q(t)y$, where the operator $Q(t)$ is uniformly bounded on the half-axis $0 \leqslant t < \infty$ and has a derivative $\dfrac{dQ}{dt}$ and an inverse $Q^{-1}(t)$ which is continuous and uniformly bounded on this half-axis, then the function y satisfies the equation

$$\frac{dy}{dt} = \left(Q^{-1}AQ - Q^{-1}\frac{dQ}{dt} \right)y,$$

for which the leading and special indices are the same as for the original equation.

An equation is called *reducible* if, by the above substitution, it can be reduced to an equation with a constant operator. The leading and special indices coincide for a reducible equation.

The magnitude of the special index depends essentially on the behavior of the operator function $A(t)$ at infinity. If the limit $A_\infty = \lim\limits_{t_i \to \infty} A(t)$ exists and the spectrum of the operator A_∞ lies in the open left-hand half-plane, then the special index is negative.

If the operators $A(t)$ $(0 \leqslant t < \infty)$ form a compact set in the space of operators, if the spectra of all limits A_∞ as $t \to \infty$ of the operators*) lie in a half-plane $\mathrm{Re}\ \lambda \leqslant -v$ $(v > 0)$ and if the derivative $A'(t)$ exists and tends to zero as $t \to \infty$, then the special index is also negative. The last condition on $A'(t)$ can be weakened by requiring that, for sufficiently large t, the norm $\|A'(t)\|$ be less than a sufficiently small number δ consistent with v. Finally, instead of the existence of the derivative we can require that the operator function $A(t)$ for sufficiently large t satisfy a Lipschitz condition $\|A(t) - A(t_1)\| \leqslant \varepsilon |t - t_1|$ with a sufficiently small coefficient ε.

In a Hilbert space we can give the following criterion for negativeness of the special index. *If there exists a Hermitian form $(V(t)x, y)$ such that*

$$0 < \alpha_1(x, x) \leqslant (V(t)x, x) \leqslant \alpha_2(x, x)$$

and

$$\frac{d}{dt}(V(t)x(t), x(t)) \leqslant -\beta(x(t), x(t)), \quad \beta > 0$$

---

*) An operator A_∞ is called a *limit* for $A(t)$ as $t \to \infty$ if a sequence $t_i \to \infty$ exists such that $\lim\limits_{t_i \to \infty} \|A(t_i) - A_\infty\| = 0$.

for an arbitrary solution $x(t)$ of the homogeneous equation, then the special index is negative.

Conversely, we can construct a form $(V(t)x, y)$ with the properties indicated for every homogeneous equation with a negative special index. The operator $V(t)$ can be obtained, for example, by the formula

$$V(t) = \int_t^\infty U^*(\tau, t) U(\tau, t) \, d\tau.$$

6. *Equations with a periodic operator.* In the equation

$$\frac{dx}{dt} = A(t)x,$$

let the operator function $A(t)$ be *periodic with a period ω*:

$$A(t + \omega) = A(t) \quad (0 \leqslant t < \infty).$$

The resolving operator $U(t, 0) = U(t)$ has the property

$$U(t + \omega) = U(t) U(\omega).$$

The operator $U(\omega)$ is called the *operator of monodromy* of the equation with a periodic operator.

The leading and special indices of an equation with a periodic operator coincide and are equal to the logarithm of the spectral radius of the operator of monodromy (see ch. I, §§ 5, 6) divided by the period:

$$\sigma_s = \sigma^* = \frac{\ln r_{U(\omega)}}{\omega}.$$

In particular, *in order for the special index to be negative it is necessary and sufficient that the spectrum of the operator of monodromy lie inside the unit circle.*

If the spectrum of the operator of monodromy does not enclose zero, then the equation with a periodic operator is reducible. It can be reduced by the substitution $x = Q(t) y$ to an equation with constant coefficients by means of the operator

$$Q(t) = U(t) e^{-\frac{t}{\omega} \ln U(\omega)}.$$

The logarithm of the operator $U(\omega)$ can be determined by the Cauchy formula

$$\ln U(\omega) = -\frac{1}{2\pi i} \oint_\Gamma (U(\omega) - \lambda I)^{-1} \ln \lambda \, d\lambda,$$

where the contour Γ surrounds the spectrum of the operator $U(\omega)$ and does not contain the point $\lambda = 0$, and where $\ln \lambda$ is some single-value, branch of the logarithm.

Estimates can be given for the indices σ of exponential growth of solutions in a Hilbert space in terms of bounds of $\text{Re}(A(t)x, x)$.

If

$$\alpha_1(t)(x, x) \leqslant \text{Re}(A(t)x, x) \leqslant \alpha_2(t)(x, x),$$

then

$$\frac{1}{\omega} \int_0^\omega \alpha_1(t) \, dt \leqslant \sigma \leqslant \frac{1}{\omega} \int_0^\omega \alpha_2(t) \, dt.$$

7. *Non-homogeneous equations.* The solution of the Cauchy problem with initial condition $x(0) = x_0$ for a non-homogeneous equation $\frac{dx}{dt} = A(t)x + f(t)$ can be written, using the resolving operator $U(t, \tau)$ for the corresponding homogeneous equation, in the form

$$x(t) = U(t, 0)x_0 + \int_0^t U(t, \tau) f(\tau) \, d\tau.$$

The question about the boundedness of solutions on $(0, \infty)$ under the condition of boundedness of $f(t)$:

$$\sup_{0 \leqslant t < \infty} \|f(t)\| < \infty$$

is important for a non-homogeneous equation.

In order for the solution of the Cauchy problem, with the zero initial condition $x(0) = 0$, for a non-homogeneous equation to be bounded on the half-axis $(0, \infty)$ for every bounded function $f(t)$ it is necessary and sufficient that the leading index of the homogeneous equation be negative. With the satisfaction of this last condition, all the solutions of the non-homogeneous equation will be bounded on $(0, \infty)$.

If it is additionally known that the operator function $A(t)$ is bounded on the half-axis, then the necessary condition can be strenghened: for the boundedness on the half-axis $(0, \infty)$ of the solution of the Cauchy problem with the condition $x(0) = \theta$ for arbitrary bounded $f(t)$ it is necessary (and, of course, sufficient) that the special index of the homogeneous equation be negative. Examples exist of unbounded operator functions $A(t)$ such that all solutions of the non-homogeneous equation, for an arbitrary bounded $f(t)$, are bounded and the special index is positive.*)

Criteria for the boundedness of solutions of the Cauchy problem on the half-axis $(-\infty, 0)$ are obtained from the criteria mentioned by the replacement of the sign of the leading or special index by its opposite. Therefore the question about the boundedness of all solutions of the non-homogeneous equation on the entire axis for an arbitrary bounded $f(t)$ is meaningless. A question can be posed about the existence of even one, or of only one, bounded solution for arbitrary bounded $f(t)$. For the last question, in the case of a constant operator $A(t) \equiv A$, there is a final answer: in order that exactly one bounded solution (on the entire axis) of the non-homogeneous equation with a constant operator A corresponds to each bounded $f(t)$ $(-\infty < t < \infty)$, it is necessary and sufficient that the spectrum of the operator A not intersect the imaginary axis.

A nonlinear differential equation of the form

$$\frac{dx}{dt} = A(t)\, x + f(t, x),$$

where $f(t, x)$ for every t is, generally speaking, a nonlinear operator on x, can be considered as linear with a free member $f(t, x(t))$; then the formula for the solution of the Cauchy problem gives the equation

$$x(t) = U(t, 0)\, x_0 + \int_0^t U(t, \tau) f(\tau, x(\tau))\, d\tau.$$

If bounds of growth for the resolving operator $U(t, \tau)$ are known, in particular, if the special index σ^* is known, then bounds for the solutions of the nonlinear equation are obtained from the resulting integral equation. In this way analogies of the Lyapunov theorems about the uniform and asymptotic stability of the solutions of a nonlinear equation are obtained.

---

*) For unbounded $A(t)$ the leading and special indices are defined the same as in the bounded case.

§ 2. Equation with a constant unbounded operator.
Semi-groups

1. *Cauchy problem.* In this section we consider the differential equation

$$\frac{dx}{dt} = Ax$$

with the linear operator A having an everywhere dense domain of definition $D(A)$ in the Banach space E.

A function $x(t)$ is called a *solution of the equation* on the segment $[0, T]$ if it satisfies the conditions: 1) the values of the function belong to the domain of definition $D(A)$ of the operator A for all $t \in [0, T]$; 2) the strong derivative $x'(t)$ of the function $x(t)$ exists at every point t of the segment $[0, T]$; 3) the equation $x'(t) = Ax(t)$ is satisfied for all $t \in [0, T]$.

The *Cauchy problem* on the segment $[0, T]$ is the problem of the discovery of a solution of the equation on $[0, T]$ satisfying the initial condition

$$x(0) = x_0 \in D(A).$$

If the questions of the existence and uniqueness of the solution of the Cauchy problem and of its continuous dependence on initial data were always solved positively for a linear equation with a bounded operator and, hence, basic attention was given to the behavior of the solutions as $t \to \infty$, then the questions enumerated become central for an equation with an unbounded operator.

We say that the Cauchy problem is *correctly formulated (well posed)* on the segment $[0, T]$ if: 1) its unique solution exists for arbitrary $x_0 \in D(A)$ and 2) this solution depends continuously on initial data in the sense that if $x_{0,n} \to 0 \, (x_{0,n} \in D(A))$, then for the corresponding solutions $x_n(t)$ it follows that $x_n(t) \to 0$ for every $t \in [0, T]$.

By virtue of the constancy of the operator A, the correctness on an arbitrary segment $[0, T_1]$ $(T_1 > 0)$, that is, correctness on the entire half-axis $[0, \infty)$, follows from the correctness of the Cauchy problem on a single segment $[0, T]$.

Let $U(t)$ be the operator setting into correspondence the value of the solution $x(t)$ of the Cauchy problem $x(0) = x_0$ at time t to every element $x_0 \in D(A)$. If the Cauchy problem is well posed, then the operator $U(t)$

is defined on $D(A)$ and is linear and bounded. Therefore it can be extended by continuity to a bounded linear operator, defined on the entire space E, which we also denote by $U(t)$.

A family of bounded linear operators $U(t)$, depending on the parameter t $(0 < t < \infty)$, is called a *semi-group* if

$$U(t_1 + t_2) = U(t_1) U(t_2) \qquad (0 < t_1, t_2 < \infty).$$

The operators $U(t)$, generated by a well posed Cauchy problem, form a semi-group.

Thus, the solution of a well posed Cauchy problem is representable in the form

$$x(t) = U(t) x_0 \qquad (x_0 \in D(A)),$$

where $U(t)$ is this semi-group of operators.

If x_0 does not belong to the domain of definition of the operator A, then the function $U(t) x_0$ may not be differentiable and its values may not belong to the domain $D(A)$ of the operator A. We can call the function $U(t) x_0$ a *generalized solution of the equation* $x' = Ax$.

2. *Uniformly correct Cauchy problem.* A correctly formulated Cauchy problem is called *uniformly correct* if it follows from $x_{0,n} \to \theta$ that the solutions $x_n(t)$ tend to θ uniformly on every finite segment $[0, T]$.

If the operator A is closed and its resolvent $R_A(\lambda)$ (see ch. I, § 5, no. 6) exists for some λ, then uniform correctness follows from the existence and uniqueness of a continuously differentiable solution of the Cauchy problem for arbitrary $x \in D(A)$.

The semi-group $U(t)$ is *strongly continuous* for a uniformly correct Cauchy problem, that is, the function $U(t) x_0$ is continuous on $(0, \infty)$ for arbitrary $x_0 \in E$.

We say that semi-group $U(t)$ belongs to the *class* (C_0) if it is strongly continuous and satisfies the condition

$$\lim_{t \to +0} U(t) x_0 = x_0$$

for arbitrary $x_0 \in E$.

The semi-group $U(t)$, generated by a uniformly correct Cauchy problem, belongs to the class (C_0). In other words, we can say that all generalized solutions are continuous on $[0, \infty]$ in this case.

The limit

$$\lim_{t \to \infty} \frac{\ln \|U(t)\|}{t} = \omega$$

exists for an arbitrary strongly continuous semi-group $U(t)$. If the semi-group belongs to the class (C_0), then the estimate

$$\|U(t)\| \leqslant M e^{\omega t}$$

is valid for it.

Thus, for a uniformly correct Cauchy problem, the orders of exponential growth of all solutions are bounded above.

If $\omega = 0$, then the semi-group is bounded, $\|U(t)\| \leqslant M$, and the Cauchy problem is uniformly correct on $[0, \infty)$. In this case, an equivalent norm can be introduced in the space E, for example

$$\|x\|_1 = \sup_{0 \leqslant t < \infty} \|U(t) x\|,$$

in which the operators $U(t)$ have a norm not greater than one:

$$\|U(t)\|_1 \leqslant 1.$$

The semi-group is called *contractive* in this case.

The Cauchy problem for the equation of thermal conductivity:

$$\frac{\partial v}{\partial t} = \frac{\partial^2 v}{\partial x^2}, \, v(0, x) = \varphi(x) \qquad (-\infty < x < \infty, 0 \leqslant t < \infty)$$

is one of the simplest examples of a uniformly correct Cauchy problem. Let the space $C(-\infty, \infty)$, consisting of all continuous bounded functions on the x-axis, serve as the space E. Here the operator A is the second derivative operator with respect to x defined on the set $D(A)$, dense in $C(-\infty, \infty)$, consisting of all twice continuously differentiable functions $v(x)$ for which v and $v'' \in C(-\infty, \infty)$.

The Cauchy problem is uniformly correct, that is, a unique solution $v(t, x)$ of the thermal conductivity equation, having the property that $\lim_{t \to +0} v(t, x) = \varphi(x)$ uniformly with respect to x, exists for an arbitrary function $\varphi \in D(A)$. Furthermore, if $\varphi_n(x) \in D(A)$ converge uniformly to $\varphi(x) \in D(A)$, then the solutions $v_n(t, x) \to v(t, x)$ uniformly with respect to x and t on every finite segment $[0, T]$ of variation of t.

The corresponding semi-group $U(t)$ of bounded operators is given by the *integral formula of Poisson*

$$[U(t)\,\varphi]\,(x) = \frac{1}{2\sqrt{\pi t}} \int\limits_{-\infty}^{\infty} e^{-\frac{(x-s)^2}{4t}}\,\varphi(s)\,ds \qquad (t > 0)$$

and consists of contractive operators.

3. *Generating operator and its resolvent.* For the semi-group $U(t)$ the question is raised: for which elements x_0 will the function $U(t)\,x_0$ be differentiable? Differentiability of this function for arbitrary t follows from its differentiability for $t=0$.

The linear operator

$$U'(0)\,x_0 = \lim_{h \to +0} \frac{U(h)\,x_0 - x_0}{h}$$

is defined on the elements x_0 for which $U(t)\,x_0$ is differentiable at zero. The operator $U'(0)$ is called the *generating operator of the semi-group* $U(t)$.

If the semi-group belongs to the class (C_0), then the domain of definition D of the generator $U'(0)$ is everywhere dense; it is closed and commutes with the semi-group on its domain of definition:

$$U'(0)\,U(t)\,x_0 = U(t)\,U'(0)\,x_0 \qquad (x_0 \in D).$$

If the Cauchy problem is uniformly correct for the equation $x' = Ax$, then the operator A allows closure. This gives the generator of the corresponding semi-group $U(t)$:

$$\bar{A} = U'(0).$$

The Cauchy problem is uniformly correct for the equation $x' = Ax$ where A is the generating operator of a semi-group of class (C_0).

Thus, if we restrict ourselves to equations with closed operators, then *the class of equations for which the Cauchy problem is uniformly correct coincides with the class of equations for which A is a generator for a semi-group of class (C_0).* This explains the role which the study of semi-groups and their generating operators plays in the theory of differential equations.

The spectrum of the generator of a semi-group of the class (C_0) lies alway sin some half-plane Re $\lambda \leqslant \omega$.

The class of generating operators can be characterized by the behavior of the resolvents $R_A(\lambda)$ of the operators: *in order for the operator A to be the generator of a semi-group in (C_0), it is necessary and sufficient that a real ω and a positive M exist such that*

$$\|R_A^k(\lambda)\| \leqslant \frac{M}{(\lambda - \omega)^k} \quad \text{for} \quad \lambda > \omega \quad (k = 0, 1, 2, \ldots).$$

If the operator A in the equation $x' = Ax$ is closed, then the conditions mentioned are necessary and sufficient for the uniform correctness of the Cauchy problem.

The estimate $\|U(t)\| \leqslant Me^{\omega t}$ for the corresponding semi-group is valid under the satisfaction of the indicated list of conditions.

The verification of the necessary and sufficient conditions is difficult since all the powers of the resolvent appear in them. They will be obviously satisfied if

$$\|R_A(\lambda)\| \leqslant \frac{1}{\lambda - \omega} \quad (\lambda > \omega).$$

Such a condition is satisfied, for example, for the equation of thermal conductivity (see example, no. 2).

In the presence of the last estimate, the inequality

$$\|U(t)\| \leqslant e^{\omega t}$$

is valid for the semi-group.

In particular, if $\omega = 0$, then $\|U(t)\| \leqslant 1$ and the semi-group is contractive. It should be emphasized that the satisfaction of the condition

$$\|R_A(\lambda)\| \leqslant \frac{M}{\lambda - \omega} \quad (\lambda > \omega)$$

with $M > 1$ is not sufficient for the correctness of the Cauchy problem.

The semi-group $U(t)$ can be constructed in terms of the resolvent $R_A(\lambda)$ by the formula

$$U(t)x = -\lim_{v \to \infty} \frac{1}{2\pi i} \int_{\mu - iv}^{\mu + iv} e^{\lambda t} R_A(\lambda) x \, d\lambda,$$

which is valid for $x \in D(A)$, $t > 0$ and sufficiently large positive μ. The integral converges uniformly on every interval $0 < \varepsilon \leqslant t \leqslant \frac{1}{\varepsilon}$. The limit of the integral as $t \to 0$ is equal to $x/2$.

The resolvent of the operator A is the Laplace transform (with opposite sign) of the semi-group:

$$R_A(\lambda) x = - \int_0^\infty e^{-\lambda t} U(t) x \, dt.$$

The integral converges when the real part of λ is sufficiently large.

A uniformly correct Cauchy problem is always the limit of Cauchy problems with bounded operators in the following sense: a sequence of bounded operators A_n exists such that the solutions of the problems $\frac{dx_n}{dt} = A_n x_n$, $x_n(0) = x_0 \in D(A)$ converge to the solution of the problem $\frac{dx}{dt} = Ax$, $x(0) = x_0$. Moreover, the convergence is uniform on every finite interval $[0, T]$. The operators A_n can be constructed by the formula $A_n = -nI - n^2 R_A(n) = -nAR_A(n)$. The semi-group $U(t)$ is the limit of the operator-functions $e^{-nAR_A(n)t}$ where the convergence is uniform on an arbitrary finite interval $[0, T]$.

4. *Weakened Cauchy problem.* In no. 1 it was required that the solution of the equation satisfy the equation for $t = 0$ also. This requirement must often be weakened.

A function $x(t)$, continuous on $[0, T]$, strongly differentiable and satisfying the equation on $(0, T]$, is called a *weak solution of the equation* $x' = Ax$ on the segment $[0, T]$.

We understand by the *weakened Cauchy problem* on $[0, T]$ the problem of finding a weak solution satisfying the initial condition $x(0) = x_0$. Here the element x_0 does not have to belong to the domain of definition of the operator A.

If we leave aside the question of the existence of a solution of the Cauchy problem, then rather general conditions for its uniqueness can be pointed out. The condition

$$\varlimsup_{\lambda \to \infty} \frac{\ln \|R_A(\lambda)\|}{\lambda} = h < T$$

(this limit is always non-negative) is sufficient for the uniqueness of the solution of the weakened Cauchy problem. The solution is unique on $[0, T-h]$ and can branch for $t > T-h$. If $h=0$, then the solution of the weakened Cauchy problem is unique on the entire half-axis $(0, \infty)$. On the other hand, a differential equation $x' = Ax$ with an operator A for which $\|R_A(\lambda)\| < \varrho(\lambda)$, having a nontrivial solution with the initial condition $x(0) = \theta$ exists for every function $\varrho(\lambda) > 0$ satisfying the condition $\dfrac{\ln \varrho(\lambda)}{\lambda} \to \infty$ $(\lambda - \infty)$.

If a real ω, a positive M and a $\beta, 0 < \beta \leqslant 1$, exist such that for the resolvent of the operator A

$$\|R_A(\sigma + i\tau)\| \leqslant \frac{M}{\sigma - \omega + |\tau|^\beta}$$

in the half-plane $\sigma > \omega$, then the weakened Cauchy problem has a unique solution on $[0, \infty)$ for an arbitrary x_0 from the domain of definition of the operator $A (x_0 \in D(A))$. The weakened Cauchy problem, in this connection, will be correct but not uniformly correct.

When the last condition on the resolvent is satisfied, then the solution of the weakened Cauchy problem is given by the formula $x(t) = U(t) x_0$, where $U(t)$ is a strongly continuous semi-group.

If $x_0 \notin D(A)$, then the generalized solution $U(t) x_0$ can be discontinuous at the point $t = 0$. However, it is always Abel summable to its initial value:

$$\lim_{\lambda \to \infty} \lambda \int_0^\infty e^{-\lambda t} U(t) x_0 \, dt = x_0.$$

If the estimate given above with index $\beta > \frac{1}{2}$ is valid for the resolvent of the operator A, then we have

$$\int_0^T \|U(t)\| \, dt < \infty$$

for the operator $U(t)$.

The weakened Cauchy problem in this case will be correct in the mean: if $x_{0,n} \to \theta$, then the integrals $\int_0^T \|x_n(t)\| \, dt$ tend to zero for the solutions.

The derivative of the solution of the weakened Cauchy problem can be discontinuous at the point $t = 0$, but its norm is summable on an arbitrary finite interval:

$$\int_0^T \|x'(t)\| \, dt < \infty .$$

5. *Abstract parabolic equation. Analytic semi-groups.* An equation $x' = Ax$, for which the weakened Cauchy problem with arbitrary initial condition $x_0 \in E$ is correct, is called an *abstract parabolic equation.* Thus, a unique solution of the weakened Cauchy problem exists for an abstract parabolic equation for arbitrary $x_0 \in E$, and this solution continuously depends on initially given data.

If the operator A is closed and has a resolvent for sufficiently large positive λ, then the Cauchy problem is uniformly correct for the abstract parabolic equation, and the operator A coincides with the generating operator of the semi-group $U(t)$ of class (C_0) by means of which the solution $x(t) = U(t) x_0$ is given.

If the Cauchy problem is uniformly correct for the equation $x' = Ax$, then, in order for the equation to be abstract parabolic, it is sufficient that

$$\overline{\lim_{\tau \to \infty}} \left(\ln \tau \, \| R_A(\sigma + i\tau) \| \right) = 0$$

for some real σ.

Every generalized solution of the abstract parabolic equation is weak and, hence, differentiable for $t > 0$. It follows from the commutivity of the generating operator and the operators of the semi-group that every generalized solution is infinitely differentiable. The operators $A^k U(t)$ $(t > 0, k = 0, 1, 2, \ldots)$ are linear operators, bounded for every $t > 0$. The norms of the operators $A^k U(t)$, generally speaking, are not bounded as $t \to 0$.

An important class of abstract parabolic equations is formed from equations for which all generalized solutions are analytic functions of t and can be analytically extended to some (fixed for a given equation) sector of the complex plane containing the positive real half-axis. The semi-group $U(t)$ is itself analytically extended to some operator-function $U(z)$ which is analytic in the sector. In the sequel such semi-groups are called *analytic.*

In order for a semi-group to be analytic it suffices that the estimate

$$\| R_A(\lambda) \| \leqslant \frac{M}{|\lambda - \omega|}$$

for the resolvent $R_\lambda(A)$ of the operator A be satisfied in the half-plane
Re $\lambda > \omega$.

The angle of the sector of analyticity can be defined in the following manner: it follows from the indicated estimate for the resolvent that the analogous estimate

$$\|R_A(\lambda)\| \leqslant \frac{M_1}{|\lambda - \omega|}$$

holds in some sector $-\varphi < \arg(\lambda - \omega) < \varphi \, (\varphi > \frac{\pi}{2})$; then the semi-group $U(z)$ is analytic in the sector $-\psi < \arg z < \psi$, where

$$\psi = \varphi - \frac{\pi}{2}.$$

If the semi-group belongs to the class (C_0), then the indicated estimate for the resolvent is necessary for its analyticity.

The estimate

$$\|x'(t)\| \leqslant \frac{Ce^{\omega t}}{t} \|x_0\| \qquad (0 < t < \infty)$$

holds for the derivatives of all generalized solutions of the equation $x' = Ax$.

Conversely, if the estimate

$$\|AU(t)\| \leqslant \frac{Ce^{\omega t}}{t}$$

is valid for a semi-group of the class (C_0), then this semi-group has an analytic extension $U(z)$ in some sector containing the positive half-axis. The inequality

$$\|A^k U(t)\| \leqslant \left(\frac{kC}{t}\right)^k e^{\omega t}.$$

holds for the norms of the operators $A^k U(t)$.

It remains to remark that it follows from the satisfaction of the estimate for the resolvent for some ω that it will be satisfied (possibly with another constant M) for an arbitrary ω lying to the right of all the points of the spectrum of A.

6. *Reverse Cauchy problem.* The problem of finding a solution

on the interval $[0, T]$ with a given terminal value

$$x(T) = x_T \in D(A).$$

is called the *reverse Cauchy problem*.

By a replacement of the independent variable, the reverse Cauchy problem for the equation $x' = Ax$ can be reduced to the Cauchy problem for the equation $x' = -Ax$.

If the Cauchy problem is well posed for an equation, then the reverse Cauchy problem, generally speaking, is not well posed: for some x_T the solution will, in general, not exist, for others it will cut off, not reaching zero; the solution will not depend continuously on initially given data.

If the direct and reverse Cauchy problems are uniformly correct for an equation, then the operators $U(t)$ are defined and bounded for all $t \, (-\infty < t < \infty)$ and form a group where $U(-t) = U^{-1}(t)$.

In order for the direct and reverse Cauchy problems with a closed operator A to be uniformly correct, it is necessary and sufficient that constants M and $\omega > 0$ exist such that

$$\|R_A^n(\lambda)\| \leqslant \frac{M}{(|\lambda| - \omega)^n} \qquad (n = 1, 2, 3, ...)$$

for all real λ with $|\lambda| > \omega$. The spectrum of the operator A, in this connection, lies inside the strip $|\lambda| < \omega$.

We digress from the question of the existence of a solution to the reverse Cauchy problem and consider only the question of the uniqueness of the solution and its continuous dependence on terminal values. The reverse Cauchy problem is called *correct (well posed)* on $[0, T]$ *in the class of bounded solutions* if a number $\delta = \delta(\varepsilon, M, t_0)$ can be found for every positive M, ε and $t_0 \in (0, T)$ such that the inequality $\|x(t_0)\| \leqslant \delta$ is satisfied for an arbitrary weak solution on $[0, T]$ satisfying the conditions $\|x(t)\| \leqslant M \, (0 \leqslant t \leqslant T)$ and $\|x(T)\| \leqslant \varepsilon$.

If all the generalized solutions of the equation $x' = Ax$ are analytic in some sector, then the reverse Cauchy problem is well posed in the class of bounded solutions on an arbitrary segment $[0, T]$. This fact follows from the inequality

$$\|x(t_0)\| \leqslant M \|x(0)\|^{1 - \omega(t_0)} \|x(T)\|^{\omega(t_0)},$$

where the continuous function $\omega(t)$, $0 \leqslant \omega(t) \leqslant 1$ does not depend on the choice of the generalized solution.

7. *Equations in a Hilbert space.* The differential equation $x' = -Bx$ in a Hilbert space H with an unbounded self-adjoint positive definite operator B is the simplest example of an abstract parabolic equation.

The Cauchy problem is uniformly correct for this equation, and the semi-group $U(t)$ corresponding to it can be written in the form

$$U(t) = e^{-Bt},$$

where the function e^{-Bt} is defined by means of the spectral decomposition of the operator A (see ch. II, § 3. no. 4):

$$e^{-Bt} = \int_0^\infty e^{-\lambda t} \, dE_\lambda.$$

The semi-group $U(t)$ will be contractive, $\|U(t)\| \leqslant 1$.

All generalized solutions $x(t) = e^{-Bt} x_0 (x_0 \in H)$ are solutions of the weakened Cauchy problem and allow analytic extension to the right hand half-plane $\operatorname{Re} z > 0$.

The behavior of the weak solutions can be studied in more detail as $t \to 0$. An arbitrary (non-integral) power of the operator B is defined by means of the spectral expansion

$$B^\alpha = \int_0^\infty \lambda^\alpha \, dE_\lambda.$$

If $x(0) \in D(B^\alpha)$, $\alpha > 0$, then the inequality

$$\|x'(t)\| \leqslant \frac{C}{t^{1-\alpha}} \|B^\alpha x(0)\|$$

is valid for the solution of the weakened Cauchy problem.

In the case $\alpha = \frac{1}{2}$ the solutions allow a more precise characterization: in order for the derivative $x'(t)$ of the solution $x(t)$ to have a square-integrable norm on $[0, T]$,

$$\int_0^T \|x'(t)\|^2 \, dt < \infty,$$

it is necessary and sufficient that $x(0) \in D(B^{\frac{1}{2}})$.

It remains to note the useful inequality

$$\|x(t)\| \leqslant \|x(0)\|^{1-\frac{t}{T}} \|x(T)\|^{\frac{t}{T}},$$

which is valid for an arbitrary generalized solution.

A linear operator C is called an *operator of fractional order* with respect to the self-adjoint positive definite operator B if it is defined on $D(B)$ and the operator $CB^{-\alpha}$ is bounded on $D(B)$ for some $\alpha \in (0, 1)$. The greatest lower bound γ of the numbers α is called the *order* of the operator C with respect to B.

In order for the operator C to be an operator of fractional order γ with respect to the operator B it is necessary, and if C allows closure it is sufficient, that the inequality

$$\|Cx\| \leqslant K_\alpha (\delta^{1-\alpha} \|Bx\| + \frac{1}{\delta^\alpha} \|x\|),$$

where K_α does not depend on x and δ, be satisfied for $x \in D(B)$, $\alpha > \gamma$ and sufficiently small δ.

If the operator C is an operator of fractional order with respect to a self-adjoint positive definite operator B, then the equation $x' = -(B+C)x$ is abstract parabolic. Generalized solutions will be analytic in some sector containing the positive half-axis. The behavior of the solutions as $t \to 0$ will be the same as for the equation $x' = -Bx$.

In a Hilbert space the generating operators for several important classes of semi-groups can be completely described. The description is made in terms connected with the operator itself and not with its resolvent.

1) *In order for the operator A to generate a strongly continuous contractive semi-group of operators it is necessary and sufficient that it be a maximal dissipative operator with an everywhere dense domain of definition* (see ch. II, § 4, no. 4).

This criterion can be formulated differently as follows: *in order for a closed operator with an everywhere dense domain of definition to be the generating operator of a strongly continuous contractive semi-group of operators it is necessary and sufficient that the conditions*

$$\mathrm{Re}(Ax, x) \leqslant 0 \quad (x \in D(A)) \quad \textit{and} \quad \mathrm{Re}(A^*x, x) \leqslant 0 \quad (x \in D(A^*))$$

be satisfied.

2) *If the operators A and A^* have the same everywhere dense domain of definition and the operator* $\operatorname{Re}\ A = \dfrac{A + A^*}{2}$ *is bounded, then the operator A is the generating operator of a semi-group $U(t)$ of class (C_0) for which the estimate $\|U(t)\| \leqslant e^{\omega t}$ is valid, where ω is an upper bound of the operator* $\operatorname{Re}\ A$.

3) *In order for the operator A to be the generating operator of a strongly continuous semi-group of isometric transformations $\left(\|U(t)\| = 1\right)$, it is necessary and sufficient that it be a maximal dissipative conservative operator with a dense domain of definition. In this case $A = iB$ where B is a maximal symmetric operator.*

4) *In order for the operator A to be the generating operator of a strongly continuous group of unitary operators, it is necessary and sufficient that $A = iB$ where B is a self-adjoint operator. The group of operators $U(t)$ is representable in terms of the spectral decomposition of the operator B in the form*

$$U(t) = e^{iBt} = \int\limits_{-\infty}^{\infty} e^{i\lambda t}\, dE_\lambda.$$

The direct and reverse Cauchy problems are uniformly correct on the entire axis. Generalized solutions $U(t)\, x_0$ are differentiable only when $x_0 \in D(A)$.

5) *If the condition*

$$|\operatorname{Im}(Ax, x)| \leqslant \beta\, |\operatorname{Re}(Ax, x)|$$

is satisfied for a maximal dissipative operator, then it is the generating operator for a semi-group allowing analytic extension to the sector $|\arg z| < \dfrac{\pi}{2} - \arctan \beta$.

The estimate

$$\|R_A(\lambda)\| \leqslant \frac{M_\varepsilon}{|\lambda|}$$

is valid for the resolvent of the operator in the sector $|\arg \lambda| \leqslant \pi - \arctan \beta - \varepsilon$ for arbitrary $\varepsilon > 0$.

The equation $x' = Ax$ is an abstract parabolic equation; the properties of its solutions are analogous to the properties of the solutions of the equation $x' = -Bx$ with a self-adjoint positive definite operator B.

Recently, many of the above-enumerated criteria that an operator be generating for semi-groups of one class or another have been extended to certain classes of Banach spaces.

8. *Examples of well posed problems for partial differential equations.*
1. Cauchy problem for the diffusion equation. Let $b(x)$ be a continuous bounded function on the entire axis $-\infty < x < \infty$. We consider the problem

$$\frac{\partial u}{\partial t} = \frac{\partial^2 u}{\partial x^2} + b(x)\frac{\partial u}{\partial x}, \qquad u(0, x) = \varphi(x) \qquad (-\infty < x < \infty, t > 0).$$

The domain of definition of the operator $\dfrac{\partial^2}{\partial x^2}$ is described in the example in no. 2. The domain of definition of the operator $\dfrac{\partial^2}{\partial x^2} + b(x)\dfrac{\partial}{\partial x}$ is assumed to be the same. The Cauchy problem is uniformly correct in the space $C(-\infty, \infty)$.

2. Boundary value problems for strongly parabolic systems of equations. Let G be a bounded region in an n-dimensional space with a sufficiently smooth boundary Γ; $x = (x_1, x_2, ..., x_n)$ a point of the region G; $L(x, \dfrac{\partial}{\partial x})$ an N-dimensional matrix with each element $L_{ij}(x, \dfrac{\partial}{\partial x})$ a linear differential operator of order $2m$ of the form

$$L_{ij}\left(x, \frac{\partial}{\partial x}\right) = \sum_{|\alpha| \leqslant 2m} a_{ij}^{(\alpha)}(x)\, D^\alpha \qquad (i, j = 1, 2, ..., N),$$

where $\alpha = (\alpha_1, ..., \alpha_n)\ (\alpha_i \geqslant 0)$, $|\alpha| = \alpha_1 + \alpha_2 + \cdots + \alpha_n$, $D^\alpha = \dfrac{\partial^{\alpha_1 + \cdots + \alpha_n}}{\partial x_1^{\alpha_1}...\partial x_n^{\alpha_n}}$. The coefficients $a_{ij}^{(\alpha)}(x)$ are assumed to be real and sufficiently smooth. The matrix $L(x, \dfrac{\partial}{\partial x})$ is assumed to be *strongly elliptic*, that is,

$$(-1)^m \sum_{i, j = 1}^{N} \sum_{|\alpha| = 2m} a_{ij}^{(\alpha)}(x)\, \xi_1^{\alpha_1} ... \xi_n^{\alpha_n} \eta_i \eta_j > 0$$

for arbitrary $x \in G$ and real ξ_k and η_l with $\sum \xi_k^2 \neq 0$ and $\sum \eta_i^2 \neq 0$.
For the system of equations

$$\frac{\partial u}{\partial t} = -L\left(x, \frac{\partial}{\partial x}\right) u \quad (x \in G, t > 0),$$

one formulates the first boundary value problem of finding a solution $u(t, x) = (u_1(t, x), \ldots, u_N(t, x))$ satisfying the initial condition

$$u(0, x) = \varphi(x)$$

and the boundary conditions

$$u\Big|_\Gamma = \frac{\partial u}{\partial n}\Big|_\Gamma = \cdots = \frac{\partial^{m-1} u}{\partial n^{m-1}}\Big|_\Gamma = 0,$$

where n is the direction of the normal to the boundary Γ.

The strongly elliptic expression $L(x, \frac{\partial}{\partial x})$ generates a linear operator in the space $\bar{L}_2(G)$ of vector functions $u(x)$ whose modulus squared is summable, defined on the smooth functions satisfying the boundary conditions. The operator allows the closure L, having the properties

$$\mathrm{Re}(Lu, u) \geqslant q(u, u) \, (u \in D(L))$$

and

$$\mathrm{Re}(L^*u, u) \geqslant q(u, u) \, (u \in D(L^*)).$$

Thus, the operator $qI - L$ will be a maximal dissipative operator. From this follows the uniform correctness of the first described boundary value problem for the equation $u'_t = -Lu$ in the space $\bar{L}_2(G)$. The estimate $\|U(t)\| \leqslant e^{qt}$ is valid for the corresponding semi-group.

An analogous problem can be considered in the space $\bar{L}_p(G)$ of functions for which

$$\|u\|_{L_p} = \{ \int |u(x)|^p \, dx \}^{\frac{1}{p}} < \infty \qquad (p > 1).$$

It turns out that for arbitrary $p > 1$, the estimate

$$\|(L + \lambda I)^{-1}\| \leqslant \frac{C}{|\lambda + q|}$$

is valid in some sector $|\arg \lambda| < \varphi \, (\varphi > \frac{\pi}{2})$ for the resolvent of the operator $-L$ generated by a strongly elliptic matrix.

Thus, the semi-group generated by a strongly parabolic system for the first boundary value problem in $\bar{L}_p(G)$ is analytic. The solution of the problem exists for an arbitrary initial function $\varphi(x) \in \bar{L}_p(G)$ and is an analytic function of t. The initial condition is satisfied in the sense that

$$\|u(t, x) - \varphi(x)\|_{L_p} \to 0 \quad \text{as} \quad t \to 0.$$

The estimate of the derivatives

$$\|u'_t\|_{L_p} \leqslant \frac{C}{t} \|\varphi\|_{L_p},$$

where C does not depend on φ, is valid for all solutions.

For the simpler case of one equation with a scalar function $u(t, x)$, the concept of a strongly elliptic differential expression coincides with the concept of an elliptic expression (for real coefficients!) (see ch. II, § 6, no. 1). The indicated estimate is valid in an arbitrary sector $|\arg \lambda| \leqslant \varphi$ with $\varphi < \pi$ for the resolvent of the corresponding operator $\bar{L}$. Therefore, the solutions are analytic in the entire right-hand half-plane.

Recently, operators were studied which are generated by a strongly elliptic expression $L(x, \frac{\partial}{\partial x})$ on functions satisfying the boundary conditions not of the first boundary value problem but some general boundary conditions (the so-called Lopatinskiĭ conditions). In the spaces $L_p (p > 1)$, the estimates obtained for the resolvents of operators are analogous to those mentioned above. Thus, for the solutions of corresponding boundary value problems for a strongly parabolic system, the same conclusions are valid as for the first boundary value problem.

3. **Cauchy problem for an equation with constant coefficients.** We consider the equation

$$\frac{\partial^m u}{\partial t^m} = P\left(\frac{\partial}{\partial t}, \frac{\partial}{\partial x}\right) u = \sum_{k=1}^m P_k\left(\frac{\partial}{\partial x}\right) \frac{\partial^{k-1} u}{\partial t^{k-1}},$$

where $x = (x_1, \ldots, x_n)$ is a point of the n-dimensional space R_n and $P(\frac{\partial}{\partial t}, \frac{\partial}{\partial x})$ is a linear differential expression with constant coefficients containing derivatives with respect to t of order not greater than $m - 1$.

With the substitution $u = u_1$, $u'_t = u_2, \ldots, u_t^{(m-1)} = u_m$ the equation can be reduced to the system

$$\frac{\partial u_1}{\partial t} = u_2, \ldots, \frac{\partial u_{m-1}}{\partial t} = u_m, \frac{\partial u_m}{\partial t} = \sum_{k=1}^m P_k\left(\frac{\partial}{\partial x}\right) u_k.$$

The system thus obtained can be treated as the equation $u'_t = Au$ in the Hilbert space of vector functions $L_2(R_n)$. The operator A is the closure of the operator defined by the matrix

$$\begin{pmatrix} 0 & 1 & 0 & \dots & 0 \\ 0 & 0 & 1 & \dots & 0 \\ \cdot & \cdot & \cdot & \cdot & \cdot & \cdot \\ 0 & 0 & 0 & \dots & 1 \\ P_1 & P_2 & P_3 & \dots & P_k \end{pmatrix}$$

on sufficiently smooth functions such that $P_k(\dfrac{\partial}{\partial x}) u_k \in L_2(R_n)$.

The question is raised: when will the equation $u'_t = Au$ be abstract parabolic? It turns out that it is necessary and sufficient that these conditions be satisfied:

a) the equation is *correct according to Petrovskiĭ*, that is, for all roots of the equation

$$s^m = P(s, i\xi)$$

for an arbitrary real vector $\xi = (\xi_1, \dots, \xi_n)$, the inequality

$$\operatorname{Im} s < C$$

is satisfied, where the constant C does not depend on the choice of the vector ξ;

b) for arbitrary $a > 0$ a b can be found such that

$$|\operatorname{Im} s| \leqslant a \ln |\operatorname{Re} s| - b.$$

If the stronger inequality

$$|\operatorname{Im} s| \leqslant a_1 |\operatorname{Re} s|^h - b_1$$

is satisfied for some real b_1, $a_1 > 0$ and $h \geqslant 1$, then the semi-group corresponding to the problem will allow analytic extension to some sector containing the positive real half-axis.

Finally, if $h > 1$, then the semi-group is analytic in the right-hand half-plane.

4. Schrödinger equation. In the 3-dimensional space R_3, the equation

$$i \frac{\partial \psi}{\partial t} = -\Delta \psi + v(x) \psi$$

is considered, where Δ is the Laplace operator.

Under certain conditions on the function $v(x)$, the operator H, obtained by closure of the operator defined by the differential expression $-\Delta + v(x)$ on finitary functions, will be a self-adjoint operator in the Hilbert space $L_2(R_3)$. The direct and reverse Cauchy problems are uniformly correct on the entire axis. The equation generates a group of unitary operators $U(t) = e^{-iHt}$ (see ch. VII, § 1, no. 4).

5. Symmetric hyperbolic systems. We consider the system of equations

$$\frac{\partial u}{\partial t} = \sum A_i(x) \frac{\partial u}{\partial x_i} + Bu,$$

where $x = (x_1, ..., x_n)$; $u(t, x)$ is an m-dimensional vector function; $A_i(x)$ $(i = 1, 2, ..., n)$, $B(x)$ are m-dimensional square matrices depending smoothly on x; $A_i(x)$ are symmetric matrices.

If, for example, it is assumed that $A_i(x)$ and $B(x)$ are periodic functions in all variables, and the operator $A_i \dfrac{\partial}{\partial x_i} + B$ is considered on periodic differentiable functions, then the real part of its closure will be the bounded operator $B + B^* - \dfrac{\partial A_i}{\partial x_i}$ (case 4, no. 7). The problem of the determination of periodic (with respect to spacial coordinates) solutions of the system will be uniformly correct on an arbitrary segment $[0, T]$ in the space $\bar{L}_2(Q)$, where Q is a period parallelepiped.

If the condition

$$B + B^* - \frac{\partial A_i}{\partial x_i} \leqslant 0 \qquad (i = 1, 2, ..., n)$$

is satisfied, that is, the matrices $B + B^* - \dfrac{\partial A_i}{\partial x_i}$ are negative definite, then the corresponding operator will be a maximal dissipative operator and the corresponding semi-group will consist of contractive operators.

Upon satisfaction of the condition $B + B^* - \dfrac{\partial A_i}{\partial x_i} \leqslant 0$ in some region G, the operator $A_i \dfrac{\partial}{\partial x_i} + B$ will be dissipative on the functions satisfying the condition

$$\int_\Gamma \left(\sum A_i n_i u, u \right) dx \leqslant 0$$

on the boundary Γ of the region, where n_i are the components of the normal to Γ. For example, this condition is satisfied on finitary functions.

Recently, the forms of boundary conditions defining maximal dissipative extensions of the corresponding operator were studied in detail.

9. *Equations in a space with a basis. Continual integrals.* If a basis $\{e_i\}$ exists in the Banach space E, then solutions of differential equations can be sought in the form of expansions with respect to the basis. In this subsection, several formulas for solutions of a formal nature will be mentioned. Questions about the convergence of the expansions will not be considered.

If a basis $\{e_i\}$ of eigenvectors of the operator A exists in the space E, then the solution of the Cauchy problem $x' = Ax$, $x(0) = x_0$ can be written in the form

$$x(t) = \sum_i c_i e^{\lambda_i t} e_i,$$

where the λ_i are eigenvalues of the operator A and the coefficients c_i can be found from the expansion $x_0 = \sum_i c_i e_i$.

If we now consider the equation

$$\frac{dx}{dt} = (A + B) x,$$

then the solution of the Cauchy problem for it can also be sought in the form $x(t) = \sum a_i(t) e_i$. An infinite system of differential equations

$$\frac{da_i}{dt} = \lambda_i a_i + \sum_k b_{ik} a_k$$

is obtained for the functions $a_i(t)$, where (b_{ik}) is the matrix which is assigned to the operator B in the basis (e_i).

The method described of finding solutions in the form of their expansions with respect to a basis of eigenvectors of the operator A is called the *Fourier method.*

Now let an arbitrary basis $\{e_i\}$ be given in the space E, forming a biorthogonal system with the collection of linear functionals $\{f_i\}$. Let $x_i(t)$ be the solution of the Cauchy problem

$$\frac{dx}{dt} = Ax, \qquad x_i(0) = e_i.$$

The system $\{x_i(t)\}$ is called a *fundamental system of solutions with respect to the basis* $\{e_i\}$.

The matrix-function with elements $s_{ij}(t)=f_i(x_j(t))$ is called the *fundamental matrix of the equation in the basis* $\{e_i\}$.

If x_0 is the element having the expansion $x_0=\sum_i c_i e_i$ in the basis, then the solution of the Cauchy problem with the initial condition $x(0)=x_0$ can be written in the form

$$x(t) = \sum_{i,\,j} s_{ij}(t)\, c_j e_i.$$

The functions $s_{ij}(t)$ satisfy the identity

$$\sum_k s_{ik}(t)\, s_{kj}(\tau) = s_{ij}(t+\tau).$$

If the operator A is such that $s_{ij}(t)\geqslant 0$ and $\sum_j s_{ij}(t)=1$, then the numbers $s_{ij}(t)$ can be treated as the probabilities of the passage of some system from the state e_i to the state e_j. In this case, the matrix $(S_{ij}(t))$ describes the so-called Markov process with a denumerable number of states.

Now let the operator B be such that the functionals $f_i(x)$ are eigenvectors for the operator B^* with eigenvalues α_i:

$$f_i(Bx) = \alpha_i f_i(x).$$

We formulate the problem of finding the fundamental matrix $(S_{ij}(t))$ for the equation

$$\frac{dx}{dt} = (A + B)\, x.$$

It turns out that this matrix can be evaluated by the formula

$$S_{ij}(t) = \lim \sum_{i_1=1}^{\infty} \sum_{i_2=1}^{\infty} \cdots \sum_{i_n=1}^{\infty} e^{\sum_{k=1}^{n+1} \alpha_{i_k}\Delta t_k}\, s_{ii_1}(\Delta t_1)\, s_{i_1 i_2}(\Delta t_2)\cdots s_{i_n j}(\Delta t_{n+1}),$$

where $\Delta t_k>0$ and $\sum_{k=1}^{n+1}\Delta t_k=t$; the limit is taken with respect to partitions of the segment $[0, t]$ as $\max \Delta t_k\to 0$. The last limit is often written in the form of a so-called *continual integral*

$$S_{ij}(t) = \int e^{\int_0^t \alpha(\tau)\,d\tau}\, d\mu_{ij}.$$

In this connection, we understand the integral

$$\int \Phi(\alpha(\tau))\,d\mu_{ij}$$

of the functional Φ, defined on the set of all step-functions $\alpha(\tau)$ with values in the set of eigenvalues of the operator B^*, to be

$$\lim_{\max \Delta t_k \to 0} \sum_{i_1=1}^{\infty} \sum_{i_2=1}^{\infty} \cdots \sum_{i_n=1}^{\infty} \Phi(\alpha_{i_1,\dots,i_n}(\tau))\, s_{ii_1}(\Delta t_1) \dots s_{i_n j}(\Delta t_{n+1}),$$

where the function $\alpha_{i_1,\dots,i_n}(\tau)$ is equal to μ_{ik} on the segment Δt_k.

If we consider the equation

$$\frac{dx}{dt} = (A + f(B))\,x,$$

then its fundamental matrix is written in the form

$$S_{ij}(t) = \int e^{\int_0^t f(\alpha(\tau))\,d\tau}\, d\mu_{ij}.$$

Up to now the case of a discrete basis was considered. Sometimes the solutions are expanded in terms of continual bases. Thus, for example, we take the collection of all generalized eigenfunctions of a self-adjoint operator in a Hilbert space with a continuous spectrum as such a basis (see ch. II, § 3, no. 8). Then, in all the preceding formulas, the infinite sums are replaced by integrals. In particular, the continual integral is no longer the limit of n-multiple sums, but the limit of n-multiple integrals.

Example. The *heat conduction* equation:

$$\frac{\partial u}{\partial t} = \frac{\partial^2 u}{\partial x^2}.$$

The continual basis consists of functions

$$e_i = \delta(x - i)\,(-\infty < i < \infty)$$

(see ch. VIII, § 1, no. 1), the eigenfunctions of the operator of multiplication by the variable. The fundamental system of solutions (see no. 2) is

$$u_i(t, x) = \frac{1}{2\sqrt{\pi t}} \int_{-\infty}^{\infty} e^{\frac{-(x-s)^2}{4t}} \delta(s - i) \, ds = \frac{1}{2\sqrt{\pi t}} e^{\frac{-(x-i)^2}{4t}}.$$

The fundamental matrix is

$$s_{ij}(t) = \int_{-\infty}^{\infty} u_j(t, x) \delta(x - i) \, dx = u_j(i) = \frac{1}{2\sqrt{\pi t}} e^{\frac{-(i-j)^2}{4t}}.$$

The continual integral

$$\int \Phi(\alpha(\tau)) \, d\mu_{ij}$$

on the interval $[0, T]$ is defined as follows: a partitioning of the interval $[0, T]$ is made by the points $0 = t_0 < t_1 < t_2 < \cdots < t_n < t_{n+1} = T$. We consider all continuous functions $\alpha(t)$ which are linear on each of the subintervals Δt_k such that $\alpha(0) = i$ and $\alpha(T) = j$. Let $\alpha(t_k) = \alpha_k$ $(k = 1, 2, ..., n)$; then the functional $\Phi(\alpha(\tau))$ becomes some function of n variables $\Phi(\alpha(\tau)) = \Phi(\alpha_1, \alpha_2, ..., \alpha_n)$ on the class of functions considered. Then

$$\int \Phi(\alpha(\tau)) \, d\mu_{ij} = \lim_{\max \Delta t_k \to 0} \frac{1}{\left(2\sqrt{\pi}\right)^n} \overbrace{\int_{-\infty}^{\infty} \cdots \int_{-\infty}^{\infty}}^{n} \Phi(\alpha_1, \alpha_2, ..., \alpha_n) \times$$

$$\times \frac{1}{\sqrt{\Delta t_1 ... \Delta t_{n+1}}} e^{-\frac{1}{4} \sum_{s=1}^{n+1} \frac{(\alpha_s - \alpha_{s-1})^2}{\Delta t_s}} \, d\alpha_1 ... d\alpha_n,$$

where $\alpha_0 = i$ and $\alpha_{n+1} = j$.

The continual integral obtained is called a *Wiener integral*.

The fundamental matrix for the equation

$$\frac{\partial u}{\partial t} = \frac{\partial^2 u}{\partial x^2} + V(x) u$$

can be written in the form

$$s_{ij}(t) = \int e^{\int_0^t V(\alpha(\tau)) \, d\tau} \, d\mu_{ij}.$$

§ 3. Equation with a variable unbounded operator

1. *Homogeneous equation.* When considering the homogeneous equation

$$\frac{dx}{dt} = A(t)\, x \qquad (0 \leqslant t \leqslant T)$$

with an operator $A(t)$ which depends on t, we generally assume that for every t this operator is the generating operator for a semi-group having certain properties. Moreover, the dependence of $A(t)$ on t is assumed to be smooth. In order to formulate conditions of smoothness for an unbounded operator-function $A(t)$, it is natural to assume that this function is defined for different t on the same elements of the space. In connection with this, the following assumption is made in this subsection.

The domain of definition of the operator $A(t)$ is not dependent on t:

$$D\big(A(t)\big) = D.$$

We can now formulate two sets of conditions which guarantee the correctness of the statement of the Cauchy problem for an equation with a variable operator:

I.1) The operator $A(t)$ is the generating operator of a contractive semi-group for every $t \in [0, T]$ and, moreover, the inequality

$$\|R_{A(t)}(\lambda)\| \leqslant \frac{1}{1 + \operatorname{Re} \lambda} \quad \text{for} \quad \operatorname{Re} \lambda \geqslant 0$$

is valid for the norm of its resolvent.

2) The bounded operator-function $A(t)\, A^{-1}(s)$ is strongly continuously differentiable with respect to t for arbitrary s.

II.1) The operator $A(t)$ is the generating operator of an analytic semi-group for every $t \in [0, T]$ and the estimate

$$\|R_{A(t)}(\lambda)\| \leqslant \frac{C}{1 + |\lambda|} \quad \text{for} \quad \operatorname{Re} \lambda \geqslant 0,$$

where C does not depend on t, is valid for the norm of its resolvent.

2) The operator-function $A(t)\, A^{-1}(s)$ satisfies the Hölder condition

$$\|[A(t) - A(\tau)]\, A^{-1}(s)\| \leqslant C_1 |t - \tau|^\gamma,$$

where C_1 does not depend on s, t and τ and $0 < \gamma \leqslant 1$.

When the sets of conditions I or II are satisfied, then there exists a resolving operator $U(t, s)$ $(t \geqslant s)$, bounded and strongly continuous with respect to the variables t and s, for $0 \leqslant s \leqslant t \leqslant T$. In this connection $U(s, s) = I$. The solution of the Cauchy problem for the equation $x' = A(t)x$ with the initial condition $x(0) = x_0 \in D$ is unique and is given by the formula

$$x(t) = U(t, 0)x_0.$$

If the initial condition is given for $t = s$, $x(s) = x_0$, then the solution has the form $x(t) = U(t, s)x_0$.

The operators $U(t, s)$ have the property

$$U(t, \tau) = U(t, s)U(s, \tau) \qquad (0 \leqslant \tau \leqslant s \leqslant t \leqslant T).$$

Generalized solutions $U(t, 0)x_0$ of the equation $x' = A(t)x$ for arbitrary $x_0 \in E$ are continuous functions on $[0, T]$. If the conditions II are satisfied, then every generalized solution is a solution of the weakened Cauchy problem, i.e., is differentiable and satisfies the equation for all $t \in (0, T]$. The estimate $\|x'(t)\| \leqslant \dfrac{C_1}{t}\|x_0\|$ holds for derivatives of the weak solutions.

For the conditions I, the resolving operator $U(t, s)$ can be constructed as a "multiplicative integral"

$$U(t, s) = \lim \prod_{i=1}^{n} U_{A(\tau_i)}(\Delta t_i).$$

Here $s = t_0 < t_1 < \cdots < t_n = t$ is a partitioning of the segment $[s, t]$, $\Delta t_i = t_i - t_{i-1}$, τ_i are interior points of the segments $[t_{i-1}, t_i]$; $U_{A(\tau)}$ is the semi-group generated by the operator $A(\tau)$. The limit is taken for max $\Delta t_i \to 0$ and exists in the strong sense.

The resolving operator $U(t, s)$ can be obtained for the conditions II as a solution of the integral equation

$$U(t, s) = U_{A(s)}(t) + \int_{s}^{t} U(t, \tau)[A(\tau) - A(s)]U_{A(s)}(\tau - s)\, d\tau.$$

If the operator $A(t)$ allows an analytic extension to a region containing the interval $(0, T)$, then, with the satisfaction of the conditions II, all generalized solutions $U(t, 0)x_0$ will be analytic in some region containing $(0, T)$.

Remark.   In the conditions I and II, λ can be replaced everywhere by $\lambda - \omega$ and the operator $A(t)\,A^{-1}(s)$ by the operator $(A(t) - \omega I) \times (A(s) - \omega I)^{-1}$. This case reduces to consideration of the replacement $x = e^{\omega t} y$.

2. *Case of an operator $A(t)$ with a variable domain of definition.*   In the case when $A(t)$ is a differential operator, its domain of definition usually consists of sufficiently smooth functions satisfying certain boundary conditions. The independence of the domain of definition $D(A(t))$ of t assumed in the preceding subsection means, in the applications, the independence of t of the coefficients in the boundary conditions; therefore, the removal of this condition is of considerable interest. It turns out that some fractional power $A^{\alpha}(t)$ of the operator $A(t)$ (for the definition of fractional powers of operators, see no. 4) has, in several cases, a domain of definition consisting of functions not restricted by boundary conditions and, hence, the domain of definition does not depend on t.

Condition 2) of the set II can be replaced by the following:

2') For some $\alpha \in (0,1)$, the operators $A^{\alpha}(t)$ have a domain of definition independent of t, where the operator $A^{\alpha}(t)\,A^{-\alpha}(s)$ satisfies the Hölder condition with the index $\gamma > 1 - \alpha$:

$$\|[A^{\alpha}(t) - A^{\alpha}(\tau)]\,A^{-\alpha}(s)\| \leqslant C_1 |t - \tau|^{\gamma}.$$

If conditions 1) of group II and 2') are satisfied, then a resolving operator $U(t,s)$ exists which has the same properties as in the satisfaction of the conditions of II.

It remains to remark that the verification of condition 2') is difficult, since there are no explicit formulas for fractional powers of concrete, for example differential, operators. This condition can be verified for operators in a Hilbert space which satisfy conditions 5), § 2, no. 7. Let $A(t)$ be a maximal dissipative operator for every t for which the form $(A(t)\,x, x)$ has the properties

$$|\mathrm{Re}\,(A(t)\,x, x)| \geqslant \delta(x, x)$$

and

$$|\mathrm{Im}\,(A(t)\,x, x)| \leqslant \beta\,|\mathrm{Re}\,(A(t)\,x, x)|,$$

where the constants $\delta > 0$ and β do not depend on t. Then the operator $A(t)$ satisfies condition 1) of II. The form $(A(t)\,x, x)$ can be extended by means of closure to a form $A_t(x, x)$, defined on a wider set of elements than $D(A(t))$. It is assumed that the domain D of the form $A_t(x, x)$ does not depend on t and the form satisfies a Hölder condition of the form

$$|A_t(x, x) - A_\tau(x, x)| \leqslant M|t - \tau|^\gamma(x, x)$$

on D. The domain of definition of the operator $A^\alpha(t)$ for $0 < \alpha < \frac{1}{2}$ does not depend on t under these conditions, and

$$\|[A^\alpha(t) - A^\alpha(\tau)]A^{-\alpha}(s)\| \leqslant C_1|t - \tau|^\gamma.$$

Thus, if $\gamma \geqslant \frac{1}{2}$, then the operator $A(t)$ satisfies condition 2').

3. *Non-homogeneous equation.* The method of variation of parameters can be applied to solve a non-homogeneous equation

$$\frac{dx}{dt} = A(t)x + f(t).$$

Then the solution of the Cauchy problem with the initial condition $x(0) = x_0$ can be formally written as

$$x(t) = U(t, 0)x_0 + \int_0^t U(t, s)f(s)\,ds.$$

If the resolving operator $U(t, s)$ is strongly continuous with respect to t and s and the function $f(s)$ is continuous, then the above integral exists and is a continuous function of t. However, generally speaking, it will not be a differentiable function of t; therefore, we must assume that the above formula gives a *generalized solution* of the non-homogeneous equation.

If the conditions of set I, no. 1, are satisfied, then the generalized solution will be a true solution of the Cauchy problem under the conditions that $x_0 \in D$ and the function $f(t)$ is continuously differentiable. The last condition can be replaced by the following: $f(t) \in D$ for all $t \in [0, T]$ and the function $A(0)f(t)$ is continuous.

If the conditions of set II, no. 1 are satisfied, then the generalized solution will be a solution of the weakened Cauchy problem (see §2, no. 4) for an arbitrary $x_0 \in E$ and functions $f(t)$, satisfying the Hölder condition

$$\|f(t) - f(\tau)\| \leqslant C|t - \tau|^\delta \qquad (0 < \delta \leqslant 1).$$

Moreover, if $x_0 \in D$, then the above formula provides a true solution of the Cauchy problem.

In the investigation of a non-homogeneous equation this fact is useful:

the operator-function $A(t) U(t, s) A^{-1}(0)$ is bounded and is a strongly continuous function of the variables t and s under conditions I or II. Moreover, if the function $A(t) A^{-1}(0)$ has a second strongly continuous derivative, then the operator-function $A^2(t) U(t, s) A^{-2}(0)$ is bounded and is a strongly continuous function of t and s.

4. *Fractional powers of operators.*   For self-adjoint positive operators in a Hilbert space, fractional powers are defined by means of a spectral expansion (see § 2, no. 7).

Let A be a closed linear operator with an everywhere dense domain $D(A)$, having a resolvent on the negative half-axis, and satisfying the condition

$$\|(A + \lambda I)^{-1}\| < \frac{M}{|\lambda|} \qquad (\lambda > 0).$$

The operator

$$I_\alpha x = \frac{\sin \pi\alpha}{\pi} \int_0^\infty \lambda^{\alpha-1}(A + \lambda I)^{-1} x \, d\lambda$$

is defined for $0 < \alpha < 1$ and $x \in D(A)$.

The operator I_α allows a closure which is called a *fractional power of the operator A* and is denoted by A^α.

Moreover, if the operator A has a bounded inverse operator A^{-1}, then we can directly define bounded operators which are fractional powers of the operator A^{-1}:

$$A^{-\alpha} = \frac{\sin \pi\alpha}{\pi} \int_0^\infty \lambda^{-\alpha}(A + \lambda I)^{-1} \, d\lambda.$$

The operator $A^{-\alpha}$ is the inverse of the operator A^α. A negative power $A^{-\alpha}$ can be defined by the formula

$$A^{-\alpha} = \frac{\sin \pi\alpha}{\pi} \frac{(n - 1)!}{(1 - \alpha)(2 - \alpha)\dots(n - 1 - \alpha)} \int_0^\infty \lambda^{n-1-\alpha}(A + \lambda I)^{-n} \, d\lambda$$

for the indices $\alpha > 1$ which makes sense for fractional α contained between 0 and n. If α tends to an integer $k < n$, then $\lim_{\alpha \to k} A^{-\alpha} = A^{-k}$ where the limit is understood in terms of the operator norm.

The operators $A^{-\alpha} (0 \leqslant \alpha < \infty)$ form a semi-group of bounded operators of class (C_0). This semi-group is uniformly continuous (with respect to the operator norm) for $t > 0$.

If the operator B is the generating operator of a strongly continuous semi-group of operators $U(t)$ for whose norm the estimate

$$\|U(t)\| \leqslant M e^{-\omega t} \qquad (\omega > 0)$$

is valid, then the estimate

$$\|(B - \lambda I)^{-1}\| \leqslant \frac{M}{\lambda - \omega} \qquad (\lambda > \omega)$$

is valid for the resolvent of the operator and, hence, fractional powers can be defined for the operator $A = -B$. These powers can be expressed by a semi-group:

$$A^{-\alpha} = (-B)^{-\alpha} = \frac{1}{\Gamma(\alpha)} \int\limits_0^\infty \tau^{\alpha-1} U(\tau) \, d\tau \qquad (0 < \alpha < 1).$$

An important "*inequality of moments*" holds for fractional powers of operators: if α and β have the same sign and $|\alpha| < |\beta|$, then

$$\|A^\alpha x\| \leqslant K(\alpha, \beta) \|A^\beta x\|^{\frac{\alpha}{\beta}} \|x\|^{1-\frac{\alpha}{\beta}},$$

where the constant $K(\alpha, \beta)$ does not depend on the choice of the element x (if $\beta > 0$, then $x \in D(A^\beta)$).

The resolvent is defined for $\lambda < 0$ for the operator A^α $(0 < \alpha < 1)$ and it can be found by the formula

$$(A^\alpha + \lambda I)^{-1} = \frac{\sin \pi \alpha}{\pi} \int\limits_0^\infty \frac{\mu^\alpha (A + \mu I)^{-1}}{\lambda^2 + 2\lambda \mu^\alpha \cos \pi \alpha + \mu^{2\alpha}} \, d\mu.$$

It follows from this formula that the inequality

$$\|(A^\alpha + \lambda I)^{-1}\| \leqslant \frac{M}{\lambda} \qquad (\lambda > 0),$$

with the same constant M as for the analogous inequality for the operator A, is valid for the operator A^α.

It follows from the inequality $\|\lambda(A+\lambda I)^{-1}\| \leqslant M$ for all $\lambda > 0$ that the operator $\lambda(A+\lambda I)^{-1}$ is uniformly bounded in every sector of the complex plane $|\arg \lambda| \leqslant \varphi$ for φ not greater than some number $\pi - \psi$ $(0 < \psi < \pi)$. Then the operator $\lambda(A^\alpha + \lambda I)^{-1}$ will be uniformly bounded in every sector $|\arg \lambda| \leqslant \varphi$ for $\varphi < \pi - \alpha \psi$. In particular, the operator $(-A)^{\frac{1}{2}}$ is the generating operator of an analytic semi-group. If the operator $-A$ is a generating operator of class (C_0), then the operator $(-A)^\alpha$ will be the generating operator of an analytic semi-group for $0 < \alpha < 1$.

If $0 < \alpha, \beta < 1$, then $(A^\alpha)^\beta = A^{\alpha\beta}$.

The following fact is very important in applications.

If A and B are two positive self-adjoint operators in a Hilbert space and $D(B) \supset D(A)$, then the inclusion

$$D(B^\alpha) \supset D(A^\alpha)$$

holds for $0 < \alpha < 1$.

More generally, let A and B be two positive self-adjoint operators acting in the Hilbert spaces H and H_1, respectively. If Q is a bounded linear operator from H into H_1 such that $QD(A) \subset D(B)$ and

$$\|BQu\| \leqslant M \|Au\| \qquad (u \in D(A)),$$

then $QD(A^\alpha) \subset D(B^\alpha)$ and

$$\|B^\alpha Qu\| \leqslant M_1 \|A^\alpha u\| \qquad (u \in D(A^\alpha)).$$

We can set $M_1 = M^\alpha \|Q\|^{1-\alpha}$.

An analogous statement with another constant M_1 is valid for the case when A and B are maximal dissipative operators in the spaces H and H_1 respectively.

For Banach spaces, the last inequality is proven for the replacement, in the right hand side of the inequality, of the norm $\|A^\alpha u\|$ by the norm $\|A^\beta u\|$ with $\beta > \alpha$. The validity of the inequality for identical indices is not established.

Fractional powers of operators play an essential role in the investigation of nonlinear differential equations.

CHAPTER IV

NONLINEAR OPERATOR EQUATIONS

Introductory remarks

In this chapter the equation

$$x = Ax$$

is considered, where A is an operator (generally speaking, nonlinear) defined in some Banach space E with range of values in the same space.
The operator

$$Ax(t) = \int_0^1 K[t, s, x(s)] \, ds$$

and, in particular, the operator

$$Ax(t) = \int_0^1 K(t, s) f[s, x(s)] \, ds$$

can serve as examples of the operator A.

The first of these is usually called a *Urysohn* operator while the second is called a *Hammerstein* operator.

The first question occurring in the study of the indicated equation involves the existence of a solution. This question is often formulated in this form: is there a *fixed point* for the transformation A?

The operator A can be defined on a part T of the space E; then we talk about fixed points of the operator which belong to T.

If we are required to find a solution having an additional property, then a subset $T_0 \subset T$ of elements having this property is selected, and a fixed point is sought in T_0. For example, in the problems where non-negative solutions are sought, we set $T_0 = T \cap K$ where K is the corresponding cone of the non-negative elements of E (see ch. V).

The second question involves the uniqueness of a solution; that is, the uniqueness (in T_0) of a fixed point of the transformation A.

Theorems of non-uniqueness are of basic interest for non linear equations in many cases, that is, existence theorems for two or more solutions. For example, in various problems of stability theory and the theory of waves, a (trivial) solution of the problem is known in advance and the determination of other (non-trivial) solutions is a fundamental goal.

In those cases when the solution is not unique, the problem concerns itself with the number of solutions or with upper and lower bounds for this number.

We often consider the equation

$$x = A(x; \lambda),$$

where the operator $A(x; \lambda)$ depends on a numerical parameter λ (in some problems the parameter can be an element of some space).

For equations with a parameter, several new problems arise which are connected with a change in the number of solutions for a change of the parameter.

Those critical values of the parameter λ for which the solutions branch or merge are of particular interest.

The simplest example of an equation with a parameter is given by the problem involving eigenvalues and eigenvectors of a linear operator, that is, the problem concerning solutions of the equation

$$x = \frac{1}{\lambda} Ax,$$

where A is a linear operator. Here the trivial solution $x = \theta$ occurs for all $\lambda \neq 0$. Those values of λ for which other solutions appear are called eigenvalues. Analogously, λ is called an eigenvalue for a nonlinear operator if the equation $Ax = \lambda x$ has a solution $x \neq \theta$; x is called an eigenvector of the operator A.

§ 1. Nonlinear operators and functionals

1. *Continuity and boundedness of an operator.* Let the operator A be defined on the set T of the Banach space E and let its values belong to the Banach space E_1. The operator A is called *continuous at the point* $x_0 \in T$ if $\|Ax_n - Ax_0\| \to 0$ follows from $\|x_n - x_0\| \to 0 (x_n \in T)$.

An operator A is called *weakly continuous at the point* x_0 if the weak convergence of Ax_n to Ax_0 follows from the weak convergence of the sequence x_n to x_0.

Sometimes we consider operators A which transform a weakly convergent sequence of elements into a strongly convergent sequence, and operators A which transform a strongly convergent sequence of elements into a weakly convergent sequence.

If the space E_1 is the number line, then the operators with values in the space E_1 are called *functionals*.

An operator A is *bounded* on T if

$$\sup_{x \in T} \|Ax\| < \infty .$$

In contrast to linear operators, the boundedness of a nonlinear operator on some sphere does not imply its continuity. The continuity of the operator A on some set T implies its boundedness on a neighborhood of each point (*local boundedness*); but, an operator A which is continuous at every point of a closed sphere might not be bounded on the entire sphere (if the space E is infinite-dimensional). The operator defined on the entire space l_2 by the equality

$$Ax = (\xi_1, \xi_2^2, ..., \xi_n^n, ...) \quad (x = (\xi_1, \xi_2, ..., \xi_n, ...))$$

can serve as an example of such an operator. This operator is continuous at every point of the space l_2, but it is not bounded on any sphere $S(\theta, r)$ for $r > 1$.

If the set T is compact, then a continuous operator A is bounded on the set.

An operator A is *completely continuous* on T if it is continuous and transforms every bounded part of the set T into a (relatively) compact set of the space E_1.

An operator A satisfies a *Lipschitz condition* on T if

$$\|Ax_1 - Ax_2\| \leqslant L \|x_1 - x_2\| \qquad (x_1, x_2 \in T).$$

An operator A, satisfying a Lipschitz condition, is continuous.

2. *Differentiability of a nonlinear operator.* Operators, defined on subsets of the real line, are called *abstract functions*. Let $x(t)$ $(a \leqslant t \leqslant b)$ be an

abstract function with values in the Banach space E. The derivative $x'(t)$ of the function $x(t)$ is defined as the limit as $\Delta t \to 0$ of the difference quotient:

$$x'(t) = \lim_{\Delta t \to 0} \frac{x(t + \Delta t) - x(t)}{\Delta t}.$$

If the limit is considered with respect to the norm of the space E, then the derivative is *strong*; if the limit is considered in the sense of weak convergence in the space E, then the derivative is called *weak*.

An operator A, acting from the Banach space E into the Banach space E_1, is *differentiable in the sense of Fréchet* at the point x_0 if a bounded linear operator $A'(x_0)$ exists, acting from E into E_1, such that

$$A(x_0 + h) - A(x_0) = A'(x_0) h + \omega(x_0; h),$$

where

$$\lim_{\|h\| \to 0} \frac{\|\omega(x_0; h)\|}{\|h\|} = 0.$$

The operator $A'(x_0)$ is called the *Fréchet derivative* of the operator A at the point x_0. We say that the operator is *uniformly differentiable* on the sphere $T = \{\|x\| \leqslant a\}$ if

$$\lim_{\|h\| \to 0} \frac{1}{\|h\|} \|\omega(x_0; h)\| = 0$$

uniformly with respect to $x \in T$.

A bounded linear operator $A'(x_0)$ is called the *Gâteaux derivative* of the operator A at the point x_0 if

$$A'(x_0) h = \lim_{t \to 0} \frac{A(x_0 + th) - A(x_0)}{t}$$

for all $h \in E$. In other words, $A'(x_0)$ is called the Gâteaux derivative if $A'(x_0) h$ is a strong derivative at the point $t = 0$ of the function $A(x_0 + th)$ of the variable t with values in the space E_1:

$$A'(x_0) h = \frac{d}{dt} A(x_0 + th) \Big|_{t=0}.$$

The Fréchet derivative of the operator A, if it exists, is also the Gâteaux derivative. If the Gâteaux derivative $A'(x)$ exists in a neighborhood of the point x_0 and is continuous at the point x_0 (as an operator on x), then it is the Fréchet derivative.

We call the expression $A'(x_0)h$ the *Fréchet differential* (the *Gâteaux differential*, respectively) of the operator A at the point x_0.

If the operator A is completely continuous, then its Fréchet derivative $A'(x_0)$ is a completely continuous linear operator.

If the operator A has a Gâteaux derivative $A'(x)$ on the convex set T, then the equality

$$l\big(A(x+h) - Ax\big) = l\big(A'(x+\tau h)h\big),$$

where $\tau = \tau(l) \in (0, 1)$, holds for every pair of points x, $x+h \in T$ and every linear functional l from the space E_1^* conjugate to E_1. This equality is called the *Lagrange formula*.

An operator A is called *asymptotically linear* if it is defined on all the elements x with a sufficiently large norm and if a linear operator $A'(\infty)$ exists such that

$$\lim_{\|x\| \to \infty} \frac{\|A(x) - A'(\infty)x\|}{\|x\|} = 0.$$

The operator $A'(\infty)$ is called the *derivative at infinity* of the operator A.

3. *Integration of abstract functions.* The *Riemann integral* of an abstract function $x(t)$ $(a \leqslant t \leqslant b)$ with values in a Banach space E is defined as the limit of Riemann sums:

$$\int_a^b x(t)\,dt = \lim_{\max \Delta t_k \to 0} \sum_{k=1}^n x(\tau_k)\,\Delta t_k \quad (\Delta t_k = t_k - t_{k-1};\; t_{k-1} \leqslant \tau_k \leqslant t_k).$$

If this limit exists for an arbitrary sequence of partitionings of the interval $[a, b]$ and does not depend on the choice of this sequence and the choice of points τ_k, then the function $x(t)$ is called *integrable according to Riemann*. The integral is called *strong* if the sums converge to it with respect to the norm of the space E; if the sums are weakly convergent, then the integral is called *weak*.

A strongly continuous abstract function is strongly integrable according to Riemann. It remains to remark that the norm of an abstract function, strongly integrable according to Riemann, can be a scalar function which is not integrable according to Riemann.

The usual properties of an integral hold for the integral of an abstract function. In particular, if the abstract function $x(t)$ has a continuous derivative $x'(t)$ on $[a, b]$, then the formula

$$\int_a^b x'(t)\, dt = x(b) - x(a)$$

is valid.

The integral representation of the increment of an operator A having a continuous Gâteaux derivative:

$$A(x + h) - A(x) = \int_0^1 A'(x + th)\, h\, dt$$

stems from this formula.

The *Bochner integral* is a generalization of the Lebesgue integral to abstract functions. The abstract function $x(t)\,(a \leqslant t \leqslant b)$ is called *integrable according to Bochner* if it is strongly measurable and $\|x(t)\|$ is a scalar function summable according to Lebesgue. In this connection, an abstract function $x(t)$ with values in the space E is called *strongly measurable* if it is a uniform limit of a sequence of finite-valued functions. If the space E is separable, then the concept of strong measurability of an abstract function coincides with the concept of weak measurability. A function $x(t)$ is called *weakly measurable* if the scalar function $f(x(t))$ is Lebesgue measurable for every linear functional $f \in E^*$.

The Bochner integral is defined in the following manner: first let $x(t)$ be a finite-valued function

$$x(t) = x_i\,(t \in \varDelta_i, \quad \varDelta_i \cap \varDelta_j = 0 \quad (i \neq j), \quad \bigcup_1^k \varDelta_i = [a, b]).$$

Then

$$(B) \int_a^b x(t)\, dt = \sum_{i=1}^k x_i \operatorname{mes} \varDelta_i.$$

Now let $x(t)$ be an arbitrary function which is integrable according to Bochner, and let $x_n(t)$ be a sequence of finite-valued functions converging to $x(t)$. Then, by definition,

$$(B) \int_a^b x(t)\, dt = \lim_{n \to \infty} (B) \int_a^b x_n(t)\, dt.$$

The Bochner integral has many of the usual properties of the Lebesgue integral. In particular,

$$\left\| (B) \int_a^b x(t)\, dt \right\| \leqslant \int_a^b \| x(t) \|\, dt .$$

If $[a, b] = \cup \varDelta_i$, $\varDelta_i \cap \varDelta_j = 0$, $i \neq j$, then

$$(B) \int_a^b x(t)\, dt = \sum (B) \int_{\varDelta_i} x(t)\, dt .$$

If A is a bounded linear operator acting from the space E into the space E_1, and the abstract function $x(t)$ with values in E is integrable according to Bochner, then the function $Ax(t)$ with values in E_1 is also integrable according to Bochner and

$$(B) \int_a^b Ax(t)\, dt = A \left((B) \int_a^b x(t)\, dt \right).$$

If A is an unbounded closed linear operator, if the values of the function $x(t)$ belong to its domain of definition and the function $Ax(t)$ is integrable according to Bochner, then the preceding equality is also valid. In this connection, the integral on the right belongs to the domain of definition of the operator A.

4. *Urysohn operator in the spaces* C *and* L_p. Let the function $K(t, s, x)$ be continuous with respect to the collection of variables $(0 \leqslant t, s \leqslant 1, |x| \leqslant r)$. Then the Urysohn operator

$$Ax(t) = \int_0^1 K[t, s, x(s)]\, ds$$

is defined on all the functions of the sphere $S(\theta, r)$ of the space C, and its values belong to C. The operator A is completely continuous on $S(\theta, r)$.

If the continuous derivative $K_x'(t, s, x)$ exists, then the operator A is differentiable in the sense of Fréchet at every interior point $x_0(t)$ of the sphere $S(\theta, r)$. Its derivative $A'(x_0)$ is defined by the formula

$$A'(x_0)\, h(t) = \int_0^1 K'_x[t,\, s,\, x_0(s)]\, h(s)\, ds.$$

For the operator A to be asymptotically linear, it suffices that the function $K(t, s, x)$ satisfy the condition

$$|K(t, s, x) - K_\infty(t, s)\, x| \leqslant \varphi(x)\,(0 \leqslant t, s \leqslant 1, -\infty < x < \infty),$$

where

$$\lim_{|x| \to \infty} \frac{\varphi(x)}{|x|} = 0.$$

The operator $A'(\infty)$ is expressed by the formula

$$A'(\infty)\, h(t) = \int_0^1 K_\infty(t, s)\, h(s)\, ds.$$

It is necessary for the consideration of the Urysohn operator on the entire space C or in the space L_p that the function $K(t, s, x)$ be defined for all values $0 \leqslant t, s \leqslant 1, -\infty < x < \infty$. If this function increases faster than an arbitrary power with respect to the variable x (for example, contains exponential nonlinearities), the Urysohn operator will not be defined on any space L_p. In connection with this, restrictions are imposed on the growth of the function $K(t, s, x)$ with respect to x.

Let

$$|K(t, s, x)| \leqslant R(t, s)\,(a + b\,|x|^{\alpha_0})$$
$$(0 \leqslant t, s \leqslant 1, -\infty < x < \infty),$$

where $\alpha_0 \geqslant 0$ and the function $R(t, s)$ is summable, with respect to both variables, to some power $\beta_0 > 1$:

$$\int_0^1 \int_0^1 |R(t, s)|^{\beta_0}\, dt\, ds < \infty.$$

If

$$\alpha_0 \leqslant \beta_0 - 1,$$

then the Urysohn operator acts, and is completely continuous, in every space

L_p, where $p > 1$ and

$$\frac{\alpha_0 \beta_0}{\beta_0 - 1} \leqslant p \leqslant \beta_0 .$$

In some cases it is convenient to consider the Urysohn operator as an operator acting from one space L_{p_1} into another space L_{p_2}.

The Urysohn operator acts from L_{p_1} into L_{p_2} and is completely continuous if

$$p_1 > 1, \; p_1 \geqslant \frac{\alpha_0 \beta_0}{\beta_0 - 1}, \quad 1 < p_2 \leqslant \beta_0 .$$

Here, the condition $\alpha_0 \leqslant \beta_0 - 1$ is, naturally, not assumed to be satisfied.

If the function $K(t, s, x)$ contains essentially non-power nonlinearities, then in several cases the Urysohn operator is completely continuous in some Orlicz space.

As in the case of the space C, it is natural to look for the derivative of the Urysohn operator, acting in the space L_p, in the form of an integral operator with a kernel $K'_x(t, s, x_0(s))$. However, the differentiability of the Urysohn operator as an operator acting in L_p does not follow from the continuous differentiability of the function $K(t, s, x)$ with respect to the variable x. For example, the operator

$$Ax(t) = \int\limits_0^1 \sin(e^{x(s)}) \, ds \qquad (0 \leqslant t \leqslant 1)$$

acts and is completely continuous in any L_p. However, the operator

$$\int\limits_0^1 e^{x_0(s)} \cos e^{x_0(s)} \, h(s) \, ds$$

is not defined on L_p if the function $e^{x_0(t)}$ is not summable.

In order for the integral operator with kernel $K'_x[t, s, x_0(s)]$ to be the Fréchet derivative of a Urysohn operator acting in the space $L_p \, (p > 1)$, it suffices that the function $K'_x(t, s, x)$ be continuous with respect to x and that the inequality

$$|K'_x(t, s, x)| \leqslant a + b|x|^{p-1} \qquad (0 \leqslant t, s \leqslant 1, -\infty < x < \infty)$$

be satisfied.

5. *Operator f.* Let the function $f(t, x)$ be defined for $0 \leqslant t \leqslant 1$, $-\infty < x < \infty$. It is assumed everywhere in the sequel that the function $f(t, x)$ is continuous with respect to x and measurable with respect to t for every x. The equality

$$fx(t) = f[t, x(t)]$$

defines an operator f.

If $f(t, x)$ is continuous in both variables, then the operator f acts in the space C and is continuous and bounded on every sphere.

If the operator f acts from the space L_{p_1} into the space L_{p_2}, then it is continuous and bounded on every sphere. In order for the operator f to act from L_{p_1} into L_{p_2} it is necessary and sufficient that the inequality

$$|f(t, x)| \leqslant a(t) + b|x|^{\frac{p_1}{p_2}}$$

be satisfied, where $a(t) \in L_{p_2}$.

One must keep in mind that the operator f does not have the property of complete continuity (except for the trivial case when $f(t, x)$ does not depend on x).

If the function $f(t, x)$ and its derivative $f'_x(t, x)$ are continuous in both variables, then the operator f, considered as an operator in the space C, is differentiable in the sense of Fréchet. Its Fréchet derivative has the form

$$f'(x_0) h(t) = f'_x[t, x_0(t)] h(t).$$

Let f act from L_{p_1} into L_{p_2}. The differentiability in the sense of Fréchet of the operator f does not follow from the existence of the continuous derivative $f'_x(t, x)$. The inequality

$$|f'_x(t, x)| \leqslant a_1(t) + b_1|x|^{\frac{p_1}{p_2} - 1},$$

where $a_1(t) \in L_{\frac{p_1 p_2}{p_1 - p_2}}$, serves as a sufficient condition for which the operator of multiplication by $f'_x[t, x_0(t)]$ is the Fréchet derivative of the operator f acting from L_{p_1} into L_{p_2}, where $p_1 > p_2$.

In some cases the operator f is differentiable at certain points of the space L_{p_1} and not differentiable at others.

6. *Hammerstein operator.* If the kernel $K(t, s)$ is continuous, then the linear operator

$$Kx(t) = \int_0^1 K(t, s) x(s) ds$$

acts from an arbitrary space L_p and from the space C into every space L_{p_1} and into the space C and is a completely continuous operator. *In order for the operator to act from L_{p_1} into L_{p_2} and be completely continuous, it suffices that the inequality*

$$\int_0^1 \int_0^1 |K(t, s)|^r \, dt \, ds < \infty$$

be satisfied, where $r = max \left\{ p_2, \dfrac{p_1}{p_1 - 1} \right\}$.

The Hammerstein operator

$$Ax(t) = Kfx(t) = \int_0^1 K(t, s) f[s, x(s)] \, ds$$

is a particular case of the Urysohn operator considered above. The possibility of the representation of the Hammerstein operator in the form of a product Kf allows the search for less restrictive conditions for complete continuity of this operator in the spaces L_p.

For the complete continuity of the Hammerstein operator in the space L_p it suffices that the operator f act from L_p into some space L_{p1} and that the linear operator K be a completely continuous operator acting from L_{p1} to L_p.

If the operator f, considered as an operator from L_p into L_{p_1}, is differentiable at the point x_0 and the operator K acts from L_{p_1} to L_p, then the Hammerstein operator $A = Kf$ is also differentiable at the point x_0; moreover,

$$A'(x_0) h(t) = Kf'(x_0) h(t) = \int_0^1 K(t, s) f'_x[s, x_0(s)] h(s) \, ds.$$

7. *Derivatives of higher order.* Derivatives of higher order of abstract functions are defined in the usual manner. The situation is more complicated with the derivatives of higher order of operators.

An operator $B(x_1, x_2)$ $(x_1, x_2 \in E)$ with values in the space E_1 is called *bilinear* if it is a bounded linear operator with respect to each variable. A bilinear operator $B(x_1, x_2)$ is called *symmetric* if

$$B(x_1, x_2) = B(x_2, x_1).$$

If we set $x_1 = x_2 = x$ in the symmetric bilinear operator $B(x_1, x_2)$, then the operator $B_2(x) = B(x, x)$ is called a *quadratic* operator.

An operator A, acting from the Banach space E into the Banach space E_1, is called *twice differentiable in the sense of Fréchet* at the point x_0 if

$$A(x_0 + h) - A(x_0) = B_1(h) + \tfrac{1}{2}B_2(h) + \omega_2(x_0; h),$$

where $B_1 = B_1(x_0)$ is a linear operator with respect to h, $B_2 = B_2(x_0)$ is a quadratic operator with respect to h and

$$\lim_{\|h\| \to 0} \frac{\|\omega_2(x_0; h)\|}{\|h\|^2} = 0.$$

The quadratic operator $B_2(x_0)$ is called the *second Fréchet derivative* of the operator A at the point x_0:

$$B_2(x_0) = A''(x_0).$$

The expression $B_2(x_0)(h)$ is called the *second Fréchet differential* of the operator A at the point x_0.

We sometimes consider the *second successive Fréchet derivative* of the operator A, the Fréchet derivative of the first Fréchet derivative. The second successive Fréchet derivative is a bilinear operator with respect to the increments h and h_1.

If a function $K(t, s, x)$, which is continuous with respect to the collection of variables, is twice continuously differentiable with respect to x, then the Urysohn operator, acting in the space C, has a second Fréchet derivative

$$A''(x_0) h = \int_0^1 K''_{x^2}[t, s, x_0(s)] h^2(s) \, ds.$$

The second successive Fréchet derivative of the Urysohn operator is given by the formula

$$A''(x_0)(h_1, h_2) = \int_0^1 K''_{x^2}[t, s, x_0(s)] h_1(s) h_2(s) \, ds.$$

A quadratic operator $B(x_0)$ is called the *second Gâteaux derivative* of the operator A at the point x_0 if

$$\frac{d^2}{dt^2} A(x_0 + th)\big|_{t=0} = B(x_0)(h).$$

for arbitrary $h \in E$. The expression $B(x_0)(h)$ is called the *second Gâteaux differential* of the operator A at the point x_0.

The *second successive Gâteaux derivative* of an operator A is defined as the Gâteaux derivative of the first Gâteaux derivative. It is a bilinear operator.

One must keep in mind that, generally speaking, the existence of the second Gâteaux derivative does not follow from the existence of the second Fréchet derivative of the operator A. For example, the scalar function $f(t) = t^3 \cos \dfrac{1}{t^2}$ has a second Fréchet derivative which is equal to zero at the point $t = 0$, but it does not have a second Gâteaux derivative at this point.

8. *Potential operators.*   Differentiable functionals are a particular case of differentiable operators. If $\Phi(x)$ is a nonlinear functional, defined on the Banach space E, which is differentiable in the sense of Fréchet, then

$$\Phi(x + h) - \Phi(x) = l(h) + \omega(x; h),$$

where l is a linear functional depending on x and

$$\lim_{\|h\| \to 0} \frac{|\omega(x; h)|}{\|h\|} = 0.$$

If the functional $\Phi(x)$ is differentiable at every point x of some set $T \subset E$, then the equality

$$\Gamma(x) = l$$

defines an operator acting from $T \subset E$ into the conjugate space E^*. This operator is called the *Fréchet gradient of the functional* $\Phi(x)$:

$$\Phi(x + h) - \Phi(x) = \Gamma(x)(h) + \omega(x; h).$$

The functional $\Phi(x)$ is called *uniformly differentiable* on the set T if the ratio $\dfrac{|\omega(x; h)|}{\|h\|}$ tends uniformly to zero with respect to $x \in T$.

The norm

$$\Phi(x) = \|x\| = \sqrt{(x, x)}$$

is an example of a differentiable functional in a Hilbert space H. In this case

$$\operatorname{grad} \Phi(x) = \frac{x}{\|x\|}$$

for $\|x\| \neq 0$.

The norm is a differentiable functional in the spaces $L_p (p>1)$: if

$$\Phi(x) = \|x\| = \left(\int\limits_0^1 |x(s)|^p \, ds \right)^{\frac{1}{p}},$$

then

$$\text{grad } \Phi(x) = \frac{|x(s)|^{p-1} \text{ sgn } x(s)}{\|x\|^{p-1}}$$

for $\|x\| \neq 0$.

The Gâteaux gradient of the functional $\Phi(x)$ is defined by the equality

$$\frac{d}{dt} \Phi(x + th) \bigg|_{t=0} = \Gamma(x)(h).$$

An operator which is the gradient of a functional is called a *potential* operator. A bounded linear self-adjoint operator A acting in a Hilbert space H can serve as an example of a potential operator. It is the gradient of the functional $\Phi(x) = \frac{1}{2}(Ax, x)$.

The operator

$$fx(t) = f[t, x(t)],$$

acting from L_p into $L_{p'}$ $\left(\dfrac{1}{p} + \dfrac{1}{p'} = 1 \right)$, is another example of a potential operator. It is the gradient of the so-called *Hammerstein-Golomb functional*

$$\Phi(x) = \int\limits_0^1 \left[\int\limits_0^{x(t)} f(t, u) \, du \right] dt,$$

defined on the space L_p.

Let B be a bounded linear operator acting from the Banach space E into the Banach space E_1. Let $\Phi(y)$ be a differentiable functional defined on E_1. Then the functional

$$F(x) = \Phi(Bx),$$

defined on E, is also differentiable and

$$\text{grad } \Phi(x) = B^* \text{ grad } \Phi(Bx),$$

where B^* is the adjoint of B.

§ 2. Existence of solutions

1. *Method of successive approximations.*   Let the equation

$$x = Ax$$

be given, where A is some nonlinear operator. The method of successive approximations remains a basic method for the proof of the existence of solutions of this equation, that is, the proof of the existence of fixed points of the operator A. It consists of forming the sequence

$$x_n = Ax_{n-1} \qquad (n = 1, 2, ...)$$

from some initial element x_0 and proving that this sequence converges to some element x^*, thus establishing the equality $x^* = Ax^*$.

Example

(EXISTENCE OF A SOLUTION FOR THE VOLTERRA EQUATION).   In a nonlinear Volterra equation

$$x(t) = \int_0^t K[t, s, x(s)] \, ds$$

let the functions $K(t, s, x)$ and $K'_x(t, s, x)$ be continuous with respect to the collection of variables $t, s \geqslant 0$, $-\infty < x < \infty$, and let

$$|K(t, s, x)| \leqslant \varphi(x),$$

where $\varphi(x)$ is a non-decreasing function. If the differential equation

$$\frac{du}{dt} = \varphi(|u|)$$

has a solution on the segment $[0, \omega]$ satisfying the condition $u(0) = 0$, then the Volterra equation has a solution $x^*(t)$, defined on $[0, \omega]$. If we set $x_0(t) \equiv 0$, then the successive approximations

$$x_n(t) = \int_0^t K[t, s, x_{n-1}(s)] \, ds \qquad (n = 1, 2, ...)$$

will converge uniformly to some function $x^*(t)$ on $[0, \omega]$ which is a solution of the Volterra equation. This is implied by the fact that all $x_n(t)$

are in the region $-u(t) \leqslant x(t) \leqslant u(t)$ and satisfy the relation

$$|x_n(t) - x_{n-1}(t)| \leqslant \frac{M}{L} \frac{(Lt)^n}{n!} \qquad (n = 1, 2, \ldots),$$

where M and L are constants such that

$$|K(t, s, 0)| \leqslant M \qquad (0 \leqslant t, s \leqslant \omega)$$

and

$$|K(t, s, x_1) - K(t, s, x_2)| \leqslant L |x_1 - x_2|$$
$$(0 \leqslant t, s \leqslant \omega, \; -u(s) \leqslant x_1, x_2 \leqslant u(s)).$$

2. *Principle of contractive mappings.* In most cases, the method of successive approximations reduces to the verification of the conditions of the following general principle.

PRINCIPLE OF CONTRACTIVE MAPPINGS. *Let T be a closed set of the Banach space E. Let the operator A map T into itself and be an operator of contraction, that is, satisfy the Lipschitz condition*

$$\|Ax_1 - Ax_2\| \leqslant q \|x_1 - x_2\| \qquad (x_1, x_2 \in T)$$

with a constant $q < 1$. Then the equation $x = Ax$ has a unique solution x^* in T which is the limit of the successive approximations $x_n = Ax_{n-1}$ for an arbitrary initial approximation $x_0 \in T$.

In the conditions of this principle, the role of T is usually played by either the entire space E or some sphere $S(\theta, r)$. In some cases, the set T must be constructed in a special way.

Example. In the integral equation

$$x(t) = \int_0^1 K[t, s, x(s)] \, ds + f(t)$$

let the functions $K(t, s, x)$ and $f(t)$ be continuous and $K(t, s, x)$ satisfy a Lipschitz condition with a constant $q < 1$ with respect to the variable x.

This integral equation can be considered as an operator equation in the space C of functions which are continuous on $[0, 1]$. The operator defined by the right side of the equation satisfies a Lipschitz condition with a constant $q < 1$. Therefore, by virtue of the principle of contractive mappings, the integral equation has a continuous solution which is a limit of the successive approximations

$$x_n(t) = \int_0^1 K[t, s, x_{n-1}(s)] \, ds + f(t) \qquad (n = 1, 2, \ldots).$$

It is sometimes convenient to use a corollary of the principle of contractive mappings: *let the operator B also map the closed set T of the space E into itself and commute with the operator A, satisfying the conditions of the principle of contractive mappings, i.e.,*

$$BAx = ABx \qquad (x \in T).$$

Then the fixed point of the operator A is a fixed point (possibly not unique) of the operator B.

In particular, if some iterate B^n of the operator B satisfies the conditions of the principle of contractive mappings on the set T, then the fixed point x^* of the operator B^n is also a fixed point of the operator B. In this case x^* is the unique fixed point of the operator B.

It remains to stress that all the cases when the solution of a nonlinear equation can be obtained as the limit of successive approximations are not exhausted by the principle of contractive mappings.

3. *Uniqueness of a solution.* Under the conditions of the principle of contractive mappings, the solution of the equation $x = Ax$ is unique in T. However, the uniqueness of the solution does not generally follow from the uniqueness of the solution in T. For example, the equation

$$x(t) = \int_0^1 x^2(t) \, dt$$

satisfies the conditions of the principle of contractive mappings in the sphere $\|x\| \leqslant \frac{1}{4}$ of the space C and has a unique solution $x_0(t) \equiv 0$ in it; however this equation has a second continuous solution $x_1(t) \equiv 1$.

One must keep in mind that the uniqueness of the solution of some operator equation in the Banach space E does not imply the uniqueness of the solution of this equation considered in a wider space. Examples of linear integral Volterra equations exist which have, besides unique continuous solutions, non-summable solutions.

4. *Equations with completely continuous operators. Schauder principle.*
The principle of contractive mappings imposes on a continuous operator
the rigid restriction of strict contraction. If we consider completely
continuous operators, then this condition can be weakened considerably.

SCHAUDER PRINCIPLE. *Let the operator A be completely continuous and
map a closed bounded convex set T into itself. Then the equation $x = Ax$
has at least one solution (uniqueness of the solution is not guaranteed) in T.*

If the set T is compact, then it is sufficient that the operator A be
continuous.

In the application of the Schauder principle to the study of concrete
equations, we first construct a space E in which the operator A is completely continuous. For the convex set T we take some sphere of the space
E. Furthermore, it is necessary to select the radius and center of the
sphere so that the operator A maps the sphere into itself.

For example, let the completely continuous operator A have the property

$$\|Ax\| \leqslant a + b \|x\|^{\alpha} \qquad (x \in E, \alpha, a, b > 0).$$

If a number $r > 0$ satisfying the condition

$$a + br^{\alpha} \leqslant r$$

exists, then we apply the Schauder principle to the operator A in the
sphere $S(\theta, r)$. Such a number r always exists for $\alpha < 1$ and for $\alpha = 1$ and
$b < 1$. If $\alpha > 1$, then r exists under the condition that

$$\min_{0 \leqslant s < \infty} (bs^{\alpha} - s) \leqslant - a.$$

The Schauder principle only states the existence of a solution and does
not give a method for finding it. In the case when the operator A is
linear, we can indicate a method for finding solutions. Starting with
some $x_0 \in T$, we construct the successive approximations $x_n = Ax_{n-1}$
$(n = 1, 2, ...)$. Under the conditions of the Schauder principle, the sequence of elements

$$z_N = \frac{1}{N} \sum_{n=0}^{N-1} x_n$$

is compact and all of its limit points are solutions of the equation $x = Ax$.

We can formulate an assertion for which both the Schauder principle
and the principle of contractive mappings are special cases.

COMBINED PRINCIPLE. *Let the operator, defined in a closed bounded convex set T, allow the representation $A = A_1 + A_2$, where A_1 is completely continuous and A_2 satisfies a Lipschitz condition with constant $q < 1$. If the condition*

$$A_1 x + A_1 y \in T \qquad (x, y \in T)$$

is satisfied, then the equation $x = Ax$ has at least one solution in T.

The Schauder principle is proved by topological methods. These same methods (see no. 5) allow it to be strengthened.

THE STRENGTHENED SCHAUDER PRINCIPLE. *If the completely continuous operator A does not have eigenvectors with eigenvalues greater than 1, on the boundary Γ of the closed convex set T containing θ as an interior point, then a solution of the equation $x = Ax$ exists in T.*

Thus, it is possible not to require that the boundary Γ of the region be mapped by the operator A into T. It suffices that there be no vectors on it which the operator A "expands" $(Ax = \varrho x, \varrho > 1)$. Frequently the last condition is verified significantly more easily. For example, if a linear functional $f_0(x)$ exists such that $f_0(x_0) > 0$ and $f_0(Ax_0) \leqslant f_0(x_0)$ for every point x_0 of the boundary Γ, then the condition of the strengthened Schauder principle is satisfied. In particular, if a completely continuous operator A is defined on the sphere $S(\theta, r)$ of a Hilbert space H and has the property that

$$(Ax, x) \leqslant (x, x) \qquad (\|x\| = r),$$

then the strengthened Schauder principle is valid for it.

5. *Use of the theory of completely continuous vector fields.* Let a completely continuous operator A be given on the boundary Γ of the sphere S of the Banach space E. The collection of elements of the form $x - Ax \, (x \in \Gamma)$ is called a *completely continuous vector field* on Γ. A solution of the equation $x = Ax(x \in \Gamma)$ is called a *zero of the field.*

An integer $\gamma(\Gamma)$, the so-called *degree (rotation)* of the vector field (see [23]), is set in correspondence to every completely continuous vector field $x - Ax$ without zeros on Γ. The degree can be positive, negative or zero.

PRINCIPLE OF NON-ZERO DEGREE. *If A is a completely continuous operator on the sphere S and the degree of the vector field $x - Ax$ on the boundary Γ of the sphere S is different from zero, then a solution of the equation $x = Ax$ exists in S.*

The Schauder principle and the strengthened Schauder principle are special cases of the principle of non-zero degree since the degree is equal to 1 for the conditions of these principles.

Two vector fields $x - A_0 x$ and $x - A_1 x$ are called *homotopic* on Γ if a completely continuous operator $A(x; \alpha)$ $(x \in \Gamma, 0 \leqslant \alpha \leqslant 1)$ with respect to both variables exists such that

$$A(x; 0) \equiv A_0 x, \qquad A(x; 1) \equiv A_1 x \qquad (x \in \Gamma)$$

and

$$A(x; \alpha) \neq x \qquad (x \in \Gamma, 0 \leqslant \alpha \leqslant 1).$$

An operator $A(x; \alpha)$ $(x \in \Gamma, 0 \leqslant \alpha \leqslant 1)$, continuous with respect to both variables, will, in particular, be completely continuous if it is completely continuous for every fixed α and uniformly continuous with respect to α relative to $x \in \Gamma$.

The degrees of homotopic completely continuous vector fields are identical. This fact allows the application of the following method for the proof of the existence of a solution of the equation $x = Ax$ with a completely continuous operator A. We introduce the parameter λ so that the operator $A(x; \lambda)$ is completely continuous, $A(x; 1) \equiv Ax$ and $A(x; \lambda) \neq x$ $(x \in \Gamma)$, $0 \leqslant \lambda \leqslant 1$. If it is now known that the degree of the vector field $x - A(x; 0)$ on Γ is different from zero (for example, if $A(x; 0)$ satisfies the conditions of the Schauder principle), then it is immediately implied by the equality of the degrees of the fields $x - A(x; 0)$ and $x - Ax$ that the initial equation $x = Ax$ has at least one solution on S. This method of the proof of the existence of solutions is called the *Leray-Schauder method*.

The evaluation of the degree of a vector field is carried out by the methods of combinatorial topology. The character of the degree is known for some classes of vector fields. For example, if

$$\frac{x - Ax}{\|x - Ax\|} \neq \frac{- x - A(- x)}{\|- x - A(- x)\|}$$

on the sphere $\|x\| = r$ (at symmetric points of the sphere, the vectors of the field are not directed identically), then the degree of the field is different from zero (moreover, it is odd).

Let $Ax_0 = x_0$ and the equation $x = Ax$ not have solutions different from x_0 in some neighborhood of the point x_0. Then the vector field $x - Ax$ has the same degree on all the spheres $\|x - x_0\| = r$ of sufficiently small radius r. This common degree $\gamma(x_0)$ is called the *index of the fixed point* x_0 of the operator A.

If the operator A is differentiable at the point x_0, where the linear operator $A'(x_0)$ does not have 1 as an eigenvalue, then

$$\gamma(x_0) = (-1)^\beta,$$

where β is the sum of the multiplicities of the eigenvalues of the operator $A'(x_0)$ which are greater than 1.

We understand by the *multiplicity of an eigenvalue* of a bounded linear operator B the dimension of the corresponding eigenspace. The multiplicity of every eigenvalue of a completely continuous operator is finite.

If 1 is an eigenvalue of the operator $A'(x_0)$, then the evaluation of the index $\gamma(x_0)$ is complicated; this evaluation uses derivatives of higher order. Here only partial results are obtained.

Let the equation $x = Ax$ have a finite number of solutions $x_1, \ldots, x_k$ in the sphere S. Then the degree $\gamma(\Gamma)$ of the field $x - Ax$ on Γ is connected with the indices of the points $x_1, \ldots, x_k$ by the equality

$$\gamma(\Gamma) = \gamma(x_1) + \cdots + \gamma(x_k).$$

This property of the degree can be applied in proofs of uniqueness theorems. If the degree $\gamma(\Gamma)$ of the vector field $x - Ax$ on Γ is in absolute value equal to 1 and if the index of every possible solution has the same sign, then the solution is unique by virtue of the preceding.

Conversely, if the degree $\gamma(\Gamma)$ is known and the index $\gamma(x_0)$ of the known solution x_0 turns out to be different from $\gamma(\Gamma)$, then the equation $x - Ax$ has at least one more solution besides x_0 on S.

Example

(EXISTENCE OF A SECOND SOLUTION FOR AN URYSOHN EQUATION). Let the operator A, defined by the right side of the Uryson equation

$$x(t) = \int_0^1 K[t, s, x(s)] \, ds,$$

be completely continuous in L_p and differentiable at the origin of this space, where

$$A'(\theta) h(t) = \int_0^1 K_x'(t, s, 0) h(s) \, ds.$$

If the operator A satisfies the conditions of the Schauder principle on the sphere $S(\theta, r)$, then the degree $\gamma(\Gamma)$ of the field $x - Ax$ on the sphere $\|x\| = r$ is equal to 1. Let $K(t, s, 0) \equiv 0$. Then the equation has a zero solution. If 1 is not an eigenvalue of the completely continuous linear operator $A'(\theta)$ and the sum of the multiplicities of its eigenvalues greater than 1 is odd, then $\gamma(\theta) = -1$. Thus, $\gamma(\Gamma) \neq \gamma(\theta)$ and the Urysohn equation has at least one non-zero solution in $S(\theta, r)$.

6. *Variational method.* The variational method of proof of theorems about the existence of solutions consists of the construction of a solution of the operator equation as an extremal point of some functional.

A functional $F(x)$, defined on a Banach space E, is called *weakly continuous* if it is continuous in the weak topology $\sigma(E, E')$ in the space E (see ch. I, § 4, no. 3). If the space E is reflexive, then by virtue of the compactness of an arbitrary sphere of E in the weak topology, a weakly continuous functional assumes its least and greatest values on every sphere.

The gradient of a smooth weakly continuous functional on a Hilbert space is a completely continuous operator.

Let A be a potential operator in the Hilbert space H.

VARIATIONAL PRINCIPLE. *If the operator A is the gradient of a weakly continuous functional $F(x)$ and*

$$\lim_{\|x\| \to \infty} \left[\tfrac{1}{2}(x, x) - F(x) \right] = \infty,$$

then a point x_0 exists in H at which the functional $\tfrac{1}{2}(x, x) - F(x)$ assumes its least value and which is a solution of the equation $x = Ax$.

7. *Transformation of equations.* In the study of operator equations, we often find ourselves transforming equations into a form which is convenient for the application of one principle or another from which the existence of a solution follows, or for the application of some approximation method for finding a solution.

Basic forms of the transformation of operator equations are the same as for ordinary equations: a) the addition to both sides of the equation of the same element; b) the application to both sides of the equation of the same operator ("multiplication by an operator"); c) replacement of the variable.

In the first transformation, the equation changes to an equivalent one. If a bounded linear operator B is applied to both sides of the equation, then every solution of the intial equation will be a solution of the new equation. The converse will be true if the inverse operator B^{-1} exists. Thus, the change to the new equation by a transformation b) can add extraneous solutions if zero is an eigenvalue of the linear operator B.

If a replacement of the variable of the form $x = Cy$ is carried out in the equation, where C is some operator, and the solutions y^* of the new equation can be found, then in order to obtain a solution x^* of the initial equation, it is necessary to verify that y^* is in the domain of definition of the operator C, and then $x^* = Cy^*$. Moreover, in transformation c), part of the solutions can be lost. This occurs if there are solutions x^* which are not representable in the form Cy. The transformation c) happens to be especially useful in that the operator C can act from another space E_1 into the space E in which the solution x^* is sought. Therefore, it is natural to consider the new equation (with respect to y) in the space E_1. It sometimes turns out that the equation is simpler in E_1.

In the transformation of equations in infinite-dimensional spaces we encounter a specific situation: the transformed equation contains operators which are not closed but allow closure. In this connection, it is natural to study the equation with closed operators. In this case, new solutions can appear which are usually called *generalized solutions*. The basic difficulty frequently is the proof that a generalized solution belongs to the domains of definition of the operators occurring in the equation, before their closure, and, hence, is a true solution.

8. *Examples. Decomposition of operators.*

1. PREPARATION OF AN EQUATION FOR THE APPLICATION OF THE METHOD OF SUCCESSIVE APPROXIMATIONS.   In the equation

$$Bx = f,$$

let the operator B be linear, bounded and have a spectrum lying inside the right half-plane Re $\lambda > 0$ of the complex plane. After multiplication of both sides of the equation by the number k and the addition to both sides of the element x it reduces to the form

$$x = (I - kB)\, x + kf.$$

For sufficiently small k, the operator $I-kB$ will have a spectrum lying inside the unit circle and, hence, we apply the method of successive approximations to determine the solution of this new equation (which is equivalent to the old equation).

It is sometimes convenient to apply an analogous transformation of the equation: we replace multiplication by a number k with multiplication by a suitably chosen operator K.

2. EQUATIONS WHICH ARE CLOSE TO LINEAR EQUATIONS. The equation $x=Ax$, with a completely continuous operator A, is transformed into the form $x-Bx=Ax-Bx$, where B is a completely continuous linear operator. If the number 1 is not an eigenvalue of the operator B, then this equation is equivalent to the equation

$$x = (I - B)^{-1} (A - B) x.$$

If, on the sphere $\|x\|=r$, the operator $(I-B)^{-1}(A-B)$ does not have eigenvectors corresponding to eigenvalues greater than 1, then, by virtue of the strengthened Schauder principle, the equation obtained has at least one solution in the sphere $\|x\| \leqslant r$. There will not be such eigenvectors if the operator A is close to the operator B in the sense that

$$\|Ax - Bx\| \leqslant \|x - Bx\|.$$

3. DECOMPOSITION OF OPERATORS. Let the operator B be linear in the equation

$$x = BCx$$

and allow "decomposition" into two factors: $B=B_1B_2$, where B_1 and B_2 are linear operators.

Every solution of the equation is representable in the form $x=B_1y$. This replacement reduces the equation to the equivalent equation

$$y = B_2CB_1y.$$

A special form of decomposition of an operator is often convenient: $B=B^{\alpha}B^{1-\alpha}$, where B^{α} and $B^{1-\alpha}$ are fractional powers of the operator B. In connection with this, the theory of fractional powers of linear operators was, in recent years, amply developed (see ch. III, § 3, no. 4).

The Hammerstein equation

$$x(t) = \int_0^1 K(t, s) f[s, x(s)] \, ds$$

is the simplest example of an equation of the type considered.

Let the kernel $K(t, s)$ be symmetric, bounded and positive definite. It generates a completely continuous positive definite operator B in the Hilbert space $L_2[0, 1]$. If $\{e_i(t)\}$ is a complete orthonormal system of eigenfunctions of the operator B, and λ_i are the corresponding eigenvalues, then the operator $B^{\frac{1}{2}}$ is defined by the formula

$$B^{\frac{1}{2}}x(t) = \sum_{i=1}^{\infty} \sqrt{\lambda_i} c_i e_i(t),$$

where the c_i are the Fourier coefficients of the function $x(t)$:

$$c_i = \int_0^1 e_i(s) x(s) \, ds.$$

With the replacement $x = B^{\frac{1}{2}}y$, the Hammerstein equation reduces to the form

$$y = B^{\frac{1}{2}} f B^{\frac{1}{2}} y.$$

We can show that the operator $B^{\frac{1}{2}}$ acts from the space $L_2[0, 1]$ into the space $M[0, 1]$. Therefore, if the function $f(t, x)$ is continuous, then the operator $f B^{\frac{1}{2}} y$ will be a continuous operator, acting from $L_2[0, 1]$ into $M[0, 1]$. In this case, the operator $B^{\frac{1}{2}} f B^{\frac{1}{2}}$ is completely continuous in $L_2[0, 1]$.

For the operator $B^{\frac{1}{2}} f B^{\frac{1}{2}}$ in L_2, it is convenient to verify the conditions of the strengthened Schauder principle in the form indicated at the end of no. 5. In fact,

$$(B^{\frac{1}{2}} f B^{\frac{1}{2}} y, y) = (f B^{\frac{1}{2}} y, B^{\frac{1}{2}} y).$$

If the function $f(t, x)$ does not increase too quickly with respect to x, then

$$(f B^{\frac{1}{2}} y, B^{\frac{1}{2}} y) < (y, y)$$

on a sphere of sufficiently large radius $r : \|y\| = r$. Therefore the equation $y = B^{\frac{1}{2}} f B^{\frac{1}{2}} y$ has, by virtue of the indicated principle, at least one solution

y^* inside the sphere $\|y\| \leqslant r$. Hence $x^* = B^{\frac{1}{2}}y^*$ will be a solution of the Hammerstein equation. Moreover, $y^* \in L_2[0, 1]$ and, hence, $x^* \in M[0, 1]$.

The proof mentioned for the existence of a bounded solution of the Hammerstein equation can be successfully carried out if, for example, the function $f(t, x)$ satisfies the inequality

$$xf(t, x) \leqslant ax^2 + b(t)|x|^{2-\gamma} + c(t),$$

where $0 < \gamma < 2$, $b(t) \in L_\gamma^2[0, 1]$, $c(t) \in L_1[0, 1]$ and $a < \dfrac{1}{\lambda_1}$.

In the example considered, the operator $B^{\frac{1}{2}}$ acts from L_2 to M. In more general cases it acts from L_2 to $L_p (p > 2)$ and is completely continuous. Moreover, it can be naturally extended to the operator $(B^{\frac{1}{2}})^*$ which acts from $L_{p'}$ into $L_2 \left(\dfrac{1}{p} + \dfrac{1}{p'} = 1 \right)$. If the nonlinear operator f acts from L_p into $L_{p'}$, then the equation

$$y = (B^{\frac{1}{2}})^* f B^{\frac{1}{2}} y$$

will be an equation in L_2 with a completely continuous operator and the reasoning mentioned above will be applicable to it. If $y^* \in L_2$ is a solution of this equation, then $x^* = B^{\frac{1}{2}}y^* \in L_p$ is a solution of the equation

$$x = B^{\frac{1}{2}}(B^{\frac{1}{2}})^* fx.$$

The operator $B^{\frac{1}{2}}(B^{\frac{1}{2}})^*$ is an extension of the operator B, considered originally on the space L_2. Therefore the solutions of the last equation can be considered as generalized solutions of the initial equation. In the case of the Hammerstein equation they turn out to be true solutions.

§3. Qualitative methods in the theory of branching of solutions

In this section the equation

$$x = A(x, \mu)$$

is considered, where μ is real.

It is assumed everywhere that the operator A is uniformly continuous in μ relative to the elements x of an arbitrary bounded set.

If, on the basis of one of the principles studied in the preceding section, we succeed in establishing the existence of a solution of the equation for $\mu = \mu_0$, then we succeed in the majority of cases in proving the existence of a solution for close values of the parameter μ, since the conditions of the applicability of the corresponding principle are not violated for small changes of the operator $A(x, \mu_0)$. For the determination of the magnitude of the segment $[\mu_0 - a, \mu_0 + b]$ on which these conditions are retained, it is necessary to estimate $\|A(x, \mu) - A(x, \mu_0)\|$, for which *a priori* estimates for the solutions of the corresponding equations are frequently required.

1. *Extension of solutions, implicit function theorem.* If x_0 is a solution of the equation $x = A(x, \mu_0)$, then it is natural to expect that the equation $x = A(x, \mu)$ will have a solution $x(\mu)$ close to x_0 for values of the parameter μ close to μ_0. In establishing this fact, the general implicit function theorem plays an important role.

In the equation

$$F(x, u) = 0,$$

let x be an element of the Banach space E_1, u an element of the Banach space E_2, and $F(x, u)$ an operator with values in the Banach space E_3. We understand a solution of this equation to be an operator $X(u)$, defined on some set of elements $u \in E_2$ with values in E_1, such that $F(X(u), u) \equiv 0$.

An analogue of the usual theorem on the existence of an implicit function holds: *if $F(x_0, u_0) = 0$, if the operator $F(x, u)$ is continuous and is continuously differentiable according to Fréchet with respect to the variable x for $\|x - x_0\| \leqslant a$, $\|u - u_0\| \leqslant b$ and if the linear operator $F_x'(x_0, u_0)$ has a bounded inverse, then a solution $X(u)$ of the equation $F(x, u) = 0$ is defined in some neighborhood of the point u_0. This solution is unique. The operator $X(u)$ is continuous.*

If the operator $F(x, u)$ has a Fréchet derivative with respect to u of a specified order, then the operator $X(u)$ has a Fréchet derivative of the same order.

In the application of the implicit function theorem to the equation $x = A(x, \mu)$, the requirement of the invertibility of the operator $F_x'(x_0, u_0)$ is naturally replaced by the requirement that 1 not belong to the spectrum of the operator $A_x'(x_0, \mu_0)$. Under the satisfaction of this condition and the continuity of the operator $A_x'(x, \mu)$ with respect to (x, μ) in a neighborhood of the point (x_0, u_0), a unique continuous solution $x(\mu)$ of the equation $x = A(x, \mu)$ exists such that $x(\mu_0) = x_0$.

2. *Branch points.* If the operator $A(x, \mu)$ is completely continuous

for every μ in the equation $x = A(x, \mu)$, then topological methods can be applied to the investigation of the behavior of the solution as μ varies.

Let x_0 be an isolated solution of the equation $x = A(x, \mu_0)$, having a non-zero index. On a sufficiently small sphere S surrounding the point x_0, the degree of the field $x - A(x, \mu_0)$ will be different from zero. Hence the degree of the field $x - A(x, \mu)$ will be different from zero on S for μ close to μ_0. It follows from the principle of non-zero degree that at least one solution $x(\mu)$ of the equation $x = A(x, \mu)$ exists inside S. Reducing the radius of the sphere S, we can choose a solution $x(\mu)$ such that

$$\lim_{\mu \to \mu_0} \|x(\mu) - x_0\| = 0.$$

In this sense we can speak of the continuity of the solution $x(\mu)$ at the point μ_0.

The pair (x_0, μ_0) is called a *branch point* for the equation $x = A(x, \mu)$ if for every $\varepsilon > 0$ we can find a μ such that $|\mu - \mu_0| < \varepsilon$ and the equation $x = A(x, \mu)$ has at least two solutions lying in the ε-neighborhood of the point x_0.

It is implied by the arguments of the preceding subsection that the pair (x_0, μ_0) is not a branch point if the operator $A'_x(x, \mu)$ exists and is continuous with respect to (x, μ) in a neighborhood of the point (x_0, μ_0) and if 1 is not a point of the spectrum of the operator $A'_x(x_0, \mu_0)$. But if we assume only the existence of the operator $A'_x(x_0, \mu_0)$, not having 1 as a point of the spectrum, and do not assume the existence of the operator $A'(x, \mu)$ in a neighborhood of the point (x_0, μ_0), then the pair (x_0, μ_0) may turn out to be a branch point.

Let the equation $x = A(x, \mu)$ have, inside the sphere S, for every μ close to μ_0 and different from it, only a finite number of solutions, where the operator $A'_x(x, \mu)$ exists at these point-solutions and 1 is not an eigenvalue. Then the number of such solutions differs from the degree of the field $x - A(x, \mu)$ on the sphere S by an even number (the index of every solution is ± 1, the sum of the indices is equal to the degree of the field). As long as the degree of the field $x - A(x, \mu)$ on S is not changed by changing μ close to μ_0, then the number of solutions of the equation $x = A(x, \mu)$ as μ passes through μ_0 can be changed only by an even number. This statement is called the *principle of the preservation of parity of the number of solutions*.

If the index of the solution x_0 is equal to zero, then a solution $x(\mu)$ of the equation $x = A(x, \mu)$, generally, may not exist in a neighborhood of the point x_0 for μ close to μ_0. The solutions can "flow into" x_0 for $\mu \to \mu_0 - 0$ and not exist for $\mu > \mu_0$; then the pair (x_0, μ_0) is called a *point of cessation of solutions*. The solutions may not exist for $\mu < \mu_0$ and "flow out" from the point x_0 for $\mu \to \mu_0 + 0$; then the pair (x_0, μ_0) is called a *point of appearance of solutions*.

3. *Points of bifurcation, linearization principle.* The concept of a point of bifurcation is closely related to the concept of a point of branching.

Let us assume that $A(\theta, \mu) = \theta$. Then the equation $x = A(x, \mu)$ has the trivial solution $x = \theta$ for all values of the parameter μ. The number μ_0 is called a *point of bifurcation* for this equation (or for the operator $A(x, \mu)$) if to any $\varepsilon > 0$ there corresponds a value of the parameter μ in the segment $|\mu - \mu_0| < \varepsilon$ for which the equation has at least one non-zero solution $x(\mu)$ satisfying the condition $\|x(\mu)\| < \varepsilon$. Unlike the definition of a point of branching, it is assumed in the definition of a point of bifurcation that one family of solutions is known *a priori*, defined for all values of the parameter. We speak about a "branch" of solutions from the given family. In the definition of a point of bifurcation, we do not speak about those values of the parameter for which the equation has small non-zero solutions. These values can form a discrete set or even coincide with μ_0. The generality of the concept of a point of bifurcation allows us to obtain general, simple theorems concerning methods of determining these points. At the same time the concept of a point of bifurcation describes reasonably completely the occurrence of the non-zero solutions.

For a linear equation $x = \mu B x$ with a completely continuous linear operator B, the points of bifurcation coincide with the *characteristic values* of the operator B (values inverse to the eigenvalues).

If the operator $A(x, \mu)$ is continuously differentiable in the sense of Fréchet, then by virtue of the implicit function theorem its points of bifurcation can only be those values μ for which 1 is a point of the spectrum of the operator $A'_x(\theta, \mu)$. Let $A'_x(\theta, \mu) = \mu B$ where B is a completely continuous linear operator which does not depend on μ. If 1 is an eigenvalue of the operator μB, then μ is a characteristic value of the operator B. Thus, in this case, the points of bifurcation are characteristic values of the operator B. The question is raised: *is every characteristic value of the operator B a point of bifurcation?* In the general case, as examples indicate, the answer is negative.

We call the principle according to which the determination of points of bifurcation is reduced to the determination of the characteristic values of the linear operator B the *linearization principle*. The following statement serves as a basis for this principle: *if the completely continuous operator $A(x, \mu)$ $(A(\theta, \mu)=\theta)$ has a Fréchet derivative $A'_x(\theta, \mu)=\mu B$ at the point θ, then every odd-multiple (in particular, simple) characteristic value of the operator B is a point of bifurcation of the operator $A(x, \mu)$.*

If a characteristic value of the operator B has an even multiplicity, then further analysis is required which uses more than just the linear part μB of the operator $A(x, \mu)$. Let the completely continuous operator $A(x, \mu)$ allow the representation

$$A(x, \mu) = \mu B x + C(x, \mu) + D(x, \mu),$$

where B, as above, is a completely continuous linear operator; the operator $C(x, \mu)$ consists of terms of k-th order of smallness where $k > 1$ is an integer, that is,

$$C(\alpha x, \mu) = \alpha^k C(x, \mu)$$

and

$$\|C(x_1, \mu) - C(x_2, \mu)\| \leqslant q(\varrho)\|x_1 - x_2\|$$
$$(\|x_1\| \leqslant \varrho, \|x_2\| \leqslant \varrho, q(\varrho) = O(\varrho^{k-1}));$$

and the operator $D(x, \mu)$ consists of terms of a higher order of smallness

$$\|D(x, \mu)\| \leqslant L \|x\|^{k+1}.$$

Let μ_0 be a characteristic value of the operator B of even multiplicity β; let the elements $e_1, ..., e_\beta$ form a basis in the eigenmanifold of the operator B corresponding to the eigenvalue $\dfrac{1}{\mu_0}$ and let the elements $g_1, ..., g_\beta$ form a basis in the eigenmanifold of the operator B^* corresponding to the same eigenvalue (μ_0 is real).

The vector field F in a β-dimensional vector space defined by the equality

$$F\{\xi_1, ..., \xi_\beta\} = \{\eta_1, ..., \eta_\beta\},$$

where

$$\eta_i = -(C(\xi_1 e_1 + \cdots + \xi_\beta e_\beta, \mu_0), g_i) \qquad (i = 1, ..., \beta),$$

plays an important role. Let the field not be degenerate (that is, it vanishes only at the point $\xi_1 = \xi_2 = ... = \xi_\beta = 0$) and let its degree on the unit sphere be equal to γ_C. The following statement holds:

If $\gamma_C \neq 1$, then μ_0 is a point of bifurcation of the operator $A(x, \mu)$.

For the application of this statement it is necessary to be able to construct the field F, to prove that this field is not degenerate, and to be able to calculate the degree γ_C. To calculate the degree, it is useful to know that the degree of an even field $(F(x) = F(-x))$ is an even number.

For example, let $k = 2$ and μ_0 be a characteristic value of the operator B of multiplicity two $(\beta = 2)$. In this case the field F will have two components η_1 and η_2 each of which will be a quadratic form with respect to ξ_1 and ξ_2:

$$\eta_1 = a_{11}\xi_1^2 + 2a_{12}\xi_1\xi_2 + a_{22}\xi_2^2,$$
$$\eta_2 = b_{11}\xi_1^2 + 2b_{12}\xi_1\xi_2 + b_{22}\xi_2^2.$$

If one of these forms is positive or negative definite, then the field F is not degenerate and its degree is equal to zero. If neither of the forms is sign constant, then it suffices for the non-degeneracy of the field that the straight lines $\xi_1 = a\xi_2$ which one of the forms is annihilated consist of points on which the second form assumes non-zero values. The slopes a_1 ans a_2 of straight lines on which the first quadratic form becomes zero are determined by the quadratic equation $a_{11} + 2a_{12}a + a_{22}a^2 = 0$.

Example

(POINTS OF BIFURCATION OF THE HAMMERSTEIN EQUATION). In the equation

$$x(t) = \mu \int_0^1 K(t, s) f[x(s)]\, ds$$

with a bounded symmetric kernel $K(t, s)$, let the function $f(x)$ be differentiable as many times as desired, $f(0) = 0$ and $f'(0) = 1$. The linearization of this equation yields a linear integral equation with the kernel $K(t, s)$.

If the characteristic value μ_0 of the kernel $K(t, s)$ has an odd-multiplicity, then it will be a point of bifurcation for the initial equation.

Let μ_0 have multiplicity 2 and let $e_1(t)$, $e_2(t)$ be the eigenfunctions corresponding to it. If $f''(0) \neq 0$, then the operator C will have the form

$$C(x(t), \mu) = \mu \int_0^1 K(t, s) x^2(s)\, ds.$$

Therefore the components of the vector field F are given by the equalities

$$\eta_1 = \xi_1^2 \int_0^1 e_1^3(t)\, dt + 2\xi_1\xi_2 \int_0^1 e_1^2(t)\, e_2(t)\, dt + \xi_2^2 \int_0^1 e_1(t)\, e_2^2(t)\, dt,$$

$$\eta_2 = \xi_1^2 \int_0^1 e_1^2(t)\, e_2(t)\, dt + 2\xi_1\xi_2 \int_0^1 e_1(t)\, e_2^2(t)\, dt + \xi_2^2 \int_0^1 e_2^3(t)\, dt.$$

For example, if $e_1(t)=1$, $e_2(t)=\sqrt{2}\cos 2\pi t$, then

$$\eta_1 = \xi_1^2 + \xi_2^2, \qquad \eta_2 = 2\xi_1\xi_2 + \xi_2^2.$$

The first quadratic form is positive definite, the degree of the field F is equal to zero and, hence, μ_0 is a point of bifurcation for the Hammerstein equation.

The following question is interesting: for which values of the parameter μ, greater or less than μ_0, does the equation $x = A(x; \mu)$ have small non-zero solutions?

Let μ_0 be a simple characteristic value of the linearized equation $x = \mu Bx$, let e be a corresponding eigenvector, and let g be an eigenvector of the adjoint operator, where $(e, g) = 1$. The vector field F is given in this case by the number

$$\kappa = -\left(C(e, \mu_0), g\right).$$

The following statements hold:

1) *If the order k of smallness of the operator C is an even number, then the equation $x = A(x, \mu)$ has small non-zero solutions for $\mu < \mu_0$ and for $\mu > \mu_0$. If the operator $A(x, \mu)$ is sufficiently smooth, then the non-zero solution is unique for every μ (close to μ_0).*

2) *If k is odd, then small non-zero solutions exist for $\mu > \mu_0$ and are absent for $\mu < \mu_0$ in the case when $\kappa < 0$; if $\kappa > 0$, then small non-zero solutions exist for $\mu < \mu_0$ and are absent for $\mu > \mu_0$. Two non-zero solutions exist for corresponding values of μ.*

4. *Examples from mechanics.*

a) EULER PROBLEM CONCERNING STABILITY FOR BUCKLING OF A BEAM.

The deflection $y(\xi)$ of a beam of unit length with variable rigidity $\varrho(\xi) = \dfrac{1}{EJ}$

under the action of a longitudinal force μ is given by the solution of the differential equation

$$y''(\xi) + \mu\varrho(\xi)\, y(\xi)\sqrt{1 - y'^2(\xi)} = 0$$

for zero boundary conditions

$$y(0) = y(1) = 0$$

(see figure).

The function

$$K(\xi, \eta) = \begin{cases} (1 - \eta)\,\xi, & \text{if } \xi \leqslant \eta, \\ (1 - \xi)\,\eta, & \text{if } \xi \geqslant \eta, \end{cases}$$

is Green's function for the operator y'' with zero boundary conditions.

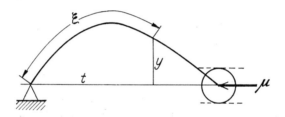

The differential equation of the buckling of the beam reduces to an integral equation with the replacement

$$y''(\xi) = -\varphi(\xi).$$

Then

$$y(\xi) = \int_0^1 K(\xi, \eta)\, \varphi(\eta)\, d\eta$$

and the equation for $\varphi(\xi)$ assumes the form

$$\varphi(\xi) = \mu\varrho(\xi) \int_0^1 K(\xi, \eta)\, \varphi(\eta)\, d\eta \sqrt{1 - \left[\int_0^1 K_\xi'(\xi, \eta)\, \varphi(\eta)\, d\eta\right]^2}.$$

This equation has the zero solution for all values of the parameter μ. For some loads μ, the equation can have non-zero solutions by which the forms of loss of stability are determined. The load μ_0 is called the *critical Euler load* if, for some loads close to μ_0, the equation has small non-zero solutions. In other words, the critical Euler load is a point of bifurcation for the equation of buckling of the beam.

The determination of the critical load is one of the important problems of the theory of elasticity.

The integral equation obtained can be regarded as an operator equation of the form $x = A(x, \mu)$ with a completely continuous operator in the space C. Linearization leads to the equation

$$\varphi(\xi) = \mu \varrho(\xi) \int_0^1 K(\xi, \eta) \, \varphi(\eta) \, d\eta \, .$$

If $e(\xi)$ is a non-zero solution of this equation for $\mu = \mu_0$, then the function

$$y(\xi) = \int_0^1 K(\xi, \eta) \, e(\eta) \, d\eta$$

will be a solution of the equation

$$y''(\xi) + \mu_0 \varrho(\xi) \, y(\xi) = 0$$

satisfying non-zero boundary conditions. It is implied that every characteristic value of the linaerized equation is simple and, hence, is a point of bifurcation. The corresponding values of μ give the critical loads.

The operator C has the form

$$C(x(\xi), \mu) = -\frac{\mu \varrho(\xi)}{2} \int_0^1 K(\xi, \eta) \, x(\eta) \, d\eta \left[\int_0^1 K'_\xi(\xi, \eta) \, x(\eta) \, d\eta \right]^2$$

for the equation considered and, hence, has a third order of smallness $(k = 3)$. Here

$$x = -\frac{1}{2} \int_0^1 e^2(\xi) \left[\int_0^1 K'_\xi(\xi, \eta) \, e(\eta) \, d\eta \right]^2 d\xi < 0 \, ;$$

therefore non-zero solutions appear for $\mu > \mu_0$. This corresponds fully with the physical meaning of the problem: loss of stability occurs then when the load exceeds the critical value.

In the literature, another equation for the buckling of a beam is some-times encountered:

$$y''(t) + \mu \varrho(t) y(t)[1 + y'^2(t)]^{\frac{3}{2}} = 0$$

for the boundary conditions

$$y(0) = y(1) = 0.$$

It corresponds to the case where the curvature is expressed not as a function of the arc length ξ but as a function of the coordinate t. In this connection it is assumed that approximately $\varrho(\xi) = \varrho(t)$; and in the boundary conditions, the change of the coordinate of the non-fixed end of the beam, representing a magnitude of the third order of smallness in comparison with the buckling of the beam, is not taken into account. However this magnitude of third order manifests itself in the sign of x. In this case, the expression

$$x = \frac{3}{2} \int_0^1 e^2(t) \left[\int_0^1 K_t'(t, s) e(s) \, ds \right]^2 dt > 0$$

is obtained for x, and non-zero solutions are obtained for $\mu < \mu_0$ for the corresponding integral equation which contradicts the physical meaning of the problem. Thus, the disregard of magnitudes of the third order of smallness in equations leads to an improper description of the problem concerning the forms of the loss of stability of a compressed beam.

b) WAVES ON THE SURFACE OF AN IDEAL INCOMPRESSIBLE HEAVY FLUID. The investigation of such waves was reduced by A. I. Nekrasov to the solution of the integral equation

$$x(t) = \mu \int_0^{2\pi} \frac{K(t, s) \sin x(s)}{1 + \int_0^s \sin x(u) \, du} \, ds,$$

where μ is a numerical parameter and

$$K(t, s) = \frac{1}{3} \sum_{n=1}^{\infty} \frac{\sin nt \sin ns}{n}.$$

This equation can be regarded as an operator equation in the space C on the segment $[0, 2\pi]$. It has the zero solution for all values of μ. Points of bifurcation of this equation correspond to the values of the parameters for which waves arise.

The linearized equation has the form

$$x(t) = \mu \int_0^{2\pi} K(t, s) \, x(s) \, ds;$$

its characteristic values are the numbers $\mu_n = 3n$ and the corresponding eigenfunctions are $e_n(t) = \sin nt$. All the characteristic values are simple; therefore they will be the only points of bifurcation for the Nekrasov equation.

The operator C has the form

$$C(x(t), \mu) = -\mu^2 \int_0^{2\pi} K(t, s) \, x(s) \int_0^s x(\tau) \, d\tau \, ds.$$

It is a magnitude of the second order of smallness ($k=2$) for small x; therefore the Nekrasov equation has small non-zero solutions for $\mu < \mu_n$ and $\mu > \mu_n$, where μ_n is an arbitrary point of bifurcation.

5. *Equations with potential operators.* For the equation

$$x = \mu A x,$$

where A is a completely continuous operator which is the gradient of a weakly continuous functional in a Hilbert space, the principle of linearization for the determination of the points of bifurcation is strengthened considerably.

If $A(\theta)=\theta$, the operator A is continuously differentiable, and its derivative $A'(\theta)=B$ is a completely continuous self-adjoint operator, then every characteristic value of the operator B, independently of its multiplicity, is a point of bifurcation of the nonlinear equation $x=\mu Ax$.

As an example we can again consider the Hammerstein equation with a bounded symmetric positive definite kernel:

$$x(t) = \mu \int_0^1 K(t, s) f[s, x(s)] \, ds, \qquad f(s, 0) \equiv 0, \qquad f'(s, 0) \equiv 1.$$

As in § 2, no. 7, we can transform it into the form

$$y = \mu B^{\frac{1}{2}} f B^{\frac{1}{2}} y.$$

The operator $B^{\frac{1}{2}} f B^{\frac{1}{2}}$ is the gradient of the functional

$$\Phi(x) = \int\limits_{0}^{1} ds \int\limits_{0}^{B^{\frac{1}{2}}x} f(s, u) \, du.$$

If the operator f is differentiable, then the Fréchet derivative of the operator $B^{\frac{1}{2}} f B^{\frac{1}{2}}$ at the point 0 is a linear integral operator B with kernel $K(t, s)$. All characteristic values of this operator are points of bifurcation for the equation $y = \mu B^{\frac{1}{2}} f B^{\frac{1}{2}} y$. The inverse replacement $B^{\frac{1}{2}} y = x$ indicates that the points of bifurcation of the last equation coincide with the points of bifurcation of the initial Hammerstein equation.

6. *Appearance of large solutions.*   In no. 2 a general pattern was described of the change of solutions for a change of the value of the parameter. This pattern is relative to the case when solutions in some sphere were considered. In a more general case, the norms of the solutions can increase indefinitely for a change of the values of the parameter. We may encounter such a case when the solutions with large norms appear for values of the parameter greater than some critical number. Here one theorem is mentioned which describes the appearance of solutions with large norms.

Let the operator $A(x, \mu)$ be asymptotically linear, where $A'_{\infty}(\infty, \mu) = \mu B$. Let μ_0 be a characteristic value of odd-multiplicity of the completely continuous linear operator B.

Then, for any $\varepsilon, R > 0$ a μ can be found which satisfies the inequality $|\mu - \mu_0| < \varepsilon$ and for which the equation $x = A(x, \mu)$ has at least one solution whose norm is greater then R.

7. *Equation of branching.*   We assume that 1 is an eigenvalue of the derivative $A'_x(x_0, \mu_0)$ of the completely continuous and continuously differentiable operator $A(x, \mu)$. For the sake of simplicity we restrict ourselves to the case where the invariant subspace E_0 corresponding to this eigenvalue consists only of eigenvectors. We denote by E^0 the invariant subspace of the operator $A'_x(x_0, \mu_0)$ which is complementary to E_0. We represent every element $x \in E$ in the form

$$x = u + v \qquad (u \in E_0, v \in E^0).$$

Let P and Q be the projectors onto E_0 and E^0 defined by the equalities $Px = u$, $Qx = v$.

The equation $x = A(x, \mu)$ can be rewritten in the form of the system

$$y = PA(x_0 + y + z, \mu) - Px_0,$$
$$z = QA(x_0 + y + z, \mu) - Qx_0,$$

where $y = P(x - x_0)$, $z = Q(x - x_0)$. If y and $\mu - \mu_0$ are sufficiently small, then the second equation has a unique small solution $z = R(y, \mu)$. Therefore the question of the solvability and the construction of a solution of the equation $x = A(x, \mu)$ is equivalent to the question of the solvability of the equation

$$y = PA(x_0 + y + R(y, \mu), \mu) - Px_0.$$

The last equation is an equation in a finite-dimensional space. It is called the *equation of branching*. Analytical and topological methods can be applied for its investigation.

8. *Construction of solutions in the form of a series.* Let x_0 be a solution of the equation $x = A(x, \mu_0)$. Let the operator $A(x, \mu)$ be analytic in a neighborhood of the point (x_0, μ_0) in the sense that it is representable in the form of a Taylor series

$$A(x, \mu) = x_0 + \sum_{i+j \geqslant 1} (\mu - \mu_0)^i C_{ij}(x - x_0),$$

where the $C_{ij}(h)$ $(i \geqslant 0, j \geqslant 0)$ are operators having j-th order of smallness with respect to h; in particular, the $C_{i0}(h) = C_{i0}$ are some fixed elements of E.

As above, the linear operator $C_{01} = A'_x(x_0, \mu_0)$ plays a special role. Let the operator $A(x, \mu)$ be completely continuous. Then the operator $C_{01} = A'_x(x_0, \mu_0)$ is also completely continuous.

If 1 is not an eigenvalue of the operator C_{01}, then the equation $x = A(x, \mu)$ has a unique solution $x(\mu)$ for values of μ close to μ_0. This solution, as it turns out, is representable by a series

$$x(\mu) = x_0 + (\mu - \mu_0) x_1 + (\mu - \mu_0)^2 x_2 + \cdots$$

To determine the elements $x_1, x_2, \ldots$, this series is substituted in the equation, then the right hand side is developed in a series in powers of $(\mu - \mu_0)$ and the coefficients of the identical powers of $(\mu - \mu_0)$ are equated. As a result we arrive at the system of equations

$$x_1 = C_{01}(x_1) + C_{10},$$
$$x_2 = C_{01}(x_2) + C_{11}(x_1) + C_{02}(x_1) + C_{20}.$$
$$\cdots \cdots \cdots \cdots \cdots \cdots \cdots$$

The linear equations which are written out can be solved successively. The series for $x(\mu)$ is convergent for $|\mu - \mu_0|$ sufficiently small. Majorizing numerical series are usually constructed to estimate the radius of convergence.

Now let 1 be an eigenvalue of the linear operator C_{01}. In this case, the question concerning the number of solutions of the equation $x = A(x, \mu)$ for values μ close to μ_0 becomes more complicated. Such solutions can sometimes be found in the form of the series

$$x(\mu) = x_0 + (\mu - \mu_0)^{\frac{1}{k}} x_1 + (\mu - \mu_0)^{\frac{2}{k}} x_2 + \cdots$$

with respect to fractional powers (k is a natural number) of the increment $\mu - \mu_0$.

To determine the elements $x_1, x_2, \ldots$, we again substitute the series for $x(\mu)$ in the equation and compare the coefficients of identical fractional powers of $\mu - \mu_0$. For example, the equations

$$x_1 = C_{01}(x_1),$$
$$x_2 = C_{01}(x_2) + C_{02}(x_1) + C_{10},$$
$$\cdots \cdots \cdots \cdots \cdots \cdots$$

are obtained in the case $k = 2$.

The first equation is a homogeneous linear equation. Its solution has the form

$$x_1 = \alpha_1 e_1 + \cdots + \alpha_s e_s,$$

where $e_1, \ldots, e_s$ is a basis for the subspace E_0 of eigenvectors corresponding to the eigenvalue 1 and $\alpha_1, \ldots, \alpha_s$ are arbitrary numbers.

To determine the numbers $\alpha_1, \ldots, \alpha_s$, conditions for the solvability of the second equation are used. These conditions can be written in the form

$$f_i[C_{02}(\alpha_1 e_1 + \cdots + \alpha_s e_s) + C_{10}] = 0 \qquad (i = 1, 2, \ldots, s),$$

where $f_1, \ldots, f_s$ is a complete system of eigenvectors (linear functionals) of the operator C_{01}^*, adjoint to C_{01}, corresponding to the eigenvalue 1.

The conditions of solvability are represented by a system of s non-linear equations with s unknowns. If it can be solved, then the element can be found. Simultaneously, we can state that the second equation can be solved (with respect to x_2). Its solution is again defined to within s arbitrary constants:

$$x_2 = x_2^0 + \beta_1 e_1 + \cdots + \beta_s e_s.$$

The coefficients $\beta_1, \ldots, \beta_2$ are determined from the conditions of the solvability of the third equation, and so on.

The determination of the elements $x_1, x_2, \ldots$ becomes more difficult if it is impossible to determine the coefficients $\alpha_1, \ldots, \alpha_s$ from the conditions of the solvability of the second equation. Here it is necessary to draw upon the conditions of solvability of the following equations.

If we do not succeed in constructing the solution in the form of a series in powers of $\mu - \mu_0$, $(\mu - \mu_0)^{\frac{1}{2}}$, then we try to construct the solution in the form of a series in powers of $(\mu - \mu_0)^{\frac{1}{3}}$, and so on.

CHAPTER V

OPERATORS IN SPACES WITH A CONE

§ 1. Cones in linear spaces

1. *Cone in a linear system.* A convex set K of elements of a real linear system is called a *cone* if this set contains, together with each element $x(x \neq \theta)$, all the elements of the form tx for $t \geqslant 0$ and does not contain the element $-x^*$).

Examples

1. The collection of all non-negative functions $x(t)(t\in[0, 1])$ of the space $C(0, 1)$ forms a cone in this space.**).

Analogously, the sets of all non-negative functions of the space $L_p(0, 1)$, the space $M(0, 1)$, and the Orlicz spaces form cones in these spaces.

2. The set of positive operators forms a cone in the space of bounded linear self-adjoint operators acting in a Hilbert space (see ch. II, § 2, no. 3).

3. The sets of elements with non-negative coordinates will be cones in the coordinate spaces l_p, m, c.

4. In function spaces, it is sometimes necessary to study cones which are narrower than the cone consisting of all non-negative functions. These cones are determined by a system of additional inequalities. Examples are the cone of non-negative non-decreasing functions:

$$x(t_1) \leqslant x(t_2) \qquad (t_1 < t_2),$$

and the cone of non-negative convex upwards functions:

$$x\left(\frac{t_1 + t_2}{2}\right) \geqslant \tfrac{1}{2}\left[x(t_1) + x(t_2)\right].$$

---

*) If the last condition is not satisfied, then the set is called a *wedge*.
**) For the definition of the spaces, see ch. I, § 2, no. 5.

The cone K in the linear system E is called *generating* if an arbitrary element $x \in E$ is representable in the form of the difference of two elements of the cone: $x = x_1 - x_2 (x_1, x_2 \in K)$.

The cone of non-negative functions of the space $C[0, 1]$ is generating. We can represent every function $x(t) \in C(0, 1)$ in the form of a difference of non-negative functions $x_+(t)$ and $x_-(t)$:

$$x(t) = x_+(t) - x_-(t),$$

where

$$x_+(t) = \begin{cases} x(t), & \text{if } x(t) \geq 0, \\ 0, & \text{if } x(t) < 0, \end{cases}$$

$$x_-(t) = \begin{cases} 0, & \text{if } x(t) \geq 0, \\ -x(t), & \text{if } x(t) < 0. \end{cases}$$

All the cones considered in examples 1–3 are generating; however, not every cone has this property. For example, the cone of non-negative non-decreasing functions (example 4) is not generating in the space $C(0, 1)$ since only functions of bounded variation can be represented in the form of a difference of non-decreasing functions.

2. *Partially ordered spaces.* The real linear system E is called a linear *partially ordered space* if a relation $x \leqslant y$ is defined for some pairs of elements $x, y \in E$ and if the sign $\leqslant$ has the usual properties of the sign of inequality. We mean the following properties:

1) it follows from $x \leqslant y$ that $tx \leqslant ty$ for $t \geqslant 0$ and $ty \leqslant tx$ for $t < 0$,
2) it follows from $x \leqslant y$ and $y \leqslant x$ that $x = y$,
3) it follows from $x_1 \leqslant y_1$ and $x_2 \leqslant y_2$ that $x_1 + x_2 \leqslant y_1 + y_2$,
4) it follows from $x \leqslant y$ and $y \leqslant z$ that $x \leqslant z$.

The relation $\leqslant$ is usually called inequality, and we say that x is *less than or equal to* y if $x \leqslant y$.

It remains to remark that the sign $\leqslant$ establishes a total ordering relation in the space of real numbers in the sense that two arbitrary numbers a and b can be united with this sign (either $a \leqslant b$ or $a \geqslant b$); the sign $\leqslant$, generally speaking, does not have this property in a linear space. The origin of the term "partial ordering" is connected with this situation.

The collection K of all elements x of a partially ordered space for which $\theta \leqslant x$ forms a cone in this space. Conversely, if the cone K is given in a linear system E, then a partial ordering can be introduced in this system by setting $x \leqslant y$ if $y - x \in K$.

Thus the consideration of linear partially ordered spaces is equivalent to the consideration of linear systems with a cone.

If K is the cone of non-negative functions in the space c (or L_p), then the partial ordering relation acquires a simple meaning: $x \leqslant y$ if $x(t) \leqslant y(t)$ for all (or almost all) values of t.

3. *Vector lattices, minihedral cones.* The partially ordered linear space E is called a *vector lattice* if the following property is satisfied:

5) for any two elements $x, y \in E$ there exists an element $z \in E$ such that $x \leqslant z$, $y \leqslant z$, and $z \leqslant \zeta$ for every element ζ having the same property

$$x \leqslant \zeta, y \leqslant \zeta.$$

The element z is called the *least upper bound* or the *supremum* of the elements x and y and is denoted by $z = \sup(x, y)$.

The existence of an *infimum* for any pair (x, y), that is, an element u which has the following properties: $u \leqslant x$, $u \leqslant y$ and if $v \leqslant x$, $v \leqslant y$, then $v \leqslant u$, follows from the existence of a supremum for arbitrary elements x, y in a vector lattice. In this connection we write $u = \inf(x, y)$.

The cone formed by the elements $x \geqslant \theta$ of a vector lattice is called *minihedral*.

If we denote the collection of elements of the form $x + u (u \in K)$ by K_x, then the minihedralness of the cone means that for any $x, y \in E$ an element z can be indicated such that $K_x \cap K_y = K_z$. In this connection $z = \sup(x, y)$.

Every element of a vector lattice allows the representation $x = x_+ - x_-$, where $x_+ = \sup(x, \theta)$ and is called the *positive part* of x and where $x_- = \sup(-x, \theta)$ and is called the *negative part* of x. The element $|x| = x_+ + x_-$ is called the *absolute value* of the element x. The relation

$$- |x| \leqslant x \leqslant |x|$$

is valid.

The cones of non-negative functions are minihedral in the spaces C, L_p, M. The cones of non-negative sequences are also minihedral in the spaces l_p, m, c. Not every cone is minihedral. The usual circular cones of three-dimensional Euclidean space are the simplest examples of non-minihedral cones.

4. *K-spaces.* Let $M \subset E$ be some subset of the partially ordered space E. If all the elements of M are less than or equal (in the sense of the operation of comparison $\leqslant$) to some element $z \in E$, then z is called an *upper bound* of the set M and the set M itself is called *bounded above*. An upper bound z of the set M is called the *least upper bound* of M (we write $z = \sup M$) if the relation $\zeta \geqslant z$ is satisfied for every other upper bound ζ of the set M. The definitions of boundedness below, lower bound and greatest lower bound of a set of elements from E are introduced analogously.

If the space E is a vector lattice, then a least upper bound exists for every finite set of elements $x_1, x_2, \ldots, x_n$ and can be defined by means of the recurrence relation

$$\sup(x_1, x_2, \ldots, x_n) = \sup\{x_1, \sup(x_2, \ldots, x_n)\}.$$

The vector lattice E is called a *K-space* if each of its bounded above non-empty subsets has a least upper bound.

In a K-space, every bounded below non-empty set of elements has a greatest lower bound.

The cone of elements $x \geqslant \theta$ in a K-space is sometimes called *strongly minihedral.*

The spaces L_p, M, l_p, m are K-spaces; the vector lattices C and c are not K-spaces.

In K-spaces, convergence with respect to the ordering can be introduced which is called (o)-*convergence.* For convenience of description, two new "non-characteristic elements" ∞ and $-\infty$ are adjoined to the K-space with respect to which we assume that $-\infty \leqslant x \leqslant \infty$ for all elements $x \in E$. Then we write $\sup M = \infty$ for a non-bounded above set $M \subset E$ and $\inf M = -\infty$ for a non-bounded below set. If we introduce ∞ into the number of elements of the set M in addition to the characteristic elements $x \in E$, then we assume that $\sup M = \infty$ and if we introduce $-\infty$, then we assume that $\inf M = -\infty$.

Let $x_n (n = 1, 2, \ldots)$ be an arbitrary sequence of elements of the K-space E. The *upper* and *lower limits* of this sequence are the elements defined by the relations

$$\overline{\lim_{n \to \infty}} \, x_n = \inf_n \left[\sup(x_n, x_{n+1}, \ldots) \right],$$

$$\underline{\lim_{n \to \infty}} \, x_n = \sup_n \left[\inf(x_n, x_{n+1}, \ldots) \right].$$

These elements can be finite or equal to ∞ or $-\infty$. If $\overline{\lim\limits_{n\to\infty}} \, x_n = \underline{\lim\limits_{n\to\infty}} \, x_n$, then the sequence x_n is called (o)-*convergent*, and the common value of its upper and lower limits is called simply the (o)-*limit* and is denoted by (o)- $\lim\limits_{n\to\infty} x_n$.

(o)-convergence is a basic form of convergence in a K-space; (o)-convergent sequences have several usual properties of convergent sequences. For example, if (o)-$\lim x_n = x$ and (o)-$\lim y_n = y$ and both limits are finite, then the sequence $x_n + y_n$ is (o)-convergent and

$$(o)\text{-} \lim_{n\to\infty} (x_n + y_n) = (o)\text{-} \lim_{n\to\infty} x_n + (o)\text{-} \lim_{n\to\infty} y_n \,.$$

Moreover, in order for (o)- $\lim\limits_{n\to\infty} x_n = x$ it is necessary and sufficient that (o)- $\lim\limits_{n\to\infty} |x_n - x| = 0$.

The property of completeness with respect to (o)-convergence is an important property of a K-space: *in order for the sequence x_n to have a finite* (o)-*limit it is necessary and sufficient that*

$$(o)\text{-} \lim_{n\to\infty} \big[\sup_{k,m \geqslant n} |x_k - x_m| \big] = 0 \,.$$

The sequence x_n is called (t)-*convergent* to the element x if from an arbitrary subsequence x_{n_i} we can choose a particular subsequence $x_{n_{i_k}}$ such that (o)-$\lim\limits_{k} x_{n_{i_k}} = x$. If a sequence is (o)-convergent, then it is (t)-convergent.

In the K-space M, (o)-convergence of a sequence of elements coincides with convergence almost everywhere, and (t)-convergence in M means convergence in measure. (o)-convergence and (t)-convergence coincide in the K-spaces $l_p (p \geqslant 1)$.

5. *Cones in a Banach space.*   If the system E is a Banach space, then a *cone in the Banach space* E is any cone of the system E which is a closed set in the space E.

If the cone K in the Banach space E is generating, then a constant M exists such that for every $x \in E$ we have the representation $x = x_1 - x_2$ $(x_1, x_2 \in K)$ in which $\|x_1\| \leqslant M \|x\|$ and $\|x_2\| \leqslant M \|x\|$.

A solid cone is a special case of a generating cone. A cone is *solid* if it contains at least one interior element. The cone of non-negative functions of the space $C[0, 1]$ can serve as an example of a solid cone. Functions with a positive minimum are interior elements of this cone. The cone from example 2, no. 1 and the cone of non-negative sequences of the space m of bounded sequences are also solid. The cones of non-negative functions of the spaces $L_p[0, 1]$ $(p \geqslant 1)$ and the non-negative sequences of the coordinate spaces $l_p (p \geqslant 1)$ do not have the property of solidity. Thus, not every generating cone is solid. Every generating cone is solid in a finite-dimensional space.

The cone K is called *normal* if there exists a $\delta > 0$ such that the inequality $\|e_1 + e_2\| \geqslant \delta$ is satisfied for all $e_1, e_2 \in K$, $\|e_1\| = \|e_2\| = 1$. If the cone K is normal, then

$$\|f_1 + f_2\| \geqslant \frac{\delta}{2} \max \{\|f_1\|, \|f_2\|\}$$

for arbitrary elements $f_1, f_2 \in K$. The set of normal cones is very important.

The cones of non-negative functions are normal in the spaces $C, L_p (p \geqslant 1), l_p, m$; 1 can be taken as the constant δ which appears in the definition of a normal cone in these examples. The cone of positive linear operators can serve as an example of a normal cone in the space of self-adjoint operators.

Not all cones are normal. For example, the cone of non-negative functions in the space $C^{(1)}[0, 1]$ of continuously differentiable functions $x(t)$ does not have the property of normality (see ch. I, § 2, no. 5).

If the cone K is normal, then the partial ordering established by this cone in E has the following property: a new norm $\|...\|_1$ can be introduced in E which is equivalent to the originally given norm and such that the inequality $\|x\|_1 \leqslant \|y\|_1$ follows from the relation $\theta \leqslant x \leqslant y$. The converse statement is valid. By virtue of the equivalence of the norms $\|...\|_1$ and $\|...\|$, this property of a normal cone can be formulated in the form of the following criterion: *the cone K is normal if and only if the inequality $\theta \leqslant x \leqslant y$ implies the scalar inequality $\|x\| \leqslant M \|y\|$ where M is a constant.*

Several further criteria for normality of a cone exist. One of these is connected with the concept of a u_0-norm. Let u_0 be some fixed non-zero element of the cone K. The element $x \in E$ is called *u_0-measurable* if

$$- t_1 u_0 \leqslant x \leqslant t_2 u_0$$

for some non-negative t_1 and t_2.

Let E_{u_0} be the set of all u_0-measurable elements, $\alpha(x)$ the lower bound of the numbers t_1, $\beta(x)$ the lower bound of the numbers t_2 for $x \in E_{u_0}$.

This set is a linear space. If we now set

$$\|x\|_{u_0} = \max \{\alpha(x), \beta(x)\}$$

for all elements x of E_{u_0}, then the set E_{u_0} becomes a normed space, and the number $\|x\|_{u_0}$ is called the u_0-*norm* of the element x.

For the normality of the cone K it is necessary and sufficient that the inequality

$$\|x\|_E \leqslant M \|x\|_{u_0} \cdot \|u_0\|_E \qquad (x \in E_{u_0}, u_0 \in K, u_0 \neq \theta),$$

where the constant M does not depend either on x or on u_0, be satisfied.

It follows from the last inequality that the normality of the cone K guarantees the completeness of the space E_u with respect to the u_0-norm.

6. *Regular cones.*   The cone $K \subset E$ is called *regular* if the partial ordering generated by it has the property that the convergence of a sequence $x_n (n = 1, 2, \ldots)$ with respect to the norm of the space E follows from the relations

$$x_1 \leqslant x_2 \leqslant \cdots \leqslant x_n \leqslant \cdots$$

and

$$x_n \leqslant u \qquad (n = 1, 2, \ldots),$$

where u is some element of the space E. In other words, the cone K is regular if every monotone (with respect to the cone) and bounded (also with respect to the cone) sequence of elements of the space converges with respect to the norm.

The cone of non-negative functions of the space $L_p (p \geqslant 1)$ serves as an example of a regular cone; the cone of non-negative functions of the space C is an example of a cone which is not regular. The cones of non-negative sequences in the spaces $l_p (p \geqslant 1)$ and numerical sequences convergent to zero in the space c_0 are regular. The cone of non-negative sequences of the space m of all bounded sequences does not have the property of regularity.

The fact that *every regular cone is normal* is important.

Analogously, we can raise the question of convergence in the norm of every monotone (in the sense of the partial ordering generated by the cone K) sequence of elements of the Banach space E bounded with respect to the norm. It is clear that such convergence occurs far from always. For example, in the space $C[0, 1]$ the monotone, with respect to the cone of non-negative functions, sequence $x_n(t)=1-t^n$ is bounded with respect to the norm and at the same time is not convergent. In this connection one more class of cones is introduced. A cone K is called *completely regular* if the partial ordering generated by it is such that for an arbitrary sequence x_n the convergence of the sequence x_n with respect to the norm of the space E follows from the relations $x_1 \leqslant x_2 \leqslant \cdots \leqslant x_n \leqslant$ and from the scalar inequality $\|x_n\| \leqslant C (n=1, 2, ...)$ where C is a constant.

The cones of non-negative functions in the spaces $L_p(p \geqslant 1)$ and the cone of non-negative sequences in $l_p(p \geqslant 1)$ can serve as examples of completely regular cones. The cone of non-negative sequences in the space c_0 is, as indicated above, regular, but it is not completely regular: the sequence

$$x_1 = (1, 0, 0, ...), \ x_2 = (1, 1, 0, 0, ...), \ ...$$
$$..., \ x_n = (1, 1, ..., 1, 0, 0, ...), \ ...$$

of elements of the space c_0 is monotone with respect to this cone, bounded with respect to the norm, but it does not converge. This example shows that not every regular cone is completely regular. However every completely regular cone is a regular cone.

A regular cone K is completely regular if it satisfies one of the following conditions:

1) the cone K is solid;
2) the space E is weakly complete.

A functional $f(x)$, not necessarily linear, is called *positive* if $f(x) \geqslant 0$ for $x \in K$; the functional is called *strictly positive* if $f(x) > 0$ for $x \in K, x \neq \theta$ The functional $f(x)$ is called *monotone* with respect to the cone K if the relation $\theta \leqslant x \leqslant y$ implies the inequality $f(x) \leqslant f(y)$. The positive functional $f(x)$ is called *strictly increasing* if for any $h_n \in K$ $(n=1, 2, ...)$ with $\|h_n\| \geqslant \varepsilon_0 > 0$ $(n=1, 2, ...)$ it follows that

$$\lim_{n \to \infty} f(h_1 + h_2 + \cdots + h_n) = \infty.$$

If a monotone strictly increasing functional can be defined on the cone K, then the cone K is regular.

If on the cone K a strictly increasing functional $f(x)$ that is bounded on every sphere can be defined, then the cone K is completely regular.

An example of a strictly increasing functional which is monotone on the cone of non-negative functions of the space $L_p[0, 1]\,(p \geqslant 1)$ is the p-th power of the norm of an element:

$$f(x) = \int_0^1 |x(t)|^p\, dt.$$

A result of a negative character is valid: a solid minihedral cone in an infinite-dimensional space can not be regular.

7. *Theorems on the realization of partially ordered spaces.* Let K be a normal cone in a separable Banach space E. Then there exists a one-to-one linear and continuous mapping of the space E into a subspace of the space $C(0, 1)$ for which the elements of K and only they map into non-negative functions.

If E is not separable, then an analogous statement with the replacement of $C(0, 1)$ by the space $C(Q)$ of continuous functions defined on some compactum Q is valid.

In the case when a solid normal minihedral cone K occurs in the Banach space E, there exists a linear one-to-one and continuous mapping of the space E onto all of the space $C(Q)$, where Q is a compactum, for which K maps onto the set K_Q of all non-negative functions on Q.

The following is a special case of the last statement: let E be an n-dimensional space and $K \subset E$ a minihedral solid cone; then a basis $(e_1, e_2, ..., e_n)$ exists in E such that the set of all vectors $x = \xi_1 e_1 + \xi_2 e_2 + \cdots + \xi_n e_n$ with non-negative coordinates $\xi_i \geqslant 0$ $(i = 1, 2, ..., n)$ coincides with K.

If a minihedral cone K is given in a separable Banach space E and $\|x + y\| = \|x\| + \|y\|$ for $x, y \in K$, then there exists a linear one-to-one and continuous mapping of the space E onto the space $L(0, 1)$, for which K maps onto the set of all almost everywhere non-negative functions of $L(0, 1)$. If E is not separable, then $L(0, 1)$ is replaced by the space $L(Q)$ on some compactum with a measure.

§ 2. Positive linear functionals

1. *Positive functionals.* The positive linear functionals are the most important class of positive functionals (i.e. functionals such that $f(x) \geqslant 0$ for $x \geqslant \theta$).

If the cone K in a Banach space is generating, then every positive linear functional is continuous. In the sequel, we understand a positive linear functional to be a continuous functional.

Positive linear functionals exist for every cone K. Moreover, for every $x \in K(x \neq \theta)$, a positive continuous linear functional f can be indicated such that $f(x) > 0$.

If the space E is separable, then a continuous linear functional $f(x)$ can be constructed such that $f(x) > 0$ for all $x \in K(x \neq \theta)$.

If the cone K is solid and u_0 is an arbitrary interior element of it, then $f(u_0) > 0$ for every positive functional f. At least one positive functional f can be found such that $f(v_0) = 0$ for an element $v_0 \in K$ which is not an interior element of the cone K. On an element not belonging to the cone K, at least one of the positive functionals assumes a negative value. Thus, every cone K is characterized uniquely by a set K' of positive linear continuous functionals.

The set $K' \subset E'$ of positive linear functionals cannot be a cone: it still does not follow from $f \in K'$ and $f \neq \theta$ that $-f \notin K'$ (that is, the set K' is, generally speaking, a wedge). However, K' is a cone if K is a generating cone. The cone K' in this case is called a *conjugate cone*.

Starting from the general form of the linear functionals in concrete function spaces, it is not difficult, as a rule, to establish the general form of the positive linear functionals in these spaces. Let K be the cone of non-negative functions. Then the general form of a positive linear functional in the space $L_p(\Omega)$ $(p \geqslant 1)$ (where Ω is some set) is defined by the formula

$$f(x) = \int_{\Omega} x(t)\, \varphi(t)\, dt,$$

where $\varphi(t)$ is a non-negative function from the conjugate space $L_q(\Omega)$ $\left(q = \dfrac{p}{p-1}\right)$, $L_\infty = M$. The general form of a positive linear functional in the space $C(0, 1)$ is described by the formula

$$f(x) = \int_0^1 x(t)\, dg(t),$$

where $g(t)$ is a non-decreasing function.

Thus, for the cone K of non-negative functions, the cone K' coincides with the cone of non-negative functions of the space L_q (the case $E = L_p$) and with the cone of non-decreasing functions (the case $E = C$).

Analogously, the general form of a positive linear functional f in the space $l_p (p \geqslant 1)$ with the cone K of non-negative sequences is given by the equality

$$f(x) = \sum_{i=1}^{\infty} x_i y_i$$
$$\left(x = (x_1, x_2, x_3, \ldots), \qquad y = (y_1, y_2, y_3, \ldots) \right),$$

where $y = (y_1, y_2, y_3, \ldots) \in l_q$ is an arbitrary non-negative sequence.

The following statement is important: *in order for the conjugate cone K' to be generating, it is necessary and sufficient that the cone K be normal.*

Thus, if the cone K is normal, then for every $f \in E'$ there exists a representation $f = f_1 - f_2$ in the form of a difference of positive functionals f_1 and f_2. This representation is not unique:

$$f = (f_1 + g) \quad (f_2 + g)$$

for arbitrary $g \in K'$.

If the cone K is solid and minihedral, then a minimal representation

$$f = f_1^0 - f_2^0 \quad (f_1^0, f_2^0 \in K')$$

exists having the property that $f_1^0 \leqslant f_1$ and $f_2^0 \leqslant f_2$ for any other representation.

2. *Extension of positive linear functionals.* Let E be a Banach space and $E_0 \subset E$ some subspace of it, and let K_0 and K be cones in E_0 and E, respectively, where $K_0 = K \cap E$. Let f be a continuous linear functional, defined on E_0 and positive with respect to the cone $K_0 (f(x) \geqslant 0$ for $x \in K_0)$. A continuous linear functional F, defined on E and positive with respect to the cone K, is called a *positive extension* of the functional f from the subspace E_0 to the space E if $F(x) = f(x)$ for $x \in E_0$.

For an arbitrary element $x \in E$, let there exist a representation $x = x_1 - x_2$ where $x_1 \in K_0$, $x_2 \in K$. Then a positive extension F to the space E exists for every continuous positive linear functional f defined on E_0. Furthermore, all the positive linear extensions F of the functional f to the space E are bounded and

$$\|f\| \leqslant \|F\| \leqslant C \|f\|,$$

where C is constant.

Another theorem about extension can be formulated for vector lattices. Let E_0 be a linear manifold in the vector lattice E, having the property that for any $x \in E$ an $x' \subset E$ can be found majorizing the absolute value of the element x:

$$|x| \leqslant x'.$$

Then every positive linear functional defined on E_0 (the cone $K_0 \subset E_0$ consists of $x \in E_0 \cap K$) can be extended to a positive linear functional defined on the entire space E.

3. *Uniformly positive functionals.* The positive linear functional $f(x)$ is called *uniformly positive* if a positive number a exists such that

$$f(x) \geqslant a \|x\| \qquad (x \in K).$$

The functional

$$f(x) = \int_{\Omega} x(t) \, dt$$

will be uniformly positive in the space $L(\Omega)$ of functions summable on some set Ω with the cone K of functions which are non-negative on Ω.

In the spaces L_p where $p > 1$, there are no uniformly positive (on the cone K of non-negative functions) linear functionals. There are also no uniformly positive linear functionals in the space C.

If a uniformly positive functional exists, then the cone is normal and even completely regular. The converse statement is not true.

In order for a uniformly positive functional to exist, it is necessary and sufficient that the cone K be included in another cone K_1 such that every non-zero element $x_0 \in K$ be contained in K_1 together with its spherical neighborhood of radius $q \|x_0\|$, where the number q does not depend on the choice of $x_0 \in K$.

The cone $K(F)$, constructed with respect to a closed bounded convex set F not containing zero in the following way: $K(F)$ consists of all elements $x \in E$ which allow the representation $x = tz$ where $t \geqslant 0$ and $z \in F$, always has the last property.

4. *Bounded functionals on a cone.* An additive and homogeneous functional $f(x)$, defined only on elements of the cone K of the Banach

space E, is called *bounded* if

$$\|f\|_+ = \sup_{x \in K, \|x\|=1} |f(x)| < \infty.$$

If an additive and homogeneous functional is defined on a solid cone and is non-negative on it, then it is bounded.

A bounded, additive and homogeneous functional, defined on a generating (in particular, on a solid) cone, is uniquely extendible to a continuous linear functional defined on the entire space E by the equality $f(x)=$ $f(x_1)-f(x_2)$ where $x=x_1-x_2$ and $x_1, x_2 \in K$. If the cone is not generating, then the indicated extension of the functional to the linear hull of the cone gives an unbounded functional. It turns out that under the condition of normality of the cone, every bounded, additive and homogeneous functional $f(x)$ on K is majorized on the cone K by some positive continuous linear functional, that is, a $\Phi(x) \in K'$ exists such that

$$|f(x)| \leqslant \Phi(x) \qquad (x \in K).$$

If a norm is introduced in E such that $\|x\| \leqslant \|y\|$ for $0 \leqslant x \leqslant y$, then the functional Φ can be chosen so that $\|\Phi\| = \|f_+\|$ for non-negative f and $\|\Phi\| \leqslant 2\|f_+\|$ in the general case.

§ 3. Positive linear operators

1. *Concept of a positive operator.* Let E be a Banach space with cone K. A linear operator A is called *positive* if it maps the cone K into itself: $AK \subset K$. A positive linear operator has the property of monotonicity: for arbitrary elements $x, y \in E$, $x \leqslant y$ implies $Ax \leqslant Ay$.

In the case of finite-dimensional spaces with a cone consisting of vectors with non-negative components, the positive linear operators are defined by matrices with non-negative elements.

The linear integral operators

$$A\varphi(t) = \int_\Omega K(t, s)\, \varphi(s)\, ds$$

with non-negative kernels $K(t, s)$ ($t, s \in \Omega$; Ω is a closed bounded set of a finite-dimensional space) are the most important examples of positive linear operators acting in different spaces of functions. If the kernel

$K(t, s)$ satisfies conditions such that the operator acts in a corresponding space, then this operator is a positive linear operator for the cone K of non-negative functions.

If the cone K is solid and an n can be found for every non-zero φ of K such that $A^n \varphi$ is an interior element of the cone, then the operator A is called *strongly positive*.

If the integral operator acts in the space C, and if some iterate of the kernel $K(t, s)$:

$$K_N(t, s) = \int_\Omega \cdots \int_\Omega K(t, s_1) K(s_1, s_2) \ldots K(s_{N-1}, s) \, ds_1 \ldots ds_{N-1},$$

is positive, then this operator will be strongly positive in the space C.

A positive linear operator A is called u_0-*bounded below* (u_0 is a fixed element of K) if for every non-zero $\varphi \in K$ a natural number n and a positive number $\alpha = \alpha(\varphi)$ can be found such that $\alpha u_0 \leqslant A^n \varphi$.

Analogously, operators are defined to be u_0-*bounded above*. It turns out that if the operator A is u_0-bounded above and below, then the relation

$$\alpha(\varphi) u_0 \leqslant A^n \varphi \leqslant \beta(\varphi) u_0$$

is satisfied for every $\varphi \in K$ for some n and $\alpha(\varphi)$, $\beta(\varphi) > 0$.

The operators, satisfying the last relation, are called u_0-*positive*.

Let the integral operator act in $L_p(\Omega)(p \geqslant 1)$. If for this kernel $K(t, s)$ the inequality

$$K(t, s) \geqslant m > 0 \qquad (t, s \in \Omega)$$

is satisfied, then the operator will be u_0-bounded below if we choose as u_0 the function $u_0(t) \equiv 1$. In this connection the operator cannot have the property of u_0-boundedness above.

2. *Affirmative eigenvalues.* An eigenvalue $\lambda_0 \neq 0$ of the positive operator A is called *affirmative* if it has a corresponding eigenvector e_0 in the cone K. This element is called a *positive eigenvector* of the operator A. An affirmative eigenvalue is always positive.

An affirmative eigenvalue has an important property: *if the cone K is generating and the operator A is u_0-positive, then an affirmative eigenvalue is simple and greater than the moduli of the remaining eigenvalues.*

This statement, generally speaking, loses force if the condition of u_0-positiveness of the operator A is dropped. If the element u_0 itself is a positive eigenvector of the operator A, then it suffices to require u_0-boundedness above of the operator A instead of u_0-positiveness.

Several theorems on the existence of affirmative eigenvalues can be formulated for completely continuous operators.

Let the closure of the linear hull of the cone K be all of the space E. *If a positive linear completely continuous operator A has eigenvalues different from zero, then it has an affirmative eigenvalue λ_0 not less than the modulus of any other eigenvalue.* The number λ_0 is always an affirmative eigenvalue for the operator A' acting in the conjugate space E' with the cone K'.

In practice it is convenient to use the following statement: for the positive linear completely continuous operator A, let an element u exist such that $u = v - w(v, w \in K)$, $-\mu \notin K$ and $A^p u \geqslant \alpha u (\alpha > 0)$ for some natural number p; then the operator A has an affirmative eigenvalue λ_0 where $\lambda_0 \geqslant \sqrt[p]{\alpha}$. The number λ_0 is also an affimative eigenvalue of the operator A'.

The preceding results obtain further development if the cone is solid.

Let A be a completely continuous linear operator, strongly positive with respect to the solid cone K. Then:

1) *The operator A has one and only one (normalized) eigenvector x_0 inside K:*

$$A x_0 = \lambda_0 x_0 (\lambda_0 > 0).$$

2) *The adjoint operator A' has one and only one normalized eigenvector ψ in K':*

$$A' \psi = \lambda_0 \psi;$$

moreover ψ is a strictly positive functional.

3) *The eigenvalue λ_0 corresponding to these elements is simple and exceeds the modulus of every other eigenvalue of the operator A.*

Conversely, if a completely continuous operator has properties 1), 2), and 3), then it is strongly positive with respect to K.

Theorems on the existence of affirmative eigenvalues can be illustrated with a Fredholm integral equation

$$\int_a^b K(t, s) \, \varphi(s) \, ds = \lambda \varphi(t)$$

with a non-negative kernel $K(t, s)$ continuous on the square $a \leqslant t$, $s \leqslant b$. If a system of points $s_1, s_2, ..., s_p$ of (a, b) exists such that

$$K(s_1, s_2) \, K(s_2, s_3) ... K(s_{p-1}, s_p) \, K(s_p, s_1) > 0,$$

then the equation has a positive eigenvalue λ_0 not less in modulus than every other of its eigenvalues. At least one non-negative solution (eigenfunction) of the integral equation corresponds to this number λ_0.

If for every continuous non-negative function $\varphi(s)$ not identically equal to zero an iterate $K_N(t, s)$ can be found such that

$$\int_a^b K_N(t, s)\, \varphi(s)\, ds > 0 \qquad (a \leqslant t \leqslant b),$$

then the Fredholm equation has a unique positive eigenfunction. The transposed equation

$$\int_a^b K(s, t)\, \psi(s)\, ds = \lambda \psi(t)$$

has a unique positive solution corresponding to the same positive eigenvalue. The eigenvalue λ_0 is in absolute value greater than all the remaining eigenvalues of the integral equation.

Now let the kernel $K(t, s)$ in the integral equation be a non-negative function measurable on the square $a \leqslant t,\ s \leqslant b$ satisfying the condition

$$\int_a^b \int_a^b |K(t, s)|^2\, dt\, ds < \infty .$$

If the inequality

$$K(s_1, s_2)\, K(s_2, s_3) \ldots K(s_p, s_1) > 0$$

is satisfied for some $p \geqslant 2$ on a set of points $(s_1, s_2, \ldots, s_p)$ of positive measure in the corresponding p-dimensional cube, then in this case the integral equation has at least one eigenvalue such that a positive eigenvalue occurs among the eigenvalues having the largest absolute value. At least one non-negative eigenfunction of L_2 corresponds to this positive eigenvalue.

3. *Positive operators on a minihedral cone.* Let K be a minihedral solid cone and A a positive completely continuous linear operator having a fixed vector v inside K:

$$Av = v .$$

Then the eigenvalues of the operator A, equal in absolute value to one, are roots of an integral power of one. The sets of fixed vectors of the operators A and A' have bases $v_1, v_2, ..., v_r$ and $\psi_1, \psi_2, ..., \psi_r$, respectively, having the properties:

1) The systems $v_1, v_2, ..., v_r$ and $\psi_1, \psi_2, ..., \psi_r$ are biorthogonal: $\psi_i(v_j) = \delta_{ij}(i, j = 1, 2, ..., r)$.

2) For every pair $i \neq j(i, j = 1, 2, ..., r)$

$$\inf(\psi_i, \psi_j) = \theta .$$

3) Linear combinations $\sum c_i v_i$ (or $\sum c_i \psi_i$) are non-negative if, and only if all the coefficients are non-negative.

In the linear manifold M_1 of all eigenvectors and associated vectors of the operator A, corresponding to all eigenvalues equal in absolute value to one, we can choose a basis which always has property 3). The operator A allows the expansion $A = U_1 + A_1$ where the operator U_1 maps all the space E onto M_1 and permutes the elements of the basis; the operator A_1 has spectral radius <1 (see ch. I, § 5, no. 6).

In the finite-dimensional case the statements mentioned are applicable to the so-called *stochastic matrices*, i.e., to the matrices (a_{ik}) with non-negative elements satisfying conditions

$$\sum_{k=1}^{n} a_{ik} = 1 \qquad (i = 1, 2, ..., n).$$

For the integral equation

$$\int_a^b K(t, s) \, \varphi(s) \, ds = \lambda \varphi(t)$$

with a continuous non-negative kernel, satisfying the condition

$$\int_a^b K(t, s) \, ds \equiv 1 \qquad (a \leqslant t \leqslant b),$$

the following statements are obtained:

a) All eigenvalues in absolute value equal to one are natural roots of unity.

b) The set of eigenfunctions, corresponding to the value $\lambda = 1$, has a basis consisting of non-negative functions $\varphi_1(s), \varphi_2(s), ..., \varphi_r(s)$ and having the following properties:

1. At least one point exists in (a, b) for every function $\varphi_j(s)(j=1, 2, ...,$ $r)$ at which the given function is positive and the remaining functions of the basis are equal to zero.

2. For every point of the interval (a, b), at least one function of the basis can be found which is positive at this point.

c) The set of eigenfunctions of the transposed equation corresponding to the value $\lambda = 1$ has a basis consisting of non-negative functions $\psi_1(s), \psi_2(s), ..., \psi_r(s)$ biorthogonal with the basis $\varphi_1(s), \varphi_2(s), ..., \varphi_r(s)$ and having the property that

$$\psi_i(s)\,\psi_j(s) \equiv 0 \qquad (a \leqslant s \leqslant b; \, i \neq j; \, i, j = 1, 2, ..., r).$$

4. *Non-homogeneous linear equation.* For the non-homogeneous equation

$$\lambda\varphi = A\varphi + f$$

with a bounded positive linear operator A, we consider for what conditions this equation has a solution in the cone K if f belongs to K.

Let r be the spectral radius of the operator A. If $\lambda > r$, then the non-homogeneous equation has a unique solution $\varphi \in K$ for $f \in K$. If $\lambda \leqslant r$ and the operator A is u_0-positive, then the non-homogeneous equation does not have a solution in the cone K for arbitrary $f \in K$.

5. *Invariant functionals and eigenvectors of conjugate operators.* A continuous linear functional $f(x)$ is called *invariant* with respect to the bounded operator A if

$$f(x) = f(Ax).$$

In other words, an invariant functional is a fixed vector for the conjugate operator:

$$A'f = f.$$

The following statement is very important: *if the collection of bounded positive operators $\{A_h\}$, commuting with one another, has a common fixed element inside the solid cone K, then a positive functional $F(x)$ exists which is invariant with respect to all the operators A_h.*

Example

Let G be a commutative group, E be the space of functions $x(g)$ bounded on G, and the operators $A_h(h \in G)$ be defined by the equality

$A_h x(g) = x(g+h)$. The function $u(g) \equiv 1$ is an interior element of the cone of all non-negative functions of E, which is fixed for the transformations A_h. There exists an invariant functional

$$F(x(g+h)) = F(x(g)) \qquad (h \in G).$$

If a topology is introduced in the group G, then for the defined conditions the functional $F(\alpha)$ can be represented in the form of an integral, that is an invariant integral exists on the group (see ch. VI, § 2, no. 1).

The following statement is more general than the existence theorem for an invariant functional: *a positive functional φ, which is a common eigenvector of all the adjoint operators:*

$$A_h' \varphi = \lambda_h \varphi \qquad (\lambda_h > 0)$$

exists for every collection $\{A_h\}$ of pairwise commutative bounded linear operators mapping the interior of a solid cone K into itself.

6. *Inconsistent inequalities.* If $y - x \notin K$, then we write $x \nleq y$.
Let the positive operator A be u_0-bounded below where

$$Au_0 \geqslant \lambda_0 u_).$$

Then

$$Ax \nleq \lambda x$$

for arbitrary non-zero $x \in K$ and $\lambda < \lambda_0$, and it follows from

$$Ax \leqslant \lambda_0 x (x \in K)$$

that

$$x = ku_0 \quad \text{and} \quad Au_0 = \lambda_0 u_0.$$

Let the operator A be u_0-bounded above where

$$Au_0 \leqslant \lambda_0 u_0;$$

then

$$Ax \ngeq \lambda x$$

for arbitrary non-zero $x \in K$ and $\lambda > \lambda_0$.
Let the operator A be u_0-positive where

$$Au_0 = \lambda_0 u_0;$$

then the elements $\lambda_0 x$ and Ax are incomparable for arbitrary non-zero $x \in K(x \neq cu_0)$:

$$\lambda_0 x \nleq Ax \quad \text{and} \quad \lambda_0 x \ngeq Ax.$$

The theorem formulated here is applicable to the comparison of the eigenvalues of two operators.

Let A_1 and A_2 be two linear operators, $A_1 x \leq A_2 x$ for $x \in K$, A_1 is u_0-bounded below where $A_1 u_0 \geq \lambda_0 u_0$. Then every affirmative eigen-value λ_2 of the operator A_2 is less than λ_0.

§ 4. Nonlinear operators

1. *Basic concepts.* Positiveness and monotonicity for nonlinear operators are defined the same as for linear operators. An operator A is *positive* if $AK \subset K$ and *monotone* if $Ax \leq Ay$ follows from $x \leq y$. Unlike linear operators, in the case considered, monotonicity does not follow from the positiveness of the operator.

The operator A is *strongly differentiable with respect to the cone K* at the point x_0 if

$$A(x_0 + h) - Ax_0 = A'(x_0) h + \omega(x_0, h)$$

for all $h \in K$, where $A'(x_0)$ is a linear operator and

$$\lim_{h \in K, \, \|h\| \to 0} \left\| \frac{\omega(x_0, h)}{\|h\|} \right\| = 0.$$

The linear operator $A'(x_0)$ is called the *strong derivative with respect to the cone of the operator A at the point x_0*. In the case where $\dfrac{\omega(x_0, h)}{\|h\|}$ weakly tends to θ for $\|h\| \to 0 (h \in K)$, we speak of the weak derivative with respect to the cone.

It turns out that the strong derivative $A'(x_0)$ of a completely continuous operator with respect to the cone K transforms every bounded set $T \subset K$ into a compact set, and the strong derivative with respect to a generating cone K of a completely continuous operator is a completely continuous operator.

Along with the derivative with respect to a cone, the derivative at infinity plays an important role in the investigation of nonlinear operators. An operator A is called *strongly differentiable at infinity with respect to the cone K* if a continuous linear operator $A'(\infty)$ exists for which

$$\lim_{R \to \infty} \sup_{\|x\| \geqslant R, \, x \in K} \frac{\|Ax - A'(\infty)\,x\|}{\|x\|} = 0.$$

In this connection $A'(\infty)$ is called a *strong asymptotic derivative with respect to the cone K*. Analogously, the concept of *weak differentiability at infinity* is developed.

2. *Existence of positive solutions.* Here the equation

$$x = Ax$$

with a positive operator A is considered. The solutions of this equation will be fixed elements of the operator.

Let the positive continuous operator A have a strong asymptotic derivative $A'(\infty)$ with respect to a cone and let the spectral radius of the operator $A'(\infty)$ be less than one. *It suffices that one of the following conditions be satisfied for the existence of at least one fixed point in the cone:*

a) *the operator A is completely continuous,*

b) *the operator A is monotone and the cone K is completely regular,*

c) *the space E is reflexive and the operator A is weakly continuous.*

For a completely continuous operator the condition of existence of $A'(\infty)$ can be replaced by the following: an R exists such that for all $\varepsilon > 0$, $Ax \not\geqslant (1+\varepsilon)x \ (x \in K, \ \|x\| \geqslant R)$.

The collection of elements x for which $x_0 \leqslant x \leqslant u_0$ is called the *conical interval* $\langle x_0, u_0 \rangle$.

It suffices for the existence, for an operator monotone on the interval $\langle x_0, u_0 \rangle$, of at least one fixed point that the operator transform $\langle x_0, u_0 \rangle$ into itself and that one of the following conditions be satisfied:

a) *the cone K is strongly minihedral,*

b) *the cone K is regular, the operator A is continuous,*

c) *the cone K is normal, the operator A is completely continuous,*

d) *the cone K is normal, the space E is reflexive, the operator A is weakly continuous.*

In the satisfaction of conditions b)–d), the fixed point of the operator A can be obtained as the limit of the sequence $x_n = Ax_{n-1} \ (n = 1, 2, ...)$. If it is additionally known that a unique fixed point of the operator A lies in $\langle x_0, u_0 \rangle$, then the successive approximations $y_n = Ay_{n-1} (n = 1, 2, ...)$

converge with respect to the norm to the solution for any $y_0 \in \langle x_0, u_0 \rangle$ in cases b)–c).

3. *Existence of a non-zero positive solution.* When $A\theta = \theta$, then not unfrequently the question arises of the existence in a cone of a second (different from θ) fixed point for a positive A. In several cases the answer to this question can be obtained.

We say that the positive operator $A(A\theta = \theta)$ is a *contraction of the cone* on the part from r to $R(0 < r < R)$ if

$$Ax \nleq x \, (x \in K, \|x\| \leqslant r, x \neq \theta)$$

and

$$Ax \ngeqslant (1 + \varepsilon) x \qquad (x \in K, \|x\| \geqslant R).$$

for all $\varepsilon > 0$.

The operator $A(A\theta = \theta)$ is called an *expansion of the cone* on the part of the cone from r to $R(0 < r < R)$ if

$$Ax \ngeqslant (1 + \varepsilon) x \, (x \in K, \|x\| \leqslant r, x \neq \theta)$$

for all $\varepsilon > 0$ and

$$Ax \nleq x \, (x \in K, \|x\| \geqslant R).$$

Let the positive completely continuous operator A be an operator of contraction (expansion) on some part of the cone K; then the operator A has at least one non-zero fixed point on K.

The verification of the conditions of contraction or expansion is facilitated if two linear operators A_- and A_+ exist for which

$$A_- x \leqslant Ax \leqslant A_+ x \qquad (x \in K).$$

The conditions of contraction will automatically hold if

$$A_- x \nleq x \qquad (x \in K, \|x\| \leqslant r),$$
$$A_+ x \ngeqslant x \qquad (x \in K, \|x\| \geqslant R).$$

Analogously we can verify the conditions for expansion. In this connnection it suffices to construct the operator A_- only on the elements of small norm, and the operator A_+ on elements with a large norm. In the construction of the operator A_- we often use the derivative $A'(\theta)$ with respect to the cone and in the construction of A_+, the derivative $A'(\infty)$ at infinity with respect to the cone (see [25]).

4. *Concave operators.* Let u_0 be a fixed non-zero element of K. The operator A is called u_0-concave on K if it is positive and monotone and positive numbers α and β exist for arbitrary non-zero $x \in K$ such that

$$\alpha u_0 \leqslant Ax \leqslant \beta u_0$$

and $\eta = \eta(x, a, b) > 0$ can be found such that

$$A(tx) \geqslant (1 + \eta) t \, Ax \qquad (a \leqslant t \leqslant b)$$

for arbitrary $x \in K$ with the condition $x \geqslant \gamma u_0 (\gamma > 0)$ and for every segment $[a, b] \subset (0, 1)$.

The set of those λ for which the equation

$$Ax = \lambda x$$

with a completely continuous u_0-concave operator has a non-zero solution in the cone K form some interval (α, β). The equation cannot have more than one solution different from θ in the cone K for every $\lambda \in (\alpha, \beta)$. If $\lambda_1 > \lambda_2 (\lambda_1, \lambda_2 \in (\alpha, \beta))$, then for the corresponding solutions of the equation x_1 and x_2 in K the inequality $x_1 \leqslant x_2$ is valid.

If $A\theta = \theta$ and the strong derivative $A'(\theta)$ with respect to the cone is a completely continuous operator, then the upper bound β is a positive eigenvalue of the operator $A'(\theta)$. If, in this connection, the operator $A'(\theta)$ is u_0-positive, then β coincides with the unique positive eigenvalue of the operator $A'(\theta)$.

If the operator A has a strong asymptotic derivative $A'(\infty)$ with respect to the cone and is a completely continuous u_0-positive operator, then α is a eigenvalue of the operator $A'(\infty)$.

The Uryson integral operator

$$Ax(t) = \int\limits_0^1 K(t, s, x(s)) \, ds,$$

in which the function $K(t, s, u)$ is continuous and $K(t, s, u) \geqslant 0$ for $u \geqslant 0$ can serve as an example of a nonlinear positive operator in the space $C(0, 1)$. This operator is monotone if the function $K(t, s, u)$ does not decrease as u increases. Moreover, if $K(t, s, 0) \equiv 0$ and for $u_2 > u_1$ the inequality

$$\frac{1}{u_1} K(t, s, u_1) - \frac{1}{u_2} K(t, s, u_2) > 0$$

is satisfied for every t for almost all s, then the integral operator will be u_0-concave. In this connection we take as u_0 the function identically equal to 1.

5. *Convergence of successive approximations.* Let the equation $x = Ax$ with the u_0-concave operator A on the normal cone K have a unique non-zero solution x^* in K. Then the successive approximations $x_n = Ax_{n-1}$ $(n = 1, 2, \ldots)$ converge to x^* whatever the non-zero initial approximation $x_0 \in K$ is.

Moreover, the successive approximations will converge in the u_0-norm, which, as was indicated in § 1, no. 5, is stronger than the initial norm of the space E.

The condition of u_0-concavity can be weakened, requiring only the satisfaction of the inequality $A(tx) \geqslant tAx$ $(0 \leqslant t \leqslant 1)$ for $A(tx)$. Then the convergence of the successive approximations will occur for arbitrary non-zero initial approximations of K if the cone K is regular or if the operator A is completely continuous.

CHAPTER VI

COMMUTATIVE NORMED RINGS (BANACH ALGEBRAS)

§ 1. Basic concepts

1. *Commutative normed rings.**)   A complex Banach space R, with elements x, y, ..., on which there is defined an associative and commutative multiplication xy which is commutative with multiplication by complex numbers, distributive with respect to addition, and continuous in each factor, is called a *commutative normed ring (Banach Algebra).***)

In the general theory of commutative normed rings, we can restrict ourselves to the consideration of rings with an *identity element*, that is, an element e such that $ex = x$ for every $x \in R$. If the ring does not have an identity element, then one can be formally adjoined to the ring; i.e., we consider the collection of elements of the form $\lambda e + x$, where e is an adjoined identity element and x is an arbitrary element of R, with the norm $\|\lambda e + x\| = |\lambda| + \|x\|$.

In every normed ring with an identity element, we can change the norm to an equivalent norm so that the relations $\|xy\| \leqslant \|x\| \cdot \|y\|$, $\|e\| = 1$ are satisfied for the new norm.****)

A set K of elements of the ring R is called a *system of generators* for this ring if the smallest closed subring, with an identity element, containing K is R. The identity element is not included among the generators.

2. *Examples of normed rings.*

1. Let $C(0,1)$ be the space of all complex functions, defined and con-

---

*) For definitions of a ring, group, algebra, and other algebraic definitions, see any standard text on higher algebra.

**) From point of view of the terminology of modern algebra, the term "Banach algebra" is more precise, but here the term "normed ring" introduced originally in the works of I. M. Gel'fand, is retained.

***) Thus the multiplication, which was assumed to be continuous in each factor separately, is actually continuous in x and y simultaneously [Editor].

tinuous on the segment $[0, 1]$, equipped with the norm $\|x\| = \max|x(t)|$. C is a normed ring (with the identity element $x(t) \equiv 1$) with respect to the usual multiplication.

2. Let $C^{(n)}(0, 1)$ be the space of all complex functions on the same segment $[0, 1]$ which possess a continuous n-th order derivative equipped with the norm

$$\|x\| = \sum_{k=0}^{n} \frac{\max\limits_{0 \leqslant t \leqslant 1} |x^{(k)}(t)|}{k!}.$$

$C^{(n)}$ is a normed ring (with the usual multiplication) where $\|xy\| \leqslant \|x\| \cdot \|y\|$.

3. Let $W(0, 2\pi)$ be the space of all complex functions $x(\theta)$, continuous on the circle $0 \leqslant \theta \leqslant 2\pi$ and expandable in an absolutely convergent Fourier series

$$x(\theta) = \sum_{m=-\infty}^{\infty} c_m e^{im\theta},$$

with the norm $\|x\| = \sum\limits_{-\infty}^{\infty} |c_m|$. The space W forms a normed ring (with the usual multiplication) where again $\|xy\| \leqslant \|x\| \cdot \|y\|$. We often call the ring W the *Wiener* ring.

4. Let A be the space of all functions of a complex variable ζ, defined and continuous on the disk $|\zeta| \leqslant 1$ and analytic inside this disk, equipped with the norm $\|x\| = \max\limits_{|\zeta| \leqslant 1}|x(\zeta)|$. A is a normed ring with the usual multiplication.

5. Let $L_1(-\infty, \infty)$ be the space of all absolutely summable measurable functions on the real line $-\infty < t < \infty$ with the norm

$$\|x\| = \int_{-\infty}^{\infty} |x(t)| \, dt.$$

L_1 forms a normed ring if as multiplication we take *convolution*:

$$(x * y)(t) = \int_{-\infty}^{\infty} x(t - \tau) y(\tau) \, d\tau.$$

More over, $\|x * y\| \leqslant \|x\| \cdot \|y\|$. In L_1 there is no identity element with respect to the multiplication introduced. We denote by V the normed ring obtained by means of formal adjunction of an identity element to L_1.

6. Let $V^{(b)}$ be the linear space of all complex functions $f(t)$ of bounded variation on $-\infty < t < \infty$, satisfying the condition $f(-\infty)=0$ and continuous from the right, with the norm

$$\|f\| = \operatorname*{Var}_{(-\infty,\infty)} (f).$$

$V^{(b)}$ is a commutative normed ring if multiplication is defined as convolution:

$$(f_1 * f_2)(t) = \int_{-\infty}^{\infty} f_1(t-\tau)\, df_2(\tau).$$

The function

$$\varepsilon(t) = \begin{cases} 0 & \text{if } t < 0, \\ 1 & \text{if } t \geq 0 \end{cases}$$

serves as the identity in $V^{(b)}$, and $\|\varepsilon\| = 1$.

The ring V of example 5 is isomorphically and isometrically (see ch. 1, § 2, no. 4) embeddable in the ring $V^{(b)}$. This isomorphism is attained if we set the element $\lambda e + x$ in correspondence with the function $\lambda\varepsilon(t) + \int_{-\infty}^{t} x(\tau)\,d\tau$.

7. Let $\alpha(t)$ be a positive function, defined and continuous for all real values of t and satisfying the condition

$$\alpha(t_1 + t_2) \leq \alpha(t_1)\,\alpha(t_2)$$

for all t_1 and t_2. $L^{\langle\alpha\rangle}$ is the normed space of all measurable functions $x(t)$ for which

$$\|x\| = \int_{-\infty}^{\infty} |x(t)|\,\alpha(t)\,dt < \infty.$$

$L^{\langle\alpha\rangle}$ forms a commutative normed ring with convolution

$$(x * y)(t) = \int_{-\infty}^{\infty} x(t-\tau)\,y(\tau)\,d\tau$$

as multiplication. In this ring there is no identity element. The ring obtained by the formal adjunction of an identity element to $L^{\langle\alpha\rangle}$ is denoted by $V^{\langle\alpha\rangle}$. The ring V from example 5 is a ring $V^{\langle\alpha\rangle}$ where $\alpha(t)\equiv 1$.

8. Let $L_+^{\langle\alpha\rangle}$ be the collection of all measurable complex functions $x(t)$ $(t\geqslant 0)$ satisfying the condition

$$\|x\| = \int_0^\infty |x(t)|\,\alpha(t)\,dt < \infty,$$

where $\alpha(t)$ is a function as in example 7 but considered only for $t\geqslant 0$. $L_+^{\langle\alpha\rangle}$ forms a normed ring with the multiplication defined by the formula

$$(x*y)(t) = \int_0^t x(t-\tau)\,y(\tau)\,d\tau.$$

Adjoining to $L_+^{\langle\alpha\rangle}$ the formal identity element, we obtain the ring which is denoted by $V_+^{\langle\alpha\rangle}$.

The rings of examples 7 and 8 are called *Wiener rings with weight*.

3. *Normed fields.* An element $y\in R$ is called the *inverse* of an element $x\in R$ if $xy=e$. An element of the ring, having an inverse, is called *invertible*. If $\|x-e\|<1$, then the element x is invertible and its inverse can be represented in the form of a *Neumann series*:

$$y = \sum_{k=0}^\infty (e-x)^k.$$

If every non-zero element of the ring has an inverse, then the ring is called a *normed field*. Every normed field is isomorphic and isometric to the field of complex numbers.

4. *Maximal ideals and multiplicative functionals.* A set I of elements of the ring R with an identity element is called an *ideal* if $IR\subset I$ and $I-I\subset I$ where $\{0\}\neq I\neq R$ (IR is the collection of elements of the form ab where $a\in I$ and $b\in R$, and $I-I$ is the collection of elements of the form $a-b$ where $a\in I$ and $b\in I$). Every ideal is a linear manifold of the space R. If the ideal I is closed, then it forms a subspace. The factor space R/I is itself a normed ring with the natural definitions of the operations and the norm of the coset X defined by

$$\|X\| = \inf_{x\in X} \|x\|$$

(see ch. I, § 4, no. 5). This factor space is called the *factor ring of the ring R with respect to the ideal I*.

The concept of a maximal ideal is a central concept in the theory of commutative normed rings. An ideal is called *maximal* if it is not properly contained in any other ideal. Every maximal ideal is closed.

An element $x \in R$ has an inverse if and only if it does not belong to any of the maximal ideals. The factor ring, with respect to a maximal ideal, is isomorphic and isometric to the field of complex numbers.

Among the continuous linear functionals on R the *multiplicative* functionals, that is, those non-zero functionals such that $M(xy) = M(x) M(y)$, play an important role. There is a close connection between maximal ideals and multiplicative functionals: *every maximal ideal is a hyperplane $M(x) = 0$ for some multiplicative functional, and conversely*. This last statement facilitates the description of the maximal ideals of a given ring, reducing this problem to the description of multiplicative functionals. In this way, in the case of the rings C, $C^{(n)}$, W, A (examples 1, 2, 3, 4), a multiplicative functional represents the evaluation of the functions of the rings at a fixed point of the corresponding domains of definition (segment, circle, or disk). Thus a function $x(t)$ which is not zero at any point does not belong to any of the maximal ideals of the corresponding ring, and therefore its inverse $\dfrac{1}{x(t)}$ also belongs to the ring. This statement is trivial in the case of the rings C, $C^{(n)}$ and A. However, in application to the ring W, it reduces to the proof of the well-known Wiener theorem: if the function $x(\theta)$ is expandable in an absolutely convergent Fourier series and does not vanish, then $\dfrac{1}{x(\theta)}$ is expandable in an absolutely convergent Fourier series. This idea, applied in the proof of the Wiener theorem, is applicable in many other cases.

L_1 forms a maximal ideal in the ring V (example 5). All remaining maximal ideals in V can be described in the following manner. Let s be a real number and M_s the collection of elements $\lambda e + x \in V$ for which

$$\lambda + \int_{-\infty}^{\infty} e^{ist} x(t) \, dt = 0.$$

M_s forms a maximal ideal in V. Other maximal ideals, besides L_1 and M_s ($-\infty < s < \infty$), do not exist in the ring V.

An element $\lambda e + x(t) \in V$ is invertible if and only if

$$\lambda \left(\lambda + \int_{-\infty}^{\infty} x(t) \, e^{ist} \, dt \right)$$

is different from zero for arbitrary s.

The maximal ideals in the rings $V^{\langle \alpha \rangle}$ and $V_+^{\langle \alpha \rangle}$ can be described in an analogous manner (examples 7 and 8). Thus, in the case of the ring $V^{\langle \alpha \rangle}$, besides $L^{\langle \alpha \rangle}$, maximal ideals M_s of the above indicated type still occur but s can now be an arbitrary complex number for which

$$\lim_{t \to +\infty} \frac{\ln \alpha(t)}{-t} \leqslant \operatorname{Im} s \leqslant \lim_{t \to +\infty} \frac{\ln \alpha(-t)}{t}.$$

In the case of the ring $V_+^{\langle \alpha \rangle}$, s is an arbitrary complex number for which

$$\operatorname{Im} s \geqslant \lim_{t \to +\infty} \frac{\ln \alpha(t)}{-t}.$$

The situation is more complicated for the ring $V^{(b)}$ (example 6). However, in this case all maximal ideals have also been described (see [9]).

5. *Maximal ideal space.*   The set $\mathfrak{M}$ of all multiplicative functionals on R is closed in the weak* topology $\sigma(R', R)$. Every multiplicative functional has norm 1, hence $\mathfrak{M}$ is a weak* closed subset of the unit sphere of the conjugate space which is compact in this topology (see ch. I, § 4, no. 4). Thus $\mathfrak{M}$ is a compact set in the weak* topology $\sigma(R', R)$. This set is called the *maximal ideal space.*

It often is useful to set the elements of the ring R in correspondence with continuous functions on the set $\mathfrak{M}$ by setting $x(M) = M(x)$ for $x \in R$, $M \in \mathfrak{M}$. The correspondence $x \to x(M)$ is a norm-nonincreasing homomorphism of R into the ring $C(\mathfrak{M})$ of all continuous functions on $\mathfrak{M}$ equipped with the usual norm (i.e., the norm $\|x\| = \sup_{M \in \mathfrak{M}} |x(M)|$). In the case when the correspondence is an isomorphism, R is called a *ring of functions.* We always have

$$\max_{M \in \mathfrak{M}} |x(M)| = \lim_{n \to \infty} \sqrt[n]{\|x^n\|} \leqslant \|x\|.$$

The intersection of all maximal ideals is called the *radical* of the ring. $x(M) \equiv 0$ for elements in the radical. The elements x for which $\sqrt[n]{\|x^n\|} \to 0$

are called *generalized nilpotent elements*. The radical of the ring coincides with the collection of all generalized nilpotent elements.

The set of values of the function $x(M)$ on the space $\mathfrak{M}$ coincides with the *spectrum of the element* x, i.e., with the set of those complex numbers λ for which $x - \lambda e$ does not have an inverse in R.

The ring R is, by definition, a *direct sum* of its ideals I_1 and I_2 if every element $x \in R$ can be expanded (in an unique manner) in a sum $x = x_1 + x_2$, where $x_1 \in I_1$, $x_2 \in I_2$.

The ring R is a direct sum of its ideals if and only if the space $\mathfrak{M}$ is not connected, i.e. is representable in the form of the union of two non-empty disjoint closed sets.

6. *Ring boundary of the space* $\mathfrak{M}$.   The smallest closed subset $\Gamma \subset \mathfrak{M}$ on which all the functions $|x(M)|$ $(x \in R)$ attain a maximum is called the *ring boundary* of the space $\mathfrak{M}$ (G. E. Šilov). Such a set exists and is unique. For example, in the case of the ring A (example 4), the maximal ideal space $\mathfrak{M}$ can be set into one-to-one correspondence with the disk $|\zeta| \leqslant 1$; the ring boundary Γ consists of points lying on the boundary $|\zeta| = 1$ of this disk. This and similar examples justify the name of ring boundary. If the ring R is isometrically imbedded in some larger ring R_1, then every continuous linear functional can be extended to a continuous linear functional over R_1. In this connection, the property of the functional being multiplicative is not retained. This means that not every maximal ideal of the ring R is the intersection of this ring with some maximal ideal in the larger ring R_1. As for maximal ideals corresponding to points on the boundary Γ, they all extend to maximal ideals in an arbitrary larger ring R_1. In this connection, if the norm in the initial ring is the maximum modulus on $\mathfrak{M}$, then a ring $R_1 \supset R$ exists such that only maximal ideals corresponding to points of Γ extend to the maximal ideals of this ring. The ring $C(\Gamma)$ of all continuous functions on Γ (in example 4, the ring of all continuous functions on the circle $|\zeta| = 1$) is such a ring R_1.

7. *Analytic functions on a ring*.   A function x_λ with values in a given normed ring R is called *analytic* if it is defined in some region of the complex variable λ and the ratio

$$\frac{x_{\lambda+h} - x_\lambda}{h}$$

converges in the norm to some limit $x'_\lambda \in R$ as $h \to 0$. For example, the function $(x - \lambda e)^{-1}$ is analytic in the complement of the spectrum of the element x.

For analytic functions with values in a ring, a considerable part of the elementary theory of ordinary analytic functions is extendible – in particular, the Cauchy theorem, the Cauchy integral formula, and the Liouville theorem.

If $f(\zeta)$ is an entire function, then in the ring R, for every element x, an element $f(x)$ can be defined by means of the Taylor series. This element has the property that $f(x)(M) = f(x(M))$ for all $M \in \mathfrak{M}$.

The Cauchy integral formula allows us to make this result more precise. Namely, *if x is some element of a ring and the function $f(\zeta)$ is analytic in a neighborhood of the spectrum of the element x, then the formula*

$$y = f(x) = \frac{1}{2\pi i} \int_\gamma (x - \lambda e)^{-1} f(\lambda) \, d\lambda,$$

where γ is an arbitrary rectifiable contour enclosing the spectrum of x and lying in the domain of analyticity of the function $f(\zeta)$, defines an element $y \in R$ for which $y(M) = f(x(M))$.

In this statement, if it is formulated for the ring W, is contained the generalization of the above-mentioned Wiener theorem, attributed to P. Levy.

Thus, to the element x of a normed ring, we can apply any function $f(\zeta)$ analytic in a neighborhood of the spectrum of x. In spite of the fact that this class of functions is narrow, already in the case of the ring W it, generally speaking, is impossible to extend it.

The Cauchy integral formula effects a homomorphic mapping $f(\zeta) \to f(x)$ of the ring of functions $f(\zeta)$ analytic in a neighborhood of the spectrum of the element x into the ring R.

The theory of functions of elements of a ring has many applications. For example, theorems of the Wiener-Levy type play an important role in the construction of the theory of regular and singular integral equations.

We can also apply a function $f(\zeta_1, \ldots, \zeta_n)$ of several independent variables $\zeta_1, \ldots, \zeta_n$ to the collection $x_1, \ldots, x_n$ of elements of a normed ring if this function is analytic in a neighborhood of the joint spectrum of the elements $x_1, \ldots, x_n$, i.e., in a neighborhood of the set (in n-dimensional complex space)

$$\{(\zeta_1, \ldots, \zeta_n) : \zeta_1 = x_1(M), \ldots, \zeta_n = x_n(M), M \in \mathfrak{M}\}.$$

8. *Invariant subspaces of R'*. Let R' be the conjugate space of the normed ring R. A subspace $P \subset R'$ is called *invariant* if from the relation $f \in P$ it follows that, for any $x_0 \in R$, the functional $f_{x_0}(x) = f(x_0 x)$ also belongs to P.

If I is an ideal in R, then its orthogonal complement (see ch. I, § 4, no. 5) will be an invariant subspace of R' which is closed in the weak* topology $\sigma(R', R)$. Conversely, every weak* closed invariant subspace of R' is the orthogonal complement of some ideal of the ring R.

The invariant subspace orthogonal to a maximal ideal $M \subset R$ is one-dimensional and consists of multiples of the functional $M(x)$; the converse is also true. Since every ideal is contained in a maximal ideal, every non-zero weak* closed invariant subspace $P \subset R'$ contains a one-dimensional invariant subspace.

In particular, let W' be the conjugate space of the Wiener ring W (example 3). The space W' can be identified with the space of all bounded sequences $\{..., f_{-1}, f_0, f_1, ...\}$ such that if

$$x(\theta) = \sum_{-\infty}^{\infty} c_k e^{ik\theta} \in W,$$

then

$$f(x) = \sum_{-\infty}^{\infty} f_k c_k.$$

The statement formulated above in terms of invariant subspaces in a conjugate space reduces then, in application to the space W', to the following theorem: *every weak* closed subspace of the space of sequences, invariant with respect to displacements, contains a subsequence of the form* $\{e^{ik\theta_0}\}$.

Every weak* closed invariant subspace $P \subset W'$, containing only one sequence $\{e^{ik\theta_0}\}$, is one-dimensional and consists of multiples of this sequence.

9. *Rings with involution.* A normed ring is called a *ring with involution* if an operation $x \to x^*$ is introduced in it which has the properties:

1) $(x^*)^* = x$,
2) $(\lambda x + \mu y)^* = \bar{\lambda} x^* + \bar{\mu} y^*$,
3) $(xy)^* = y^* x^*$.

The last property is written in a form in which it is extendible to non-commutative rings. A most important example of a non-commutative ring with involution is the ring of bounded operators on a Hilbert space; in this case, the operation * represents passage to the adjoint operator.

If the involution has the property

4) $\|x^*x\| = \|x^*\| \cdot \|x\|$,

then the ring is isomorphic and isometric to a subring of the ring of operators on a Hilbert space.

A commutative normed ring with involution, having property 4), is isomorphic and isometric to the ring of all continuous functions on its maximal ideal space.

If this theorem is applied to the ring of operators commuting with every operator which commutes with a given self-adjoint or normal operator, then an analogue of the spectral expansion of the operator is obtained.

The involution $x^* \to x$ is called *symmetric* if the element $e + x^*x$ is invertible for arbitrary $x \in R$.

A linear functional f, defined on a ring R with involution, is called *positive* if $f(x^*x) \geqslant 0$ for all $x \in R$.

Every positive functional on a commutative normed ring R with a symmetric involution is uniquely extendible to a positive linear functional on the space $C(\mathfrak{M})$ of all complex functions continuous on the space $\mathfrak{M}$ of maximal ideals of the ring R, and therefore it is representable in the form

$$f(x) = \int_{\mathfrak{M}} x(M) \, d\mu(M),$$

where $\mu(M)$ is a positive measure on $\mathfrak{M}$.

If, for every non-zero element $x_0 \in R$, a positive functional f_0 exists such that $f_0(x_0 x_0^*) \neq 0$, then the ring R does not have a (non-trivial) radical.

§ 2. Group rings. Harmonic analysis

1. *Group rings.* Let G be a group with a finite number of elements and let R be the linear system consisting of formal linear combinations of elements of the group:

$$x = \sum_g x_g g \qquad (x \in R),$$

where the x_g are arbitrary complex numbers. In the system R, an operation of multiplication of elements can be introduced in a natural way:

if $x = \sum_g x_g g$ and $y = \sum_g y_g g$, then

$$xy = \left(\sum_{g'} x_{g'}\, g' \right) \left(\sum_{g''} y_{g''}\, g'' \right) = \sum_{g'} \sum_{g''} x_{g'}\, y_{g''}\, g' g''.$$

If we denote the product $g'g'' = g$, then $g'' = (g')^{-1}g$ and the formula for multiplication assumes the form

$$xy = \sum_g \left(\sum_{g'} x_{g'}\, y_{(g')^{-1}g} \right) g.$$

With the operation of multiplication thus introduced, the system R becomes a ring (algebraic) which is called the *group ring* of the finite group G.

In the sequel, commutative groups are considered, and additive notation is used for the group operation. The formula for multiplication is written in the form

$$xy = \sum_g \left(\sum_{g'} x_{g'}\, y_{g-g'} \right) g.$$

The group ring of a commutative group is commutative. The formula for multiplication can be written in coordinate form:

$$(xy)_g = \sum_{g'} x_{g'}\, y_{g-g'}.$$

The group ring becomes a finite-dimensional commutative normed ring if we introduce the norm

$$\|x\| = \sum_g |x_g|.$$

Moreover,

$$\|xy\| \le \|x\| \cdot \|y\|.$$

The group ring of a discrete commutative group G with an infinite number of elements is constructed analogously. Here it is more convenient to consider elements of the group ring R to be complex-valued functions $x(g)$ of the elements of the group. The multiplication in R is introduced by the formula

$$(xy)(g) = \sum_{g'} x(g')\, y(g - g'),$$

and the norm by the formula

$$\|x\| = \sum_g |x(g)|.$$

The ring R consists of all functions $x(g)$ for which $\|x\| < \infty$. It is an infinite-dimensional commutative normed ring.

For the consideration of continuous groups, in the preceding formulas, sums are replaced by integrals. Furthermore, it is assumed that an *invariant integral* exists on the group, that is, an integral having the property

$$\int x(g + g_0)\, dg = \int x(g)\, dg \quad \text{for arbitrary} \quad g_0 \in G.$$

The ring R consists of the space $L_1(G)$ of summable functions $x(g)$ with the norm

$$\|x\| = \int |x(g)|\, dg$$

and with the operation of multiplication in the form of the convolution

$$(xy)(g) = \int x(g')\, y(g - g')\, dg'.$$

The group ring of a discrete group has an identity element. It is the function

$$e(g) = \begin{cases} 1, & \text{if} \quad g = 0, \\ 0, & \text{if} \quad g \neq 0. \end{cases}$$

The ring $L_1(G)$ of a non-discrete continuous group does not contain an identity element, since the function $e(g)$ is equivalent to zero in $L_1(G)$. Therefore, in this case the ring $V(G)$ obtained by means of a formal adjoining of an identity element to the ring $L_1(G)$ is called the group ring. The formally adjoined identity element can be treated as the δ-function on the group G (see ch. VIII).

The ring $V(-\infty, \infty)$ (example 5), corresponding to the group G of all real numbers, is a special case of a group ring.

We can introduce an involution in group rings. By definition, $x^*(g) = \overline{x(-g)}$ and $e^* = e$ for the identity element of the ring.

It turns out that group rings do not have (non-trivial) radicals, that is, their radicals consist only of the zero element. Thus a group ring is isomorphic to the ring of functions on the space $\mathfrak{M}$ of maximal ideals.

2. *The characters of a discrete group and maximal ideals of a group ring.* A function $\chi(g) \neq 0$ having the following two properties is called a

character of the discrete group G:

$$\chi(g + h) = \chi(g)\,\chi(h) \quad \text{and} \quad \chi^*(g) = \overline{\chi(-g)} = \chi(g).$$

It follows from the definition that $|\chi(g)| = 1$. The function $\chi(g) \equiv 1$ is a trivial character. We say that the group G has a *sufficient number of characters* if a character χ_0 can be found such that $\chi_0(g_0) \neq 1$ for every $g_0 \in G$.

From the characters of the group, we can construct multiplicative functionals on the group ring. If the group is discrete, then the multiplicative functionals are constructed according to the formula

$$M(x) = \sum_g x(g)\,\chi(g).$$

It turns out that this construction exhausts all the multiplicative functionals on the group ring. Thus, between multiplicative functionals and characters, and hence, between maximal ideals and characters, a one-to-one correspondence is established. A maximal ideal M consists of all functions orthogonal to some character:

$$\sum_g x(g)\,\chi(g) = 0.$$

The character is reconstructed from the maximal ideal by the formula

$$\chi(g_0) = M(e_{g_0}),$$

where

$$e_{g_0} = e_0(g - g_0) = \begin{cases} 1 & \text{if} \quad g = g_0, \\ 0 & \text{if} \quad g \neq g_0. \end{cases}$$

The space $\mathfrak{M}$ of maximal ideals is compact in the weak topology, and therefore the space X of characters of the group will also be compact in the corresponding topology. This topology is defined by means of a fundamental system of neighborhoods of the elements χ_0:

$$U(\chi_0, x_1, \ldots, x_n, \varepsilon),$$

where $\{x_1, \ldots, x_n\}$ is an arbitrary finite set of elements of the group ring and $\varepsilon > 0$. The character χ belongs to $U(\chi_0, x_1, \ldots, x_n, \varepsilon)$ if

$$\left| \sum_g x_k(g)\,[\chi(g) - \chi_0(g)] \right| < \varepsilon \quad (k = 1, 2, \ldots, n).$$

The topology indicated coincides with the topology introduced by L. S. Pontryagin by means of a fundamental system of neighborhoods

$U(\chi_0; F,\varepsilon)$ consisting of characters χ for which

$$|\chi(g) - \chi_0(g)| < \varepsilon$$

for $g \in F$. Here F runs through the arbitrary finite subsets of the group G.

In the set of characters χ, the operation of multiplication is introduced in the natural way: the product $\chi_1(g) \, \chi_2(g)$ of two characters of the group will again be a character of the group. The operation of multiplication is continuous in the topology of the space X.

Thus, *the space of characters of a discrete group is a compact commutative topological group.*

The group of integers $\{m\}$ is a simple example of a discrete group. The group ring consists of sequences $x = \{c_m\}$ with the law of multiplication

$$(xy)_m = \sum_{m' = -\infty}^{\infty} c_{m'} d_{m-m'}$$

and the norm

$$\|x\| = \sum |c_m|.$$

This ring is isomorphic and isometric to the Wiener ring W (example 3). The functions $e^{im\theta}$ are the characters of the group. For θ differing by an integral multiple of 2π, the characters coincide; therefore we can assume that $0 \leqslant \theta < 2\pi$, and the characters are representable as points on the unit circle. Thus the unit circle is the group of characters of the group of integers.

Multiplicative functionals are given by the formula

$$M(x) = \sum_{m=-\infty}^{\infty} c_m e^{im\theta}$$

and coincide, as noted above, with the values of the functions of the Wiener ring at the point θ.

3. *Compact groups. Principle of duality.* An invariant integral (see [38]) exists on a compact group; therefore the group ring $V(G)$ can be constructed for it. The elements of this ring have the form $\lambda e + x(g)$, where e is a formally adjoined identity element and $x(g) \in L_1(G)$. A trivial maximal ideal $M_\infty = L_1(G)$ occurs in the ring $V(G)$, and the corresponding multiplicative functional is $M_\infty(\lambda e + x) = \lambda$.

The characters of the group are defined in the same way as for a discrete group, with the additional requirement of continuity. A character defines a multiplicative functional according to the formula

$$M(\lambda e + x) = \lambda + \int x(g)\chi(g)\,dg.$$

It turns out that this construction enables us to obtain all non-trivial maximal ideals. The maximal ideal space, after rejection of the one point M_∞, becomes a discrete set. Thus, *the group of characters of a compact group is a discrete group.*

A compact group has a sufficient set of characters which form a complete orthogonal system of functions on the group G.

If we construct the group of characters G' for the group of characters X, it will be a compact group. It turns out that G' is *isomorphic to the initial group G.* The isomorphism is given by the equality

$$\chi(x) = \varphi_x(\chi),$$

where φ_x is an element of the group G'. The topology of the group of characters G' coincides with the topology of the initial group. Thus, the groups G and G' can be identified.

If we consider a discrete group G, its group of characters X, and the group of characters G' of the group X, then the groups G and G' are also isomorphic. These last statements constitute the content of the *principle of duality* of L. S. Pontryagin.

4. *Locally compact groups.* A group G is called *locally compact* if it has a compact neighborhood of zero. An invariant integral also exists on a locally compact group. The elements of the group ring $V(G)$ have the same form as in the case of a compact group.

The maximal ideal space, after removal from it of the trivial maximal ideal $M_\infty = L_1(G)$, ceases to be compact. Thus the group of characters of a locally compact group itself is a locally compact group.

A locally compact group has a sufficient number of characters, and the Pontryagin principle of duality is also valid for it.

The difference from the case of a compact group is that the characters do not belong to the ring $L_1(G)$, and we must consider them as elements of a conjugate space.

The group of all real numbers $-\infty < t < \infty$ is the simplest example of a locally compact group. The group ring is the ring $V(-\infty, \infty)$ (example 5). The functions e^{ist} are the characters of the group for an arbitrary real s.

The group of characters is also the group of real numbers. The representation of an element of the group ring in the form of a function on the space of maximal ideals

$$x(M_s) = \int_{-\infty}^{\infty} x(t) \, e^{ist} \, dt$$

corresponds to the passage from a function in $L_1(-\infty, \infty)$ to its Fourier transform.

An analogue of the Fourier transform can be constructed on an arbitrary locally compact group by means of its characters.

5. *Fourier transforms.* The passage from a function on the group G to a function on its group of characters X according to the formula

$$Tx(\chi) = \int_G x(g) \, \chi(g) \, dg$$

is called a *Fourier transform*.

A uniqueness theorem holds: *if the Fourier transforms of two functions coincide, then the functions themselves coincide for almost all $g \in G$.*

The operator T, considered on the intersection of the spaces $L_1(G)$ and $L_2(G)$, allows closure to an operator isometrically mapping the space $L_2(G)$ onto the space $L_2(X)$. The inverse operator on $L_1(X) \cap L_2(X)$ is given by the formula

$$T^{-1}f(g) = \int f(\chi) \, \bar{\chi}(g) \, d\chi \, .$$

Positive definite functions, that is, functions $\varphi(g)$ such that the inequality

$$\sum_{k=1}^{n} \sum_{l=1}^{n} \varphi(g_k - g_l) \, \bar{\xi}_k \xi_l \geq 0$$

is satisfied for an arbitrary finite set of elements $g_1, \dots, g_n$ of the group G and arbitrary complex numbers $\xi_1, \dots, \xi_n$, form an important class of functions on the group G.

A positive linear functional on the group ring can be constructed from a positive definite function on the group by the formula

$$f_\varphi(\lambda e + x) = \lambda \varphi(0) + \int x(g) \, \varphi(-g) \, dg \, .$$

Every positive functional f on $V(G)$ is representable in the form

$$f(\lambda e + x) = \lambda \varrho + f_\varphi(\lambda e + x),$$

where φ is uniquely determined by the functional f and is a positive definite function on G, and $\varrho > 0$.

The theorem on the representation of a positive definite function on a group follows from the theorem on the representation of a positive functional (see § 1, no. 8): *a continuous function on a locally compact commutative group G is positive definite if and only if it is the Fourier transform of a non-negative measure (defined uniquely with respect to φ) on the group of characters:*

$$\varphi(g) = \int_X \bar{\chi}(g) \, d\Phi(\chi).$$

6. *Hypercomplex systems.* *Hypercomplex systems* are more general objects than group rings. In the finite-dimensional case, hypercomplex systems are linear systems with a given rule for multiplication of the elements of the basis of this system:

$$e_i e_j = \sum_k c_{ijk} \, e_k.$$

Here, in distinction to the case of the group ring, the product of elements of the basis may not be an element of the basis but rather some element of the ring. The constants c_{ijk} (so-called *structural* constants) must have properties guaranteeing the necessary properties of the operations in the ring. If $c_{ijk} = c_{jik}$, then the hypercomplex system is commutative.

The formula for the multiplication of the elements $x = \{x_i\}$ and $y = \{y_i\}$ of a hypercomplex system has the form

$$(xy)_k = \sum_{i, j} x_i y_j c_{ijk}.$$

This formula can be considered as a generalization of the formula for the convolution in a group ring.

In the continuous case, some topological space Q plays the role of a basis, elements of the hypercomplex system are functions on Q, and multiplication is given by means of a generalized convolution.

With conditions imposed on the structural constants c_{ijk} in the discrete case and on the structural measure in the continuous case, we are successful in introducing the analogue of the concept of an invariant measure and of characters, and in studying the rings of summable functions with respect to this measure in as detailed a manner as for the group rings.

The theory of this class of rings allows the study of expansions in terms of the solutions of Sturm-Liouville equations and in terms of certain classes of orthogonal polynomials.

§ 3. Regular rings

1. *Regular rings.* A ring R with the radical $\{0\}$ (i.e., a ring of functions) in which, for an arbitrary closed set $F \subset \mathfrak{M}$ and arbitrary point $M_0 \in \mathfrak{M}$ not contained in F, a function $x(M)\,(x \in R)$ can be found such that $x(M) = 0$ for all $M \in F$ and $x(M_0) = 1$ is called *regular*. Every regular ring is *normal*, i.e., an analogous property of separation is satisfied for an arbitrary pair of disjoint closed sets. Moreover, in a regular ring, for an arbitrary finite open covering $\{V_i\}$, $1 \leqslant i \leqslant n$, of the space $\mathfrak{M}$ there is a "partition of unity", i.e., a system of functions $x_1(M), \dots, x_n(M)$ for which

$$\sum_{i=1}^{n} x_i(M) \equiv 1$$

and
$$x_i(M) = 0 \quad \text{for} \quad M \notin V_i, \qquad i = 1, 2, \dots, n.$$

A function $f(M)$, by definition, *belongs locally* to the ring R if a neighborhood exists for every point $M_0 \in \mathfrak{M}$ in which this function coincides with some function of the ring. The presence of a partition of unity for every finite open covering of the space $\mathfrak{M}$ allows us to establish that every continuous function belonging locally to the ring is itself an element of this ring. An element of the ring is called real if the function $x(M)$ is real.

SUFFICIENT CONDITION FOR REGULARITY: *a ring with real generators* (see § 1, no. 1) *is regular if*

$$\int_{-\infty}^{\infty} \ln \|e^{itz}\| \, \frac{dt}{1 + t^2} < \infty$$

for each of its real generators z.

The rings $C(0, 1)$, $C^{(n)}(0, 1)$, W, V are regular. The ring A is not regular.

Among the ideals of a normed ring, the *primary ideals* are of special interest, that is, those which are contained in only one maximal ideal. In the case of a regular ring of functions, in every maximal ideal a smallest closed primary ideal occurs, and all the remaining ideals are "trapped" between it and the maximal ideal. This smallest closed primary ideal $J(M_0)$ is the closure of the ideal formed by the functions $y(M)$ equal to zero on some neighborhood (depending on y) of the point M_0.

In the ring V of absolutely convergent Fourier integrals, every maximal ideal M, among them also $M_\infty = L_2(-\infty, \infty)$, coincides with $J(M)$. In particular, the following theorem follows from this fact:

WIENER TAUBERIAN THEOREM. *If the Fourier transform $\tilde{x}_0(s)$ of the function $x_0(t) \in L_1$ does not vanish for any real value of s and if $f(t)$ is a bounded measurable function such that*

$$\int_{-\infty}^{\infty} x_0(t-\tau)f(t)\,dt \to l \int_{-\infty}^{\infty} x_0(t)\,dt \quad as \quad \tau \to \infty,$$

then

$$\int_{-\infty}^{\infty} x(t-\tau)f(t)\,dt \to l \int_{-\infty}^{\infty} x(t)\,dt \quad as \quad \tau \to \infty$$

for an arbitrary function $x(t) \in L_1$.

A ring in which there is only one maximal ideal is called a *primary ring*. The factor ring with respect to a primary ideal is a primary ring. In the ring $C^{(n)}$ (example 2), the smallest closed primary ideal $J(t_0)$ corresponding to the maximal ideal consisting of functions vanishing at the point t_0 consists of the functions equal to zero at the point t_0 together with all their derivatives up to order n. The factor ring $C^{(n)}/J(t_0)$ is finite-dimensional and isomorphic to the ring $I^{(n)}$ with one generator X, $X^{n+1} = 0$, and consists of elements of the form

$$a_0 + a_1 X + \cdots + a_n X^n$$

with norm

$$\|a_0 + a_1 X + \cdots + a_n X^n\| = \sum_{k=0}^{n} \frac{|a_k|}{k!}.$$

Let R' and R'' be regular rings of functions with the same maximal ideal space where $R' \subset R''$ and R' is dense in R''. Let J' and J'' be the smallest closed primary ideals in these rings, corresponding to a fixed point $M_0 \in \mathfrak{M}$. Then the ring of residues R'/J' allows a natural homomorphism onto the ring of residues R''/J''. In this connection, if the ring of residues R'/J' is finite-dimensional, then the ring of residues R''/J'', as a homomorphic image of the ring R'/J', is also finite-dimensional and has a dimension not greater than the dimension of R'/J'.

For example, if R is a ring of functions on a segment (or on a circle) containing all infinitely differentiable functions as an everywhere dense set, then it automatically contains the ring $C^{(n)}$ and its factor ring with respect to the smallest closed primary ideal is finite-dimensional.

2. *Closed ideals.* To an arbitrary closed ideal $I \subset R$ there corresponds a closed set $F \subset \mathfrak{M}$ on which all the functions of I become zero. If, in the case of a regular ring, we fix a closed set and consider the collection of all ideals to which this set corresponds, then among them a smallest $J(F)$ can be found; it consists of functions which are limits (in the sense of convergence in the norm) of functions of the ring which are zero in some neighborhood of the set F. This fact allows the description, in several cases, of all possible closed ideals in a regular ring of functions.

Thus, in the ring C, every closed ideal is an intersection of maximal ideals; in the ring $C^{(n)}$ every closed ideal is the intersection of primary ideals. However, the problem of the description of all closed ideals is still far from its final solution even in the case of comparatively simple normed rings. For example, up to now, the structure of the closed ideals in the ring W is not known. It turns out, even in this case, that the same closed set, if it is constructed in a sufficiently complicated way, can correspond to an infinite number of different closed ideals imbedded one in the other. All closed ideals in the ring A have been described (see [9]).

3. *The ring $C(S)$ and its subrings.* Let S be compact, and let $C(S)$ be the ring of all continuous functions on S. A subring K of the ring $C(S)$ is called *symmetric* if the functions $x(t)$ and $\overline{x(t)}$ simultaneously belong or do not belong to K and *anti-symmetric* if $x(t), \overline{x(t)} \in K$ implies that $x(t) = $ constant.

In order for a closed symmetric subring K, containing the identity element, to coincide with the entire ring $C(S)$, it is necessary and sufficient that for any two distinct points $t_1, t_2 \in S$, there exist a function $x(t) \in K$ such that $x(t_1) \neq x(t_2)$.

If for the compactum S we take the segment $[0, 1]$, then, from the formulated statement, the Weierstrass theorem concerning the approximation of a continuous function by a polynomial follows directly. If S is the unit circle, then we obtain the trigonometric form of the Weierstrass theorem. Thus, this statement is a deep generalization of the Weierstrass theorem.

An arbitrary subring K of the ring $C(S)$ can be decomposed into a continuous sum of anti-symmetric rings. By means of this decomposition, for example, the problem of the description of the closed ideals of the ring K reduces to the problem of the description of the ideals of anti-symmetric rings. A closed subring of the ring $C(S)$ can be normal on S (i.e. separate closed subsets of S) and anti-symmetric, simultaneously.

Now let S not be compact but rather some completely regular topological space (see [38]).

The ring $C(S)$ of all bounded continuous functions on S will also be a normed ring; every point $t \in S$ will generate a multiplicative functional and, hence, a maximal ideal. The space $\mathfrak{M}$ of maximal ideals of the ring $C(S)$ will contain an everywhere dense subset, homeomorphic to S. The space $\mathfrak{M}$ will be a compact extension of S.

The spaces of maximal ideals of different subrings of the ring $C(S)$ can yield other compact extensions Q of the space S. All such closed subrings K can be characterized by these conditions:

1) if $x(t) \in K$, then $\overline{x(t)} \in K$;

2) for an arbitrary set $A \subset S$ and point $t_0 \in S$ not in its closure $\bar{A}$, there is a function $x(t) \in K$, equal to zero on A and different from zero for $t = t_0$.

CHAPTER VII

OPERATORS OF QUANTUM MECHANICS

§ 1. General statements of quantum mechanics

1. *State and physical magnitudes of a quantum-mechanical system.* Here a formal discussion will be given of a logical scheme of quantum mechanics without a presentation of those physical reasons from which this scheme arises. In connection with this, questions, which will be stated, can be considered to an equal extent as problems of the spectral theory of operators on Hilbert space.

In quantum mechanics, the state of a system is described by an element ψ of a complex infinite-dimensional separable Hilbert space H which is called the *space of states*. The elements ψ are called *vectors of state*.

Experimentally defined physical magnitudes, briefly the so-called *observables*, correspond to bounded self-adjoint operators acting on the space of states. The value of the magnitude A in the state ψ is given by the quadratic form $(A\psi, \psi)$. Unbounded self-adjoint operators also have a physical meaning. Arbitrary bounded real functions of these operators are observables. On account of the self-adjointness of the operators corresponding to the observable magnitudes, their possible values are always real.

The quadratic form $(A\psi, \psi)$, having a physical meaning, remains invariant for a simultaneous unitary transformation of the vectors of state and the operators corresponding to the observable magnitudes

$$\psi' \to U\psi, \qquad A' \to UAU^*.$$

2. *Representations of algebraic systems.* Let $\mathfrak{A}$ be a system of symbols between which a series of formal relations has been established. It is assumed that all these relations can be written in the form of equalities by means of addition and multiplication signs between symbols and complex numbers. We say that a *representation* of the system $\mathfrak{A}$ is con-

structed in a Hilbert space H if a linear operator, acting on H, is set in one-to-one correspondence with each symbol, where all formal equalities between the symbols pass to identical equalities between the operators. In this connection, by addition and multiplication we understand the addition and multiplication of operators, a complex number is set in correspondence with the identity operator multiplied by this number.

If the representation is realized by means of unbounded operators, then the corresponding equalities must be satisfied identically on all those elements on which both parts of these equalities are defined.

The theory of the representation of a group in the form of a group of bounded operators on a Hilbert space is one of the highly important divisions of the theory of groups. More often, unitary representations are considered since all unitary operators themselves form a group.

Let the system of operators $A, B, C, \ldots$ realize a representation of the system $\mathfrak{A}$. Then the system of operators UAU^{-1}, UBU^{-1}, $UCU^{-1}, \ldots$, where U is an invertible bounded operator, also realizes a representation of the system $\mathfrak{A}$. The two representations so constructed are called *equivalent*. If U is a unitary operator ($U^{-1} = U^*$) then the representations are called *unitarily equivalent*.

A representation is called *irreducible* if an arbitrary bounded operator, commuting with all the operators of the representation, is a multiple of the identity operator. For example, if the space H is decomposed into a direct sum of two subspaces, invariant with respect to all the operators of the representation, then these operators commute with the projection operators onto the subspaces and, hence, the representation is reducible.

3. *Coordinates and impulses.* In quantum mechanics the symbols q_i ($i = 1, 2, \ldots, n$), the so-called *coordinates*, and the symbols p_i ($i = 1, 2, \ldots, n$), the so-called *momenta*, play an essential role. *Commutation relations* exist between coordinates and *momenta*. If we introduce the notation $[a, b] = ab - ba$, then $[q_i, q_k] = 0$, $[p_i, p_k] = 0$, $[p_i, q_k] = ih\delta_{ik}$, where h is a fundamental constant, the so-called *Planck constant*, and $\delta_{ik} = 0$ for $i \neq k$ and $\delta_{ii} = 1$.

It turns out that there exist representations of a system of coordinates and impulses in the form of a system of (unbounded) self-adjoint operators in a Hilbert space. If this space is taken as the space of states, then the coordinates and impulses which are introduced acquire a physical meaning. The bounded functions of them become observable magnitudes.

It turns out that in a given separable Hilbert space H all irreducible self-adjoint representations of coordinates and impulses are unitarily equivalent. As was remarked in no. 1, the magnitudes are not changed by a simultaneous unitary transformation of the vectors of state and the operators having a physical meaning. Therefore there are no physical reasons to prefer one representation to another. In this sense, we say that the representation of coordinates and impulses is unique.

In the case $n = 1$ (one-dimensional case), the representation of coordinates q and impulses p can be realized by the operators

$$qf = xf \quad \text{and} \quad pf = ih \frac{df}{dx}$$

in the Hilbert space $L_2(-\infty, \infty)$. This representation is *suited to the operator q* since it is the operator of multiplication by the independent variable. For brevity it is called the *coordinate* or *q-representation*. It is irreducible.

Here are more examples of representations which are common in practice. If in the q-representation we add to the operator p the operator of multiplication by an arbitrary real function $m(x)$, then the relations of commutation are not changed, i.e., the operators $q' = q$, $p'f(x) = pf(x) + m(x)f(x)$ realize the representation. A unitary transformation of the operators p and q to p' and q' respectively is realized by means of the operator of multiplication by the function $\exp\left\{\frac{i}{h} \int_{-\infty}^{x} m(t)\, dt\right\}$.

A less trivial example is obtained if we interchange the roles of the operators p and q, that is, if we consider the representation

$$\tilde{q}f(x) = ih \frac{df}{dx}, \qquad \tilde{p}f(x) = xf(x)$$

in the same space $L_2(-\infty, \infty)$. This is the so-called *momentum* or *p-representation*. The operators $\tilde{p}$ and $\tilde{q}$ are connected with p and q by a unitary transformation:

$$\tilde{p} = UpU^*, \qquad \tilde{q} = UqU^*,$$

where U is defined by means of a Fourier transformation:

$$Uf = g, \qquad g(x) = \left(\frac{1}{2\pi h}\right)^{\frac{1}{2}} \int_{-\infty}^{\infty} f(t)\, e^{\frac{itx}{h}}\, dt.$$

In the case $n>1$ the natural representation, suited to the operators $q_1, ..., q_n$, is realized in the space of square-summable functions $\psi(x_1, ... x_n)$ in the variables $x_1, ..., x_n$, $-\infty < x_i < \infty$, where q_i and p_i $(i=1, ..., n)$ are operators of coordinatewise multiplication and differentiation. In the described representation, the value of the coordinate q_i in the state characterized by the function $\psi(x_1, ..., x_n)$ is given by the expression

$$(q_i\psi, \psi) = \int x_i |\psi(x_1, ..., x_n)|^2 \, dx_1 \, ... \, dx_n \quad (i = 1, ..., n).$$

The expression $|\psi(x_1, ..., x_n)|^2$ is interpreted as the probability density that the coordinates $q_1, ..., q_n$ of the system have the values $x_1, ..., x_n$.

4. *Energy operator. Schrödinger equation.* In classical mechanics, the state of a dynamical system with n degrees of freedom is defined by the assignment to it of n generalized coordinates $q_1, ..., q_n$ and n generalized impulses $p_1, ..., p_n$. The evolution of the system is described by a *canonical system* of differential equations

$$\frac{dq_k}{dt} = \frac{\partial H}{\partial p_k}, \quad \frac{dp_k}{dt} = -\frac{\partial H}{\partial q_k} \quad (k = 1, 2, ..., n),$$

where H is the *Hamiltonian function* of the system. If the system is isolated, then H does not depend on time: $H = H(q_1, ..., q_n, p_1, ..., p_n)$.

A quantum-mechanical system is set in correspondence to every classical dynamical system. In certain representations of coordinates and impulses the function $H(q_1, ..., q_n, p_1, ... p_n)$ of the corresponding self-adjoint operators q_i and p_i is constructed. The linear operator H which is thus obtained is called the *energy operator* or also the *Hamiltonian*. Generally the Hamiltonian function consists of two terms, one of which depends only on the impulses, and the second, only on the coordinates. In this case, the question of how to define uniquely the function of non-commutative operators does not arise. For example, in the one-dimensional case, the Hamiltonian function has the form

$$H(p, q) = \frac{p^2}{2\mu} + V(q),$$

where $V(q)$ is the potential energy, μ is the mass of a particle.

Consequently, in a q-representation, the Hamiltonian is written as follows:

$$H = -\frac{h^2}{2\mu}\frac{d^2}{dx^2} + V(x).$$

The energy operator H plays an important role in the description of the development with time of a quantum-mechanical system. Two basic methods for this description exist. In the first (Schrödinger picture) the vectors of state change with time and the operators remain invariant. The evolution of the vectors of state is described by the equation

$$ih\frac{d\psi(t)}{dt} = H\psi(t),$$

which is called the *Schrödinger equation*. The solution of this equation can be written in the form

$$\psi(t) = \exp\left\{-\frac{i}{h}H(t - t_0)\right\}\psi(t_0)$$

(see ch. III, § 2, no. 7). The operator

$$U(t) = \exp\left\{-\frac{i}{h}Ht\right\}$$

is unitary and can be applied to an arbitrary state ψ.

In the second method (Heisenberg picture) the operators change with time and the vectors of state remain invariant. The evolution of the operators with time is described by the formula

$$A(t) = U^*(t - t_0)A(t_0)U(t - t_0).$$

The magnitudes $(A\psi, \psi)$. having a physical meaning, vary identically for both methods of description.

The operator $A(t)$ satisfies the following differential equation:

$$-ih\frac{dA(t)}{dt} = [H, A] = HA - AH.$$

Other methods of the description of the evolution of a system in time, for which both the operators and vectors of state vary, are equivalent to the Heisenberg and Schrödinger pictures. If the variation of the operator is described by the unitary operator $V(t)$:

$$A(t) = V^*(t)A(0)V(t),$$

then the variation of the vectors of state is given by the formula

$$\psi(t) = \tilde{U}(t)\,\psi(0) = V^*(t)\,U(t)\,\psi(0).$$

The form $(A\psi, \psi)$, having a physical meaning in all the descriptions, changes identically in time.

The described transition from a classical dynamical system to a quantum-mechanical system is called *quantization*.

5. *Concrete quantum-mechanical systems.* The basic systems which are studied in non-relativistic quantum mechanics are formed by interacting electrons in a field of nuclei or in other external fields. The classical electron is a particle with three coordinates. The classical Hamiltonian function for a system of n particles with vector coordinates x_k and momenta p_k and masses m_k has the form

$$H(x_k, p_k) = \sum_{k=1}^{n} \frac{p_k^2}{2m_k} + \sum_{i<j} V_{ij}(x_i - x_j) + \sum_{i=1}^{n} V_i(x_i).$$

Here $V_{ij}(x_i - x_j)$ is the potential of the interaction between the i-th and the j-th particles and $V_i(x_i)$ is the potential of the interaction of the i-th particle with the external field.

The energy operator of the corresponding quantum-mechanical system in a coordinate representation, i.e., in a representation suited to the coordinates $x_1, \ldots, x_n$ is defined by a differential operator in the space of functions of 3n variables of the form

$$H = - \sum_{i=1}^{n} \frac{h^2}{2m_i} \Delta_i + \sum_{i<j} V_{ij}(x_i - x_j) + \sum_{i=1}^{n} V(x_i),$$

where the Δ_i are Laplace operators. This differential expression is called the *Schrödinger operator*.

Of course, a precise study of a complicated system encounters almost insurmountable difficulties. For approximate approaches, simpler model problems play a large role. Of these the most important is the problem of a particle in a central field. The corresponding energy operator has the form

$$H_3 = - \frac{h^2}{2m} \Delta + V(x).$$

In particular, the problem of the interaction of two particles without an external field reduces to this problem.

Mathematically, only this problem has been, to any degree, extensively studied. Different concrete problems lead to different assumptions concerning the behavior of the potential $V(x)$. The problem is considerably simplified in the case when $V(x)$ allows separation of variables. The simplest cases when such a separation is possible are as follows:

1) $V(x) = V(z)$,

where z is one of the coordinates of the vector x.

The problem reduces to the study of the operator

$$H_1 = -\frac{h^2}{2m}\frac{d^2}{dz^2} + V(z)$$

in the space of functions on the axis $-\infty < z < \infty$.

2) $V(x) = V(|x|) = V(r)$, $|x| = (x_1^2 + x_2^2 + \cdots + x_n^2)^{\frac{1}{2}}$.

The problem reduces to the study of the operator

$$H_l = \frac{h^2}{2m}\left(-\frac{d^2}{dr^2} + \frac{l(l+1)}{r^2}\right) + V(r)$$

in the space of functions on the half-axis $0 \leqslant r < \infty$ for fixed integers l, $l \geqslant 0$, where the boundary condition becomes $\psi(0) = 0$ in the case $l = 0$.

In the sequel, the operators H_3, H_1 and H_l are called three-dimensional, one-dimensional and radial Schrödinger operators, respectively.

6. *Transition from quantum mechanics to classical mechanics.* Under quantization, a quantum-mechanical system corresponds uniquely to each classical system.

On the other hand, classical mechanics is a limiting case of quantum mechanics as $h \to 0$. By this we understand that in some representation the operators of quantum mechanics pass, as $h \to 0$, to the operators of multiplication by functions, corresponding to the magnitudes of classical mechanics. In this connection the coordinate operators pass to operators of multiplication by the coordinates x_i, the momentum operators pass to operators of multiplication by classical momenta $(p_i)_{cl}$, the energy operator passes to the operator of multiplication by a constant E equal to the total energy of the system. For one point in a potential field, the classical coordinates, momenta, and energies are connected by the law of conservation of mechanical energy:

$$\frac{p_{cl}^2}{2\mu} + V(x) = E.$$

Since $p_{cl} = \mu \dot{x}$, classical time is expressed in terms of p_{cl} by the formula

$$\tau = \mu \int \frac{dx}{p_{cl}}.$$

The operator $e^{\frac{i}{h}Ht}$ does not have a limit as $h \to 0$ either in the q- or in the p-representation. Those representations in which the operator $e^{\frac{i}{h}Ht}$ has a limit as $h \to 0$ are called *quasi-classical*.

In the one-dimensional case, with the assumption that the potential $V(x)$ is a bounded function, a quasi-classical representation can be obtained by means of the operator of multiplication by a function

$$M(t) = \sqrt{p_{cl}}\, e^{\frac{i}{h}\left[\int p_{cl}\, dx + Et\right]},$$

where E is a constant greater than $V(x)$.

This operator is a unitary operator acting from the space L_2 into the space $\tilde{L}_2$, where the norm is defined by the formula

$$\|f\|_{L_2} = \int\limits_{-\infty}^{\infty} |f(x)|^2 \frac{\mu}{p_{cl}}\, dx = \int\limits_{-\infty}^{\infty} |f(x(\tau))|^2\, d\tau.$$

In such a constructed quasi-classical representation, the basic operators will appear as follows:

$$M(t)\, x M^{-1}(t) = x,$$

$$M(t)\, p M^{-1}(t) = p_{cl} - 2ih \sqrt{p_{cl}}\, \frac{\partial}{\partial x} \frac{1}{\sqrt{p_{cl}}},$$

$$M(t)\, H M^{-1}(t) = E - ih\frac{\partial}{\partial \tau} + \frac{h^2}{2\mu}\sqrt{p_{cl}}\, \frac{\partial^2}{\partial x^2} \frac{1}{\sqrt{p_{cl}}},$$

$$M(t)\, e^{\frac{i}{h}Ht}\, M^{-1}(0) = \exp\left\{ t\frac{\partial}{\partial \tau} + t\frac{ih}{2\mu}\sqrt{p_{cl}}\, \frac{\partial^2}{\partial x^2} \frac{1}{\sqrt{p_{cl}}} \right\}.$$

As is seen from the formulas, the operators of quantum mechanics pass to operators of classical mechanics as $h \to 0$; in this connection, $e^{\frac{i}{h} Ht}$ passes to the operator

$$Q_t = e^{t \frac{\partial}{\partial \tau}},$$

which is the operator of displacement along a classical trajectory

$$Q_t \varphi(\tau) = \varphi(t + \tau).$$

The usual Schrödinger equation

$$ih \frac{\partial \psi}{\partial t} = -\frac{h^2}{2\mu} \Delta \psi + V(x)\, \psi, \qquad x = (x_1, \dots, x_n),$$

can be written in a quasi-classical representation if there exists a non-overlapping n-parameter family of solutions $x(\alpha, t)$ $(\alpha = (\alpha_1, \dots, \alpha_n))$ of the Newtonian equations

$$\mu \ddot{x}_i = -\frac{\partial V(x)}{\partial x_i}, \qquad i = 1, \dots, n.$$

After the substitution

$$\psi = \varphi \sqrt{J} \exp^{\frac{i}{h} S},$$

where J is the Jacobian of α with respect to x $\left(J = \dfrac{\partial(\alpha_1, \dots, \alpha_n)}{\partial(x_1, \dots, x_n)}\right)$ and S is some function playing the role of action in mechanics, the Schrödinger equation reduces to the form

$$\frac{d\varphi}{dt} = \frac{\partial \varphi}{\partial t} + \frac{1}{\mu} \operatorname{grad} S \cdot \operatorname{grad} \varphi = \frac{ih}{2\mu} J^{-\frac{1}{2}} \Delta (J^{\frac{1}{2}} \varphi).$$

Here the left member is the derivative with respect to time along the classical trajectory. For $h = 0$ the solution of the equation will be constant along the classical trajectory. In this representation, the operator $e^{\frac{i}{h} Ht}$ also passes to the operator of displacement along the classical trajectory as $h \to 0$.

If we pass to the variables α and t, then

$$\varphi(x, t) = \varphi(x(\alpha, t), t) = \tilde{\varphi}(\alpha, t)$$

and the equation takes the form

$$\frac{\partial \tilde{\varphi}}{\partial t} = \frac{ih}{2\mu} J^{-\frac{1}{2}} \Delta_\alpha [J^{\frac{1}{2}} \tilde{\varphi}(\alpha, t)],$$

where Δ_α is the Laplace operator in the coordinates α. The transformation carrying $\psi(x, t)$ into $\tilde{\varphi}(\alpha, t)$ is unitary, so that

$$\int_{-\infty}^{\infty} |\psi|^2 \, dx = \int_{-\infty}^{\infty} |\tilde{\varphi}|^2 \, d\alpha.$$

It realizes the passage from a coordinate representation to a quasi-classical. For the general case of quasi-classical representations, see [32].

§ 2. Self-adjointness and the spectrum of the energy operator

1. *Criterion for self-adjointness.* In § 1, no. 5 when examples were cited of energy operators for concrete quantum-mechanical systems, they were written out in the form of formal differential expressions (Schrödinger operators). However, in the general theory, the energy operator must be a self-adjoint operator. Therefore it is necessary to describe the domains of definition of the corresponding energy operators. The Schrödinger operator on the set D' of sufficiently smooth finitary functions defines a symmetric operator H^0. We say that the Schrödinger operator is *essentially self-adjoint* if the closure of the operator H^0 is a self-adjoint operator. This closure H will be the *energy operator*.

The following is one of the most general criteria for essential self-adjointness of the Schrödinger operator:

KATO CRITERION. *If for some constants M and R the potentials of interaction satisfy the conditions*

$$\int_{|x| \leq R} |V_{ij}(x)|^2 \, dx \leq M \quad \text{and} \quad |V_{ij}(x)| \leq M \quad \text{for} \quad |x| \geq R$$

$$(i, j = 1, 2, ..., n),$$

and the potentials of the external field satisfy analogous conditions

$$\int\limits_{|x| \leqslant R} |V_i(x)|^2 \, dx \leqslant M \quad \text{and} \quad |V_i(x)| \leqslant M \quad \text{for} \quad |x| \geq R$$

$$(i = 1, 2, ..., n),$$

then the Schrödinger operator is an essentially self-adjoint operator. The domain of definition of the energy operator will coincide with the domain of definition of the self-adjoint operator obtained by closure of the operator

$$H^0 = -\sum_{i=1}^{n} \frac{h^2}{2m_i} \Delta_i^2$$

from the set of finitary functions.

The requirements imposed in this criterion on the potentials can be weakened in several directions. Thus we can get rid of the requirement of the boundedness above of the potentials, replacing it, for example, by the condition

$$V_{ij}(x) \geqslant -M, \quad V_i(x) > -M, \quad |x| \geqslant R \quad (i, j = 1, ..., n).$$

The integral conditions of the criterion allow the potential to have, in a neighborhood of singular points, singularities of the type

$$V(x)_{x \to x_0} \cong \frac{\alpha}{|x - x_0|^{\gamma}}$$

with index γ not greater than 1.5. The result of the theorem concerning the essential self-adjointness is retained if $V(x)$ has exponential singularities with $\gamma < 2$. Of course, for these weakened conditions, we can not retain the second statement of the Kato theorem with respect to the domain of definition of the energy operator.

It is quite impossible to reject completely some restrictions on the behavior of potentials at infinity or in a neighborhood of singular points. Thus, if in the three-dimensional Schrödinger operator

$$H_3 = -\Delta^2 + V(x),$$

$V(x) \to -\infty$ (as $|x| \to \infty$) such that $V(x) \leqslant -M|x|^{2+\varepsilon}(\varepsilon > 0)$, or $V(x)$ has a singular point of type $\dfrac{\alpha}{|x - x_0|^\gamma}$, where $\alpha < 0$ and $\gamma > 2$, then the operator H_3 is not essentially self-adjoint.

2. *Nature of the spectrum of a radial Schrödinger operator.* The dependence of the nature of the spectrum of the radial Schrödinger operator

$$H_l \psi(r) = -\psi''(r) + \frac{l(l+1)}{r^2} \psi(r) + V(r) \psi(r), \qquad \psi(0) = 0$$

on the behavior of the potential $V(r)$ has been studied very extensively. Here, several simple criteria which are at the same time typical are mentioned.

a) If $V(r) \to \infty$ as $r \to \infty$, then the spectrum is simple discrete.

b) If $V(r) \to a$ as $r \to \infty$, then the interval $a < \lambda < \infty$ is filled by the continuous spectrum.

Concerning the nature of the continuous spectrum, we can say more if we define more precisely the method of tending to a limiting value.

c) Thus if

$$\int\limits_0^\infty r|V(r)|\, dr < \infty,$$

then for $\lambda > 0$ only a simple continuous Lebesgue spectrum occurs, and for $\lambda < 0$ there can only be a finite number of negative eigenvalues where the point $\lambda = 0$ can also be an eigenvalue for $l > 0$.

For condition c), the solutions of the equation

$$\psi''(r) + \lambda\psi(r) = V(r)\psi(r) + \frac{l(l+1)}{r^2}\psi(r)$$

as $r \to \infty$ behave asymptotically as solutions of the equation with $V(r) \equiv 0$, i.e., for $l = 0$ as trigonometric functions, and for $l > 0$ as cylindrical functions with a half-integer index. For example, for $l = 0$ we can take

$$f_1(r, \lambda) \cong e^{i\sqrt{\lambda}\, r}, \qquad f_2(r, \lambda) \cong e^{-i\sqrt{\lambda}\, r}, \qquad (r \to \infty).$$

as linearly independent solutions of the equation.

For $\lambda > 0$ both these solutions are bounded and only one of their linear combinations satisfies the boundary condition $\psi(0) = 0$. These facts stipulate the above-described nature of the spectrum of the corresponding radial Schrödinger operator.

d) In the preceding example the coefficient $V(r)$ decreased as $r \to \infty$, roughly speaking, faster than $\dfrac{1}{r^2}$. If $V(r)$ decreases slower than $\dfrac{1}{r^2}$ then the preceding result, concerning the continuous spectrum, can be retained; however, the discrete spectrum can become infinite. The well-known example of a Coulomb field $V(r) = -\dfrac{c}{r}$ provides an illustration of this situation.

e) If $V(r) \to -\infty$ (as $r \to \infty$), where $\displaystyle\int_1^\infty \dfrac{dr}{|V(r)|^{\frac{1}{2}}} = \infty$, then the entire axis $-\infty < \lambda < \infty$ is taken up by the continuous spectrum.

3. *Nature of the spectrum of a one-dimensional Schrödinger operator.* For the one-dimensional operator

$$H_1 \psi(z) = -\frac{d^2}{dz^2} \psi(z) + V(z)\,\psi(z)$$

many criteria concerning the structure of the spectrum can be obtained from the corresponding criteria for a radial operator for $l=0$. However, the multiplicity of the spectrum may be doubled.

a) Let

$$V(z) \to a \quad (\text{as } z \to -\infty), \qquad V(z) \to b \quad (\text{as } z \to \infty)$$

(for definiteness, it is assumed that $b > a$), where

$$\int_{-\infty}^0 (1 + |z|)\,|V(z) - a|\,dz < \infty,$$

$$\int_0^\infty (1 + |z|)\,|V(z) - b|\,dz < \infty.$$

The spectrum of the operator H_1 has the following structure: the interval $b < \lambda < \infty$ is filled by a double continuous Lebesgue spectrum; for $a < \lambda < b$ a single continuous Lebesgue spectrum occurs, and for $\lambda < a$ it is possible that a finite number of simple discrete eigenvalues occur, where, if a region occurs where $V(z) < a$, then at least one eigenvalue is necessarily present. In connection with the last fact we say that in the one-dimensional case, an arbitrary potential, no matter how small, contains without fail a discrete level.

b) $V(z) \to \infty$ as $|z| \to \infty$. The spectrum in this case is simple discrete. The typical example is a harmonic oscillator. In this case $V(z) = z^2$ and the discrete levels form an arithmetic progression.

c) The potential $V(z)$ is periodic: $V(z+1) = V(z)$. In this case the double continuous Lebesgue spectrum fills separate intervals of the positive part of the real axis. These pieces of the spectrum coincide with the so-called zones of stability for a classical motion in the periodic field being considered.

4. *Nature of the spectrum of a three-dimensional Schrödinger operator.* Concerning the structure of the spectrum of the three-dimensional Schrödinger operator

$$H_3 \psi(x) = - \Delta^2 \psi(x) + V(x) \psi(x)$$

we can form several conclusions which are similar to those which were made for a radial operator.

a) If the region where the condition $V(x) - \lambda < 0$ is satisfied is finite, then λ is either a point of the discrete spectrum of the operator H_3 or belongs to its resolvent set where only the discrete spectrum of this operator can be situated to the left of the point λ.

b) If $V(x) \to \infty$ as $|x| \to \infty$, then the spectrum is purely discrete.

c) If $V(x) \to a$ as $|x| \to \infty$, then all the half-axis $a < \lambda < \infty$ is filled with the continuous spectrum.

As earlier, we can say more about the structure of the continuous spectrum if we make more precise how the potential tends to a limiting value.

d) Thus if $V(x) \to 0$ as $|x| \to \infty$, where

$$V(x) = O\left(\frac{1}{|x|^{2+\varepsilon}}\right), \qquad \varepsilon > 0,$$

then the half-axis $0 < \lambda < \infty$ is filled only by continuous Lebesgue spectrum, and for $\lambda \leqslant 0$ there can only be a finite number of discrete eigenvalues of finite multiplicity. The continuous spectrum is infinitely singular; however, one can completely explain the structure of eigenfunctions of the continuous spectrum. Namely, in a complete set of eigenfunctions, there is exactly one function $u(x, k)$ corresponding to each plane wave $\exp\{i(k, x)\}$, $(k, x) = \sum_{j=1}^{3} k_j x_j$. Of course, we can choose different sets of eigenfunctions; in this connection, the functions of one set can be expressed as linear combinations of the functions of another set.

For the general energy operator, such simply formulated criteria of the structure of the spectrum do not exist. A mathematical reason for this fact is the essentially different asymptotic behavior of a potential term in different directions in configuration space. There exist whole series of criteria of the structure both of the discrete and continuous spectrum of the operator H either proven strictly or which are clear from a physical point of view.

§ 3. Discrete spectrum, eigenfunctions

1. *Exact solutions.* The discovery of eigenvalues and eigenfunctions of the energy operator is one of the basic problems of quantum mechanics. In the present section we shall describe basic approximation methods which are applicable to the solution of this problem.

First of all, it is necessary to remark that the problem of finding the eigenvalues and eigenfunctions of the one-dimensional equation

$$-\frac{h^2}{2\mu}\frac{d^2\psi_n}{dx^2} + V(x)\,\psi_n = E_n\psi_n$$

for certain potentials $V(x)$ is solved exactly.

1. HARMONIC OSCILLATOR: $V(x) = \dfrac{\mu\omega^2 x^2}{2}$.

$$E_n = h\omega\left(n + \frac{1}{2}\right) \qquad (n = 0, 1, 2, ...),$$

$$\psi_n(x) = 1/\sqrt{x_0}\; e^{-\xi^2/2}\, H_n(\xi),$$

where $x_0 = \sqrt{h/\omega\mu}$, $\xi = x/x_0$, and the H_n areChebyshev-Hermite polynomials.

2. MORSE POTENTIAL: $V(x) = V_0(e^{-\frac{2x}{a}} - 2e^{-\frac{x}{a}})$. The number of eigenvalues (energy levels) is finite;

$$- E_n = V_0 \left[1 - \frac{h}{a\sqrt{2\mu V_0}} \left(n + \tfrac{1}{2} \right) \right]^2, \quad n + 2 < \frac{\sqrt{2\mu V_0}\, a}{h}, \quad n \geqslant 0,$$

$$\psi_n(x) = e^{-\frac{\xi}{2}} \xi^s F(-n, 2s + 1, \xi),$$

where $\xi = \dfrac{2\sqrt{2\mu V_0}\, a}{h} e^{-\frac{x}{a}}$, $s = \dfrac{a\sqrt{-2\mu E_n}}{h}$, and F is the degenerate hyper-

geometric function (see [27]).

3. $V(x) = V_0 \cot^2 \dfrac{\pi x}{a}$.

$$E_n = \frac{\pi^2 h^2}{2\mu a^2} (n^2 + 2ns - s) \qquad (n = 0, 1, 2, \ldots),$$

where $s = \tfrac{1}{2} \left(-1 + \sqrt{1 + \dfrac{8\mu V_0 a^2}{\pi^2 h^2}} \right)$;

$$\psi_n(x) = \begin{cases} \left(\sin \dfrac{\pi x}{a} \right)^{-2s} \cos \dfrac{\pi x}{a}\, F\left(\dfrac{-n - s + 1}{2}, \dfrac{n + 1}{2}, \right. \\ \qquad\qquad\qquad\qquad\qquad\qquad \left. \dfrac{3}{2}, \cos^2 \dfrac{\pi x}{a} \right) \quad \text{for } n \text{ even,} \\[2mm] \left(\sin \dfrac{\pi x}{a} \right)^{-2s} F\left(-\dfrac{n + s}{2}, \dfrac{n}{2}, \dfrac{1}{2}, \cos^2 \dfrac{\pi x}{a} \right) \\ \qquad\qquad\qquad\qquad\qquad\qquad\qquad\qquad \text{for } n \text{ odd;} \end{cases}$$

F is the hypergeometric function (see [27]). In particular,

$$E_n = n^2 - 2, \quad \psi_n(x) = n \cos \frac{n\pi x}{a} - \sin \frac{n\pi x}{a} \cot \frac{\pi x}{a}, \quad n \geqslant 1.$$

for $a = \pi h / \sqrt{2\mu}$ and $V_0 = 2$.

4. $V(x) = -\dfrac{V_0}{\operatorname{ch}^2 \dfrac{x}{a}}$.

The number of energy levels is finite;

$$E_n = -\frac{h^2}{2\mu a^2} [s - n]^2 \qquad (0 \leqslant n < s),$$

where

$$s = \frac{1}{2}\left[-1 + \sqrt{1 + \frac{8\mu V_0 a^2}{h^2}}\right];$$

$$\psi_n(x) = \begin{cases} \operatorname{ch}^{-2s}\dfrac{x}{2}\, F\left(\dfrac{-s+\kappa}{2}, \dfrac{-s-\kappa}{2}, \dfrac{1}{2}, -\operatorname{sh}^2\dfrac{x}{a}\right) & \text{for } n \text{ even}, \\[3ex] \operatorname{ch}^{-2s}\dfrac{x}{2}\operatorname{sh}\dfrac{x}{a}\, F\left(\dfrac{-s+\kappa+1}{2}, \dfrac{-s-\kappa+1}{2},\right. \\[3ex] \qquad\qquad \left.\dfrac{3}{2}, -\operatorname{sh}^2\dfrac{x}{a}\right) & \text{for } n \text{ odd}, \end{cases}$$

where $\kappa = \dfrac{a\sqrt{-2\mu E_n}}{h}$.

In examples 1 and 2 the eigenvalues satisfy the relation

$$\int_{x_1}^{x_2} \sqrt{E_n - V(x)}\, dx = \frac{\pi h}{\sqrt{2\mu}}(n + \tfrac{1}{2}),$$

where x_1 and x_2 are zeros of the expression under the radical.

In examples 3 and 4 the eigenvalues satisfy the relation

$$\int_{x_1}^{x_2} \sqrt{2\mu[E_n - V(x)]}\, dx = \pi h(n + \tfrac{1}{2}) + \left(\sqrt{1 + \frac{8\mu V_0 a^2}{h^2}} - \sqrt{\frac{8\mu V_0 a^2}{h^2}}\right).$$

In the radialsymmetric case the problem is also sometimes solved exactly.

5. COULOMB POTENTIAL: $V(r) = -\dfrac{a}{r}$.

$$E_n = -\frac{a^2\mu}{2h^2}\frac{1}{n^2} \qquad (n = 1, 2, \ldots),$$

$$\psi_{n,l}(x) = -\left\{c_n^3\frac{(n-l-1)!}{2n[(n+1)!]^3}\right\}^{\frac{1}{2}} e^{-\frac{\varrho}{2}}\varrho^l L_{n+1}^{2l+1}(\varrho),$$

where $c_n = \dfrac{2\mu}{h^2}\dfrac{a}{n}$, $\varrho = cr$ and L_{n+1}^{2l+1} are Laguerre polynomials.

6. OSCILLATOR: $V(r) = \dfrac{\mu\omega^2 r^2}{2}$.

$$E_{n,\,l} = h\omega\,(2n + l + \tfrac{3}{2}) \qquad (n = 0, 1, 2, ...),$$

$$\psi_{n,\,l}(x) = r^l\, e^{-\frac{\mu\omega}{2h} r^2}\, F\!\left(- n,\, l + \frac{3}{2},\, \frac{\mu\omega r^2}{h}\right).$$

In the last two examples, the eigenvalues satisfy the relation

$$\int_{r_1}^{r_2} \sqrt{2\mu\!\left[E_n - V(r) - \frac{h^2\,(l + \tfrac{1}{2})^2}{2\mu r^2}\right]}\; dr = \pi h\,(n + \tfrac{1}{2}),$$

where r_1 and r_2 are zeros of the expression under the radical.

2. *General properties of the solutions of the Schrödinger equation.*   In the one-dimensional case, if the potential does not have singularities, the discrete spectrum is always simple. As is evident in the example of an oscillator (example 1), the eigenfunction $\psi_0(x)$ does not have zeros, and $\psi_n(x)$ has n zeros. In this connection, all the zeros lie in the domain $E_n \geqslant V(x)$. This property is retained for an arbitrary potential. Usually $\psi_0(x)$ is called a function of a basic state. If the potential is symmetric with respect to $x=0$, then $\psi_n'(0)=0$ for even n and $\psi_n(0)=0$ for odd n. For $n\to\infty$, when the number of zeros is enlarged, the sine

$$\psi_n(x) \approx \frac{A}{\sqrt{p_{\mathrm{cl}}}}\, \sin\!\left[\frac{1}{h}\int_{x_1}^{x} p_{\mathrm{cl}}\, dx + \frac{\pi}{4}\right]$$

(A is a normalizing constant) can serve as an asymptotic eigenfunction in the domain $E_n - V(x) \geqslant 0$. In the sequel it is assumed, both in the one-dimensional and many-dimensional cases, that this domain is finite.

It is evident from example 1 that the eigenfunctions of a harmonic oscillator as $x\to\infty$ tend to zero as $e^{-\alpha x^2}$ ($\alpha>0$ is some constant). In the general case we can say that the eigenfunctions tend to zero as $x\to\infty$ faster than

$$\exp\!\left[-\frac{1}{h}\int_{-\infty}^{x} \sqrt{2\mu\,(V(x) - E_n)}\; dx\right].$$

This estimate is also retained in the multidimensional case, only instead of $V(x)$ it is necessary to take the minimum of $V(x)$ on the sphere $|x| = r$ and the integral is taken with respect to r.

3. *Quasi-classical approximation for solutions of the one-dimensional Schrödinger equation.* The Schrödinger equation is sometimes written in the form

$$-\frac{h^2}{2\mu}\frac{d^2\psi_n}{dx^2} + V(x)\,\psi_n = E_n\psi_n,$$

where $V(x) = V_0 f\left(\dfrac{x}{a}\right)$, a is a characteristic dimension, V_0 is a characteristic potential.

The asymptotic behavior of eigenfunctions and eigenvalues of the Schrödinger equation for the conditions

$$n \gg 1, \frac{h}{\sqrt{V_0}a\sqrt{2\mu}} \ll 1,$$

$$(n + \tfrac{1}{2})\frac{h}{\sqrt{V_0}a\sqrt{2\mu}} \sim \text{const}.$$

is called a *quasi-classical approximation*. Without loss of generality, we can set $V_0 = 1$, $a = 1$ and look for the asymptotic behavior as $h \to 0$. Let the domain $E_n \geqslant V(x)$ be the segment $[x_1, x_2]$ and $V(x)$ a sufficiently smooth function. The points x_1, x_2 bounding this domain are called turning points. From here on, it is assumed that they are zeros of the first order of the function $E - V(x)$.

The first term of the quasi-classical asymptotic behavior of the eigenvalues for the Schrödinger equation is found from the *Bohr condition of quantization:*

$$\int_{x_1}^{x_2} \sqrt{2\mu[E_n^{(1)} - V(x)]}\, dx = h\pi(n + \tfrac{1}{2}),$$

$$E_n = E_n^{(1)} + O(h^2).$$

We denote the mean square of the function $\Phi(\tau)$ over the classical period

$$T = \sqrt{\frac{\mu}{2}} \int_{x_1}^{x_2} \frac{dx}{\sqrt{E_n^{(1)} - V(x)}} \quad \text{by } \tilde{\Phi}:$$

$$\tilde{\Phi}^2 = \int_0^T \Phi^2\left(x(\tau)\right) d\tau = \sqrt{\frac{\mu}{2}} \int_{x_1}^{x_2} \Phi^2(x) \frac{dx}{\sqrt{E_n^{(1)} - V(x)}}.$$

Then the quantization condition of quantification can be represented in the form

$$\frac{\tilde{p}_{cl}^2}{2\mu} = \frac{h}{4}(2n + 1)\pi.$$

The second term in the asymptotic behavior of E_n satisfies

$$E_n^{(2)} = -\frac{h^2}{24\mu} \frac{1}{T} \frac{d^2\tilde{F}^2}{dE^2},$$

where $E = E_n^{(1)}$ and $F(x) = -V'(x)$ ($F(x)$ is the force)

$$E_n = E_n^{(1)} + E_n^{(2)} + O(h^4).$$

The eigenfunction $\psi_n(x)$ converges in $L_2(-\infty, \infty)$ as $h \to 0$ and $E_n \to E$ to the function equal to

$$\psi(x) = \frac{1}{\sqrt[4]{E_n^{(1)} - V(x)}} \cos\left[\frac{\sqrt{2\mu}}{h} \int_{x_1}^{x} \sqrt{E_n^{(1)} - V(x)}\, dx - \frac{\pi}{4}\right]$$

for $x_1 \leqslant x \leqslant x_2$ and equal to zero outside this interval.

The function $\psi(x)$ is asymptotic to $\psi_n(x)$, uniformly with respect to x on any segment contained inside the interval $x_1 < x < x_2$. Outside this interval – for example, for $x \geqslant x_2 + \delta$ – the function

$$\psi_n(x) = \frac{C}{\sqrt[4]{V(x) - E_n^{(1)}}} e^{-\frac{\sqrt{2\mu}}{h} \int_{x_2}^{x} \sqrt{V(x) - E_n^{(1)}}\, dx}$$

will serve as an asymptotic, uniformly with respect to x.

If two parameters n and h are "connected rigidly", i.e. we set

$$(n_0 + \tfrac{1}{2}) \frac{h}{\sqrt{2\mu}} = \text{const} = \int_{x_1}^{x_2} \sqrt{E - V(x)}\, dx$$

(by the same token, assuming that $h = h_{n_0}$ varies discretely as $n_0 \to \infty$), then $\psi_{n_0+k}(x)$ will converge in the mean to the real part of the function*)

$$M(0)\, e^{\frac{2\pi i k \tau}{T}}, \qquad k = 0, 1, 2, \ldots,$$

where $M(t)$ and τ are defined in § 1, no. 6. In this connection, the eigenvalue can be written in the form

$$E_{n_0+k} = E + h\,\frac{2\pi k}{T} + O(h^2).$$

Sometimes asymptotics of eigenfunctions, uniform on an interval containing one or both turning points is useful. We can obtain these asymptotics by means of the following rule.

Let x_1 be a turning point in which the preceding asymptotic relations are not satisfied. We assume that for some function $V_1(x)$, the equation

$$y'' - V_1(x)\,y = 0$$

is solved exactly, and that the ratio $\dfrac{V_1(x_1 - x)}{E_n - V(x)}$ is analytic at the point x_1

and tends to one as $x \to x_1$.

Then the function

$$\varphi_n(x) = \sqrt[4]{\frac{V_1[z(x)]}{E_n - V(x)}}\; Y(z(x)),$$

where $z(x)$ is defined by the condition

$$\int\limits_0^{z(x)} \sqrt{V_1(x)}\, dx = \frac{\sqrt{2\mu}}{h} \int\limits_{x_1}^{x} \sqrt{E_n - V(x)}\, dx$$

and $Y(x)$ is some solution of the equation $y'' - V_1(x)\,y = 0$ will serve as an asymptotic eigenfunction, uniformly on the interval containing the point x_1.

In particular, if $V(x_1) \neq 0$ and $V_1(x) = x$, then $Y(x)$ is an Airy func-

---

*) The function $e^{\frac{2\pi k \tau}{T}}\, i$ is an eigenfunction of the operator Q_t of displacement with respect to the trajectory (§ 1, no. 6).

tion (see [27]) and

$$\varphi_n(x) = \sqrt[4]{\frac{\left(\dfrac{3\sqrt{2\mu}}{2h}\displaystyle\int\limits_{x_0}^{x}\sqrt{E_n - V(x)}\,dx\right)^{2/3}}{E_n - V(x)}} \times$$

$$\times Y\left[\left(\frac{3\sqrt{2\mu}}{2h}\int\limits_{x_0}^{x}\sqrt{E_n - V(x)}\,dx\right)^{2/3}\right].$$

If the domain $E_n - V(x) \geqslant 0$ is multiply connected in a neighborhood of E and consists of a finite number of simply connected segments $x_{1i} \leqslant x \leqslant$ $\leqslant x_{2i}$, $i = 1, 2, ..., k$, then $E_n^{(1)}$ will satisfy one of the equations

$$\sqrt{2\mu}\int\limits_{x_{1i}}^{x_{2i}}\sqrt{E_n^{(1)} - V(x)}\,dx = h\pi(n + \tfrac{1}{2}),$$

and $E_n^{(2)}$ will, as before, be defined by the formula

$$E_n^{(2)} = -\frac{h^2}{24\mu T}\frac{d^2\tilde{F}^2}{dE^2}, \qquad F(x) = -V'(x).$$

For definiteness let $E_0^{(1)}$ satisfy the i-th equation and not satisfy any other. Then $\psi_n(x)$ will tend exponentially to zero as $h \to 0$ in the intervals $x < x_{1i}$ and $x > x_{2i}$. Physically this means that the particle turns out to be in one of the "cavities" of the potential hole of $V(x)$.

4. *Calculation of eigenvalues in one-dimensional and radial symmetric cases.* Here a method is indicated for the discovery of eigenvalues of the one-dimensional and radial Schrödinger equation with prescribed accuracy.

If the potential $V(x)$ increases as a power of x, then the *formulas,* mentioned in no. 3, *for $E_n^{(1)}$ and $E_n^{(2)}$ serve as asymptotics with respect to only one parameter $n \gg 1$.* In this case the given formulas turn out to be convenient for the concrete calculation of eigenvalues.

Thus, for the potentials $V(x) = x^4$ and $V(x) = x^6$ the formula for $E_n^{(1)}$ for $n = 6$ already gives three correct places for the energy E_n, and together with the correction $E_n^{(2)}$ six places. For $n = 2$ the formula for $E_n^{(1)}$ yields one correct place, and together with correction, gives three places.

These formulas will not serve as asymptotics for $n \to \infty$ for a potential with singularities.

We give asymptotic formulas for the radial symmetric case. The first term of the asymptotics as $n \to \infty$ for a fixed l and a potential which increases as a power of r can be found from the relation

$$\int_0^{r_1} \sqrt{E_n^{(1)} - V(r)}\, dr = \pi \left(n + \frac{3}{4} + \frac{l}{2} \right) \frac{h}{\sqrt{2\mu}},$$

where r_1 is a zero of the expression under the radical. If $V'(0)=0$ then the second term will be equal to

$$E_n^{(2)} = -\frac{h^2}{2\mu T} \left\{ \frac{1}{12} \frac{\partial^2 \tilde{F}^2}{\partial E^2} - l(l+1) \frac{\partial F r^{-1}}{\partial E} \right\}_{E=E_n^{(1)}}.$$

The values of E_n for small n can be found with the help of a high speed electronic computer since on it the Cauchy problem for an equation of second order is easily solved.

The solution of the equation

$$-\frac{h^2}{2\mu} \frac{d^2 y}{dr^2} + \left[V(r) + \frac{h^2 l(l+1)}{2\mu r^2} \right] y - Ey = 0$$

with initial conditions

$$y(0) = 0, \qquad y'(0) = 1$$

for $E=E_n$ is equal (to within a normalizing constant) to the eigenfunction $\psi_n(r)$ of the radial equation.

The solutions of this problem for r greater than the largest root of the equation

$$E = V(r) + \frac{h^2 l(l+1)}{2\mu r^2}$$

will have different signs for the values $E=E^{(+)}$, where $E_{n-1} < E^{(+)} < E_n$, and $E=E^{(-)}$, where $E_n < E^{(-)} < E_{n+1}$.

On this fact the method of finding eigenvalues for small n (*ballistic method* or method of "firing") is based.

If after the solution of the problem for an arbitrary choice of E_1 (after "firing") the $E^{(+)}$ and $E^{(-)}$ indicated above can be found, then the segment $[E^+, E^-]$ is divided into halves and the sign of the solution of the problem for the greater r for $E = \dfrac{E^+ + E^-}{2}$ is found on the computer.

For example, suppose that the signs for $E = E^+$ and $E = \dfrac{E^+ + E^-}{2}$ are different. Then the entire process is repeated for the segment $\left[E^+, \dfrac{E^+ + E^-}{2} \right]$. Such a division is continued until a segment is obtained whose magnitude does not exceed the limits of the prescribed accuracy. The midpoint of this segment will, within the given accuracy, coincide with the unknown eigenvalue.

In an analogous manner we can find eigenvalues of a one-dimensional equation with a potential which is symmetric with respect to the point $x = 0$. In this connection it is necessary to take the initial condition

$$y(0) = 1, \qquad y'(0) = 0$$

for even n.

5. *Perturbation theory.* The equation

$$-\frac{h^2}{2\mu} \, \Delta \psi_n + \left[V_0(x) + \varepsilon V_1(x) \right] \psi_n = E_n \psi_n,$$

where ε is a small parameter, is called a *perturbed equation*. For $\varepsilon = 0$, the equation

$$-\frac{h^2}{2\mu} \, \Delta \varphi_n + V_0(x) \, \varphi_n = \lambda_n \varphi_n$$

which is called *non-perturbed* is obtained. The potential $\varepsilon V_1(x)$ is called the *perturbation*.

Here the case is considered where the spectrum of the non-perturbed equation is discrete.

If $V_0(x) > 0$ and $V_1(x)$ increase not faster than $\sqrt{V_0(x)}$, then φ_n and E_n are analytic functions of ε (see ch. II, § 3, no. 6):

$$E_n = \sum_{k=0}^{\infty} \varepsilon^k \mu_{n, k}, \qquad \psi_n = \sum_{k=0}^{\infty} \varepsilon^k \varphi_{n, k}.$$

The values $\mu_{n, k}$ and $\varphi_{n, k}$ can be found by substituting these expansions into the perturbed equation and equating the coefficients of ε^k.

One E_n corresponds to each λ_n if the eigenvalue is considered as many times as its multiplicity.

Thus if l is the multiplicity of λ_n, then it is assumed that

$$\lambda_n = \lambda_{n+i} \qquad (i = 1, ..., l-1).$$

Let P_n be a projector onto the subspace of eigenfunctions corresponding to λ_n and R_n be the operator $\displaystyle\sum_{k \neq n} \frac{P_k}{\lambda_k - \lambda_n}$ which is the resolvent of the operator H at the point λ_n. The kernel of R_n will be

$$K(x, \xi) = \sum_{k \neq n} \frac{\varphi_k(x)\, \varphi_k(\xi)}{\lambda_k - \lambda_n}.$$

The formulas for E_n and ψ_n can be rewritten as follows:

$$E_{n+i} = \sum_{k=0}^{\infty} \varepsilon^k \mu_{n+i,\,k}, \quad \psi_{n+i} = \sum_{k=0}^{\infty} \varepsilon^k \varphi_{n+i,\,k} (i = 0, 1, ..., l-1).$$

Here $\mu_{n+i,0} = \lambda_n$ and $\mu_{n+i,1}$ $(i=0,...,\ l-1)$ is equal to the i-th eigenvalue of the operator $P_n V P_n$; $\varphi_{n+i,1}$ is equal to the eigenfunction of the operator $P_n V P_n$ corresponding to $\mu_{n+s,1}$.

The following terms of the series are found from recurrence relations of the form

$$\mu_k = \int_{-\infty}^{\infty} \varphi_{k-1} V \varphi_0 \, dx,$$

$$\varphi_k = R_n \Big\{ \sum_{j=1}^{k} \mu_j \varphi_{j-1} - V \varphi_{k-1} \Big\} =$$

$$= \int_{-\infty}^{\infty} K(x, \xi) \Big\{ \sum_{j=1}^{k} \mu_j \varphi_{j-1} - V \varphi_{k-1} \Big\} \, d\xi.$$

Here for simplicity the index $n+i$ is omitted.

The above recurrence relations are often applied in physics in considerably more general cases than were indicated above.

Thus the perturbation can have the form $\varepsilon V_1(x, \varepsilon)$, where $V_1(x, \varepsilon)$ remains bounded at every point x as $\varepsilon \to 0$ and increases to infinity faster than $\sqrt{V_0(x)}$. In this connection the spectrum of the perturbed equation can be both discrete and continuous. For example, if

$$V_0(x) + \varepsilon V_1(x, a) = x^2 e^{-\varepsilon x^3},$$

then the spectrum is discrete for $\varepsilon = 0$ and continuous for $\varepsilon \neq 0$. Moreover, all the integrals in the formulas for μ_k and φ_k are divergent.

This takes place because of the following reason. In the derivation of the Schrödinger equation, terms are discarded which take into account the interaction of the given system with surrounding systems. In connection with this the potential is assumed to be unbounded. The small parameter, which we disregard when considering the system to be isolated, must be taken into account in the theory of the perturbations by unbounded operators.

For the calculation of μ_k and φ_k the integrals have to be taken with respect to the domain in which the assumption concerning the isolation of a system is valid, namely: the magnitude $\varepsilon V_1(x, \varepsilon)$ in this domain must be less than some constant not dependent on ε.

Thus the values μ_k and φ_k which are obtained are approximate only for those eigenfunctions which are appreciably different from zero in the domain being considered. The accuracy of the approximation cannot be better than the magnitude of the eigenfunction of the non-perturbed equation near the boundary of this domain (for more detail, see [32]).

§ 4. Solution of the Cauchy problem for the Schrödinger equation

1. *General information.* The solution of the equation

$$ih \frac{\partial \psi}{\partial t} = -\frac{h^2}{2\mu} \frac{\partial^2 \psi}{\partial x^2} + V(x)\, \psi = H\psi \;^*)$$

with the initial condition

$$\psi(x, t)\big|_{t=0} = \delta(x - \xi)$$

is called the *fundamental solution* or the *Green's function* and is denoted by $K(x, \xi, t)$.

The solution $\psi(x, t)$ of the equation with the initial condition

$$\psi(x, 0) = f(x),$$

---

*) For the sake of simplicity we shall consider the one-dimensional case. All the results carry over automatically to the multi-dimensional case.

where $f(x)$ is a function with a summable square, can be written in the form

$$\psi(x, t) = \int_{-\infty}^{\infty} K(x, \xi, t) f(\xi) d\xi,$$

where $\|\psi(x, t)\| = \|f(x)\|$. The kernel $K(x, \xi, t)$ satisfies the condition

$$\int_{-\infty}^{\infty} K(x, \xi', t_2) K(\xi', \xi, t_1) d\xi' = K(x, \xi, t_1 + t_2)^*).$$

In the case of a discrete spectrum,

$$K(x, \xi, t) = \sum_n \psi_n(x) \psi_n(\xi) e^{-\frac{E_n t}{h}}.$$

The solution of the equation with right hand side

$$ih \frac{\partial \psi}{\partial t} + \frac{h^2}{2\mu} \frac{\partial^2 \psi}{\partial x^2} - V(x) \psi = F(x, t),$$

with initial condition $\psi(x, t)|_{t=0} = f(x)$ can be represented in the form

$$\psi(x, t) = \int_{-\infty}^{\infty} K(x, \xi, t) f(\xi) d\xi - \frac{i}{h} \int_{-\infty}^{\infty} \int_0^t K(x, \xi, t - \tau) F(\xi, \tau) d\tau d\xi.$$

In this connection

$$\left\| \psi(x, t) - \int_{-\infty}^{\infty} K(x, \xi, t) f(\xi) d\xi \right\| \leqslant \frac{C}{h} \|F(x, t)\|,$$

where C is some constant which is independent of $F(x, t), f(x)$ and h.

2. *Theory of perturbations.* The solution $\psi(x, t)$ of the equation with perturbation $\varepsilon W(x, t)$,

$$ih \frac{\partial \psi}{\partial t} = H\psi + \varepsilon W(x, t) \psi$$

---

*) The last two formulas rise from the fact that the operator $e^{i}{}_h{}^{Ht}$ for which $K(x, \xi, t)$ serves as the kernel is unitary.

can be reduced to the solution of the integral equation

$$\psi(x, t) = \psi^0(x, t) - \frac{\varepsilon i}{h} \int\limits_0^t \int\limits_{-\infty}^\infty K(x, \xi, t - \tau) \, W(\xi, \tau) \, \psi(\xi, \tau) \, d\xi \, d\tau,$$

where $\psi^0(x, t)$ is a solution of the problem for $\varepsilon = 0$, by means of the formula for the solution of the equation with a right hand side. The method of successive approximations for this equation is called the non-stationary method of perturbation theory.

The *first term given by the theory of perturbations* is equal to

$$\frac{i\varepsilon}{h} \int\limits_0^t \int\limits_{-\infty}^\infty K(x, \xi, t - \tau) \, W(\xi, \tau) \, \psi^0(\xi, \tau) \, d\xi \, d\tau.$$

Example.

Let the spectrum of the operator H be discrete. We shall take as the initial $\psi_k(x)$ the k-th eigenfunction of the operator H (physically this means that the particle at the initial time can be found on the k-th energy level). For such an initial condition, $\psi^0(x, t)$ is equal to

$$\psi^0(x, t) = \int\limits_{-\infty}^\infty K(x, \xi, t) \, \psi_k(\xi) \, d\xi = \psi_k(x) \, e^{-\frac{iE_n t}{h}}.$$

The first term given by the theory of perturbations is equal to

$$\frac{i\varepsilon}{h} \sum_n \psi_n(x) \, e^{-\frac{iE_n t}{h}} \int\limits_0^t e^{\frac{i(E_n - E_k)\tau}{h}} \int\limits_{-\infty}^\infty \psi_n(\xi) \, W(\xi, \tau) \, \psi_k(\xi) \, d\xi \, d\tau.$$

3. *Physical interpretation.* The quantum passage of the system from the state $\varphi_1(x)$ for $t = 0$ to the state $\varphi_2(x, \tau)$ for $t = \tau$ is described by the formula

$$c_{1, 2}(\tau) = \int\limits_{-\infty}^\infty \int\limits_{-\infty}^\infty \varphi_2(x, \tau) \, K(x, \xi, \tau) \, \varphi_1(\xi) \, dx \, d\xi.$$

The probability density of this passage is equal to $|c_{1, 2}(\tau)|^2$.

Therefore in order to obtain the passage of a particle from level k of the operator H to level k' under influence of the perturbation $\varepsilon W(x, t)$ to zero-th and first orders, it is necessary to multiply from the left the last formula of no. 2 by $\psi_{k'}(x)\, e^{\frac{iE_{k'}t}{h}}$ and integrate with respect to x. Then the zero-th term given by the theory of perturbations will equal $\delta_{k'k}$ and the first term given by the theory of perturbations will equal

$$\frac{i\varepsilon}{h}\int_0^t e^{\frac{i(E_{k'}-E_k)\tau}{h}}\int_{-\infty}^{\infty}\psi_{k'}(\xi)\,W(\xi, t)\,\psi_k(\xi)\,d\xi\,dt.$$

The fundamental solution itself can be obtained from the formula for $c_{1,2}(\tau)$ if we set $\varphi_1(x)=\delta(x-x_1)$ and $\varphi_2(x, t)=\delta(x-x_2)$. Hence, the fundamental solution describes the quantum passage of a particle from the point $x=x_1$ in time t to the point $x=x_2$.

4. *Quasi-classical asymptotics of the Green's function.* According to the physical interpretation, the asymptotics of the fundamental solution reduce as $h\to 0$ to the solution of the boundary value problem for the classical Newton equation

$$\mu\frac{d^2X}{d\tau^2}=-\frac{\partial V}{\partial X}$$

under the conditions $X(0)=\xi$, $X(t)=x$ (at the initial moment the particle is at the point ξ and at the last, at the point x).

Let the solution $X(x, \xi, \tau, t)$ of such a problem be unique. The action along the trajectory $X(x, \xi, \tau, t)$ will equal

$$S(x, \xi, t)=\int_0^t\left\{\frac{\mu\dot{X}^2}{2}-V[X(x, \xi, \tau, t)]\right\}d\tau.$$

The asymptotic Green's function has the following form as $h\to 0$:

$$K(x, \xi, t)=\frac{1}{\sqrt{2\pi ih}}\sqrt{\left|\frac{\partial^2 S}{\partial x\,\partial\xi}\right|}\,e^{\frac{i}{h}S(x, \xi, t)}(1+hz(x, \xi, t, h)).$$

Here

$$z(x, \xi, t, 0)=\int_0^t\left|\frac{\partial^2 S(X, \xi, \tau)}{\partial X\,\partial\xi}\right|^{-1/2}\frac{\partial^2}{\partial X^2}\left|\frac{\partial^2 S(X, \xi, \tau)}{\partial X\,\partial\xi}\right|^{1/2}d\tau.$$

The integral is taken along the trajectory $X(x, \xi, \tau, t)$.

In the following examples the first term coincides with the Green's function.

Examples.

1. $V(x) \equiv 0$;

$$X(\tau) = a\tau + c\left(c = x, a = \frac{\xi - x}{t}\right),$$

$$S(x, \xi, t) = \frac{\mu(\xi - x)^2}{2t}, \quad \left|\frac{\partial^2 S}{\partial x \, \partial \xi}\right| = \frac{\mu}{2t},$$

$$K(x, \xi, t) = \left(\frac{\mu}{2\pi i h t}\right)^{1/2} e^{\frac{i\mu}{2ht}(x-\xi)^2}.$$

2. $V(x) = -Fx$;

$$S(x, \xi, t) = -\frac{1}{12}\frac{F^3 t^3}{h\alpha^3} + \frac{Ft}{2}(x + \xi) + \frac{\mu}{2t}(x - \xi)^2,$$

$$K(x, \xi, t) = \left(\frac{\mu}{2\pi i h t}\right)^{1/2} e^{\frac{i}{h}S(x, \xi, t)}.$$

3. $V(x) = \omega^2 x^2$;

$$K(x, \xi, t) = \sqrt{\frac{\omega}{2\pi i h \sin \omega t}} \exp\left\{\frac{i}{h}\frac{1}{\sin \omega t}[(x^2 + \xi^2)\cos \omega t - 2x\xi]\right\}.$$

For non-monotone potentials (excluding x^2), more than one curve passes through the two points, generally speaking, for large t. However, the solution of the problem for potentials which are sufficiently smooth and not increasing faster than x^2 will be unique for t less than some t_1. Hence the asymptotics given above will be valid for $K(x, \xi, t)$ for these t.

For a large (and moreover arbitrary) segment of time these asymptotics are extended by means of the formula for $K(x, \xi, t_1 + t_2)$ from no. 1. The integrals in this connection are calculated by means of the stationary phase (saddle-point) method.

Let the boundary value problem being considered have a finite number of different*) solutions $X_i(x, \xi, \tau, t)$ $(i = 1, 2, ..., k)$ and let $S_i(x_i, \xi, t)$ be

*) i.e., multiple solutions are not allowed, and $\dfrac{\partial^2 \xi}{\partial x \, \partial \xi} \neq \infty$.

the action along the i-th trajectory. Then the asymptotics of the Green's function will have the form

$$K_0(x, \xi, t) = \frac{1}{\sqrt{2\pi i h}} \sum_{j=1}^{k} e^{\frac{i\pi}{2}\sigma_j} \left| \frac{\partial^2 S_j}{\partial x \, \partial \xi} \right|^{1/2} e^{\frac{i}{h} S_j(x, \xi, t)},$$

$$K(x, \xi, t) = K_0(x, \xi, t) + O(h)$$

The quantity σ_j is equal to the number of zeros of the function $\dfrac{\partial^2 S\,(\xi, X, \tau)}{\partial \xi \, \partial X}$ for the passage of the point (τ, X) along the trajectory $X_j(x, \xi, \tau, t)$ from the point $(0, \xi)$ to the point (t, x).

Analogous formulas hold in the multidimensional case [32].

5. *Passage to the limit as $h \to 0$.* The passage to the limit as $h \to 0$ for solutions of the Schrödinger equation $\psi_h(x, t)$ satisfying the initial condition $\psi_h(x, 0) = f(x)$ has two peculiarities.

First, $\lim\limits_{h \to 0} \psi_h(x, t)$ does not exist. However the limit has the expression $\int\limits_{a \leqslant x \leqslant b} |\psi_h(x, t)|^2 \, dx$, i.e., the probability that the particle occurs on the segment $a \leqslant x \leqslant b$.

Second, it turns out that if the initial condition $\psi_h(x, 0)$ does not depend on h, then we will not obtain the entire manifold of classical motions in the limit. For example, it is necessary to give the initial condition in the form

$$\psi_h(x, 0) = \varphi(x) \, e^{\frac{i}{h} f(x)}$$

so that as $h \to 0$ the impulse

$$p\psi_h(x, 0) = f'\psi_h(x, 0) + O(h)$$

will "exist". Then if $X(x_0, t)$ is the solution of the Newton equation

$$\mu \ddot{X} = -\frac{\partial V}{\partial X}, \qquad X(0) = x_0, \qquad \mu \dot{X}(0) = f'(x_0),$$

then

$$\lim_{h \to 0} \int\limits_{a \leqslant x \leqslant b} |\psi_h(x, t)|^2 \, dx = \int\limits_{a \leqslant X(x_0, t) \leqslant b} |\psi_h(x_0, 0)|^2 \, dx_0,$$

$$\lim_{h \to 0} \int\limits_{a \leqslant p \leqslant b} |\tilde{\psi}_h(p, t)|^2 \, dp = \int\limits_{a \leqslant \mu \dot{X}(x_0, t) \leqslant b} |\psi_h(x_0, 0)|^2 \, dx_0,$$

where $\tilde{\psi}_h(p, t)$ is the Fourier transformation of the function $\psi_h(x, t)$. In particular, this means that if $\psi_h(x_0, 0)$ is different from zero only in the neighborhood of some point x^0, then the probability of finding the particle at the moment t on a phase plane in the neighborhood of the point (p, x) will be different from zero as $h \to 0$ only for the condition that $p = \mu \dot{X}(x^0, t)$ and $x = X(x^0, t)$. Hence in the limit as $h \to 0$ the quantum particle moves along a classical trajectory.

Moreover, if $F(x) \geqslant 0$ is continuous and increases not faster than some power of x, then

$$\lim_{h \to 0} \int_{-\infty}^{\infty} F(x) |\psi_h(x, t)|^2 \, dx = \int_{-\infty}^{\infty} F[X(x_0, t)] |\psi_h(x_0, 0)|^2 \, dx_0,$$

$$\lim_{h \to 0} \int_{-\infty}^{\infty} F(p) |\tilde{\psi}_h(p, t)|^2 \, dp = \int_{-\infty}^{\infty} F[\mu \dot{X}(x_0, t)] |\psi_h(x_0, 0)|^2 \, dx_0.$$

This means that the quantum mean values pass to classical values as $h \to 0$.

The initial condition for $\psi_h(x, t)$ can be given also in the form

$$\psi_h(x, 0) = \frac{1}{\sqrt{2\pi i h}} \int_{-\infty}^{\infty} e^{\frac{i}{h} px} \varphi(p) \, e^{\frac{i}{h} f(p)} dp.$$

Then if $X(p_0, t)$ is the solution of the Newton equation, satisfying the conditions

$$X|_{t=0} = f'(p_0), \qquad \mu \dot{X}|_{t=0} = p_0,$$

then

$$\lim_{h \to 0} \int_{a \leqslant x \leqslant b} |\psi_h(x, t)|^2 \, dx = \int_{a \leqslant X(p_0, t) \leqslant b} |\varphi(p_0)|^2 \, dp_0,$$

$$\lim_{h \to 0} \int_{a \leqslant p \leqslant b} |\tilde{\psi}_h(p, t)|^2 \, dp = \int_{a \leqslant \mu \dot{X}(p_0, t) \leqslant b} |\varphi(p_0)|^2 \, dp_0.$$

All these relations are valid at an arbitrary point $t > 0$ if the potential and initial data are holomorphic functions which are bounded on the real axis. For the general case, see [32].

6. *Quasi-classical asymptotics of a solution of the Dirac equation.* Let

$\alpha = (\alpha_1, \alpha_2, \alpha_3)$, β and $\sigma = (\sigma_1, \sigma_2, \sigma_3)$ be Dirac matrices of 4th order:

$$\alpha_k = \begin{pmatrix} 0 & \sigma_k^0 \\ \sigma_k^0 & 0 \end{pmatrix}, \qquad \beta = \begin{pmatrix} I & 0 \\ 0 & -I \end{pmatrix}, \qquad \sigma_k = \begin{pmatrix} \sigma_k^0 & 0 \\ 0 & \sigma_k^0 \end{pmatrix},$$

where

$$\sigma_1^0 = \begin{pmatrix} 0 & 1 \\ 1 & 0 \end{pmatrix}, \qquad \sigma_2^0 = \begin{pmatrix} 0 & -i \\ i & 0 \end{pmatrix}, \qquad \sigma_3^0 = \begin{pmatrix} 1 & 0 \\ 0 & -1 \end{pmatrix}, \qquad I = \begin{pmatrix} 1 & 0 \\ 0 & 1 \end{pmatrix}.$$

The Dirac equation has the following form:

$$ih\frac{\partial \psi}{\partial t} + e\Phi(x)\,\psi - \sum_{k=1}^{3} \alpha_k \left[ich\frac{\partial}{\partial x_k} - eA_k(x) \right]\psi + \beta mc^2 \psi = 0,$$

where $x = (x_1, x_2, x_3)$, $\psi = (\psi_1, \psi_2, \psi_3, \psi_4)$, the constants c and e are equal to the speed of light and the charge of an electron: the given functions $A(x) = (A_1, (x), A_2(x), A_3(x))$ and $\Phi(x)$ have the meaning of vector and scalar potentials in an electromagnetic field. The vectors of the electromagnetic field $E(x)$ and $H(x)$ are expressed by $A(x)$ and $\Phi(x)$ (see [28]). The solution of the equation, satisfying the initial condition

$$\psi|_{t=0} = \varphi(x) \exp\left\{ \frac{i}{h} S_0(x) \right\},$$

describes the behavior of an electron and positron.

The equation of classical relativistic mechanics has the form

$$\dot{x}_i^{\pm} = \frac{\partial H_{\pm}(p^{\pm}, x^{\pm})}{\partial p_i}, \quad \dot{p}_i^{\pm} = -\frac{\partial H_{\pm}(p^{\pm}, x^{\pm})}{\partial x_i} \quad (i = 1, 2, 3),$$

$$x^{\pm}(0) = x_0, \quad p^{\pm}(0) = \mathrm{grad}\, S_0(x_0),$$

$$\dot{x}^{\pm} = \left\{ \frac{dx_1^{\pm}(x_0, t)}{dt}, \quad \frac{dx_2^{\pm}(x_0, t)}{dt}, \quad \frac{dx_3^{\pm}(x_0, t)}{dt} \right\},$$

$$\frac{dS^{\pm}}{dt} = \sum_{k=1}^{3} p_k^{\pm} \frac{\partial H^{\pm}}{\partial p_k} - H_{\pm}; \quad S(0) = S_0(x_0),$$

where

$$H_{\pm}(p, x) = -e\Phi \pm \sqrt{(cp - eA)^2 + m^2 c^4}.$$

Let $r_{1,2}^{\pm}$ be unit vectors which the matrices

$$B_{\pm} = \pm \sqrt{(c \operatorname{grad} S_0 - cA)^2 + m^2 c^4} - \sum_{k=1}^{3} \alpha_k \left(c \frac{\partial S_0}{\partial x_k} - eA_k \right) + \beta mc^2$$

map into zero (rank of the matrices B_+ and B_- is equal to two).

The vectors $r_{1,2}^{\pm}$ form a basis. Therefore, without loss of generality, we can assume

$$\psi_{1,2}^{\pm}\big|_{t=0} = r_{1,2}^{\pm} \varphi(x) e^{\frac{i}{h} S_0(x)}$$

instead of the usual initial condition.

For simplicity, only the case $\psi_s^+(x, t)$ $(s = 1, 2)$ is considered below. In this connection we use only the solutions $x^+(x_0, t)$, $p^+(x_0, t)$ and $S^+(x_0, t)$. Therefore in the sequel the index "$+$" is omitted.

We introduce the notations:

1) $\hat{R}(x_0, t) = (\sigma, H[x(x_0, t)]) + i(\alpha, E[x(x_0, t)]) \times$

$$\times \frac{e}{2mc} \sqrt{1 - \frac{\dot{x}^2}{c^2}} (x_0, t);$$

2) $\exp\{i\int_0^t \hat{R}(x_0, t)\, dt\}$ is an operator, transferring the initial condition $f\big|_{t=0} = f_0$ to the solution f of the equation

$$\frac{\partial f}{\partial t} = i\hat{R}(x_0, t)\, f.$$

The point (x, t') is called a *focus* on the trajectory $X(x_0, t)$ if

$$J(x_0, t') = \det \left\| \frac{\partial X_i(x_0, t')}{\partial x_{0j}} \right\| = 0.$$

If (x, t) is not a focus for any trajectory coming to it, then the set of solutions $x = X(x_0, t)$ consists of no more than a finite number of points $x_{0k}(x, t)$ $(k = 1, 2, \ldots, k_0)$.

The number $m(x_0, t)$ of zeros of the Jacobian $J(x_0, \tau)$ for $0 < \tau \leqslant t$ counted according to multiplicity is called the morse index.

Let the coefficients of the Dirac equation and the initial condition be bounded together with all their derivatives. Then if the point (x, t) is not a focus, then the solution of the posed Cauchy problem for the Dirac equation *is representable in the form*

$$\psi_s = \sum_{k=1}^{k_0} |J(x_{0k}, t)|^{-1/2} \sqrt[4]{\frac{c^2 - \dot{x}^2(x_{0k}, t)}{c^2 - \dot{x}^2(x_{0k}, 0)}} \, \varphi(x_{0k}) \times$$

$$\times \exp\left\{\frac{i}{h}\left[S(x_{0k}, t) - \frac{\pi h}{2} m(x_{0k}, t)\right]\right\} \times$$

$$\times \exp\left\{i \int_0^t \hat{R}(x_{0k}, t)\, dt\right\} r_s + O(h);$$

$$x_{0k} = x_{0k}(x, t).$$

Let $S_0(x)$ and $A(x)$ be analytic, and $\varphi(x)$ be a function with a summable square. Then, for an arbitrary $t > 0$ and an arbitrary three dimensional domain D, the relation

$$\lim_{h\to 0} \iiint_{x \in D} \sum_{v=1}^{4} |\psi_{sv}|^2 \, dx = \iiint_{x(x_0, t) \in D} \varphi^2(x_0)\, dx_0, \qquad s = 1, 2.,$$

is satisfied by the solution ψ_s of the Dirac equation.

This equality, analogous to the equalities of the preceding section, means that the quantum particle in the limit as $h \to 0$ is set in motion along a classical trajectory. But the value of every component is not retained along the trajectory. The particle in the classical limit is characterized by a unit vector which varies along the trajectory[*]) according to the rule

$$\sqrt[4]{\frac{c^2 - \dot{x}^2(x_0, t)}{c^2 - \dot{x}^2(x_0, 0)}} \exp\left\{i \int_0^t \hat{R}(x_0, t)\, dt\right\} r_{1,2}.$$

For asymptotic behavior of solutions near a focus, see [32].

§ 5. Continuous spectrum of the energy operator and the problem of scattering

The continuous spectrum of the energy operator plays a basic role in the consideration of problems in collision theory. The simplest is the

[*]) That is, spinor polarization has a classical limit as $h \to 0$.

problem concerning the scattering of a particle not having an internal structure from a fixed force center. The model problem plays an important role in the consideration of more complicated concrete problems of the theory of scattering.

1. *Formulation of the problem.* The energy operator of the system being investigated has the form

$$H = H_0 + V, \qquad H_0\psi(x) = -\Delta^2\psi(x), \qquad V\psi(x) = V(x)\,\psi(x).$$

in a coordinate representation. The operator H_0 is the energy operator of a freely moving particle, and the operator V describes its interaction with a scattering center. This interaction must decrease with distance in order that we may speak of the free motion of a particle at a distance from the center. The precise mathematical condition will be formulated below.

The experiment with respect to scattering consists of two parts: first, a bundle of free particles which are emitted by a source is investigated, and, second, the scattered particles which freely move from the action of the target that they irradiate are analyzed. The vector of state, describing this free motion, satisfies the Schrödinger equation with the operator H_0 as the energy operator:

$$i\,\frac{\partial\psi(t)}{\partial t} = H_0\psi(t).$$

This equation is called a *free equation.* A solution of it has the form

$$\psi(t) = e^{-iH_0 t}\psi,$$

where ψ is an arbitrary constant vector of state.

Vectors which describe the free motion of particles in a bundle to a target and after scattering on the target, will of course be different. We denote them by ψ_- and ψ_+ respectively. The problem consists in determining ψ_+ in terms of the given ψ_-. This problem is solved by means of the Schrödinger equation:

$$i\,\frac{\partial\psi(t)}{\partial t} = H\psi(t),$$

which is satisfied by an arbitrary vector of state of the system being described. The vector ψ defines an initial condition for this equation in the following manner:

$$\lim_{t \to -\infty} \| \psi(t) - e^{-iH_0 t} \psi_- \| = 0.$$

The solution $\psi(t)$ as $t \to \infty$ must again behave like a solution of the free equation in the sense:

$$\lim_{t \to +\infty} \| \psi(t) - e^{-iH_0 t} \psi_+ \| = 0.$$

The vector ψ_+ is the desired vector describing the free motion of the scattered particles. The dependence of this vector on the initial condition ψ_- must be linear:

$$\psi_+ = S\psi_- .$$

The operator S is called a *scattering operator* or an *S-matrix*.

2. *Basis for the formulation of the problem and its solution.* The basis of the formulation of the problem of scattering, described in the preceding subsection, is implied by the following statement:
if $V(x)$ satisfies the conditions

$$\int |V(x)| \, dx < \infty, \qquad \int |V(x)|^2 \, dx < \infty,$$

then the strong limits

$$\lim_{t \to \pm\infty} e^{iHt} e^{-iH_0 t} = U^{(\pm)},$$

exist and are isometric operators. The operator

$$S = U^{(+)*}(U^{(-)})$$

is unitary and commutes with the operator H_0.

By means of this result we obtain a basis for the above formulation of the problem of scattering. In fact, for an arbitrary initial vector of state ψ-the solution of the problem is given by the formula

$$\psi(t) = e^{-iHt} U^{(-)} \psi_- ,$$

where the vector ψ_+ giving the asymptotics of the solution as $t \to +\infty$ is obtained from ψ_- according to the formula $\psi_+ = S\psi_-$.

The formulas acquire a more explicit form if the Fourier method is used. The general solution of the free equation has the form

$$\psi_-(x, t) = \int c(k) e^{-ik^2 t} e^{i(k, x)} dk \qquad (k^2 = (k, k)),$$

where $c(k)$ is an arbitrary square summable function. For the condition

$$|V(x)| \leqslant \frac{C}{(1 + |x|)^{2+\varepsilon}} \qquad (\varepsilon > 0)$$

the solution, defined by the initial condition $\psi_-(x)$, exists and has the form

$$\psi(x, t) = \int c(k) e^{-ik^2 t} u(x, k) \, dk .$$

Here $u(x, k)$ are eigenfunctions of the continuous spectrum of the operator H satisfying the *radiation condition*, that is, solutions of the equation

$$[-\Delta^2 + V(x)] u(x, k) = k^2 u(x, k),$$

defined by means of the conditions

$$u(x, k) = e^{i(k, x)} + w(x, k),$$

$$w(x, k) = O\left(\frac{1}{|x|}\right), \quad \lim_{|x| \to \infty} |x| \left(ik - \frac{\partial}{\partial |x|}\right) w(x, k) = 0 .$$

If $\int |V(x)| dx < \infty$ then $w(x, k)$ for large $|x|$ has an asymptote

$$w(x, k) = \frac{e^{i|k| \, |x|}}{|x|} f(|k|; n, \alpha) \left(n = \frac{x}{|x|}; \alpha = \frac{k}{|k|}\right).$$

In this case the operator S is an integral operator formed by means of the kernel $f(|k|; n, \alpha)$ which is called the *amplitude of scattering*. In the representation, connected with the coordinate representation in which all our formulas are given by means of a Fourier transformation:

$$\tilde{\psi}(k) = \left(\frac{1}{2\pi}\right)^{3/2} \int \psi(x) e^{i(k, x)} dx,$$

the operator S appears in the following form:

$$S\tilde{\psi}(k) = \tilde{\psi}(k) + \frac{1}{\pi i} \int f\left(|k|; \frac{k}{|k|}, \frac{k'}{|k'|}\right) \delta(k^2 - k'^2) \psi(k') \, dk' .$$

The condition of S of being unitary has the form

$$f(|k|; \alpha, \beta) - \overline{f(|k|; \beta, \alpha)} = \frac{i|k|}{2\pi} \int f(|k|; \alpha, n) \overline{f(|k|; \beta, n)} \, dn.$$

Moreover, the condition of symmetry

$$f(|k|; \alpha, \beta) = f(|k|, -\beta, -\alpha)$$

is satisfied, being a corollary of the realness of the energy operator in the coordinate representation.

3. *Amplitude of scattering and its equation.* The basic problem of the theory of scattering consists of the determination of $f(|k|; x, \beta)$. The integral equation for the function $u(x, k)$:

$$u(x, k) = e^{i(k, x)} - \frac{1}{4\pi} \int \frac{e^{ik|x-y|}}{|x-y|} V(y) u(y, k) \, dy.$$

is the basis for one of the different approaches to this problem. This equation bears the designation of the *integral equation of the theory of scattering*. It is not necessary to find the asymptote of the solution $u(x, k)$ in order to obtain $f(|k|; \alpha, \beta)$ since $f(|k|; \alpha, \beta)$ is determined more simply by means of the formula

$$f(|k|; \alpha, \beta) = -\frac{1}{4\pi} \int e^{-i(k, x)} V(x) u(x, k') \, dx,$$

$$|k| = |k'|, \qquad \alpha = \frac{k}{|k|}, \qquad \beta = \frac{k'}{|k'|}.$$

We can write out an integral equation which is equivalent to the preceding equation for $u(x, k)$ from which $f(|k|; \alpha, \beta)$ is defined more directly. For this it is necessary to consider the kernel

$$t(k, l) = -\frac{1}{4\pi} \int e^{-i(k, x)} V(x) u(x, l) \, dx.$$

The function $f(|k|; \alpha, \beta)$ is obtained from $t(k, l)$ for $|k| = |l|$. The equation has the form

$$t(k, l) = \tilde{V}(k - l) - \int \tilde{V}(k - m) \frac{1}{m^2 - l^2 - i0} t(m, l) \, dm,$$

$$\tilde{V}(k) = \left(\frac{1}{2\pi}\right)^{3/2} \int e^{i(k, x)} V(x) \, dx.$$

Here $(m^2 - l^2 - i0)^{-1}$ is a generalized function (see ch. VIII, §. 2). Formally

$$\frac{1}{m^2 - l^2 - i0} = P \frac{1}{m^2 - l^2} + i\pi\delta(m^2 - l^2),$$

where the index P shows that the integral is regarded in the sense of a principal value.

4. *Case of spherical symmetry.* The problem is considerably simplified if $V(x)$ depends only on the radius:

$$V(x) = V(|x|) = V(r).$$

In this case the solution $u(x, k)$ and the amplitude $f(|k|; \alpha, \beta)$ can be expanded in series with respect to the Legendre polynomials:

$$u(x, k) = \sum_{l=0}^{\infty} (2l + 1) \frac{R_l(r, s)}{r} P_l(\cos(k, x)),$$

$$f(|k|; \alpha, \beta) = \sum_{l=0}^{\infty} (2l + 1) f_l(s) P_l(\cos(\alpha, \beta)),$$

$$r = |x|, \quad s = |k|.$$

Here the functions $R_l(r, s)$ are eigenfunctions of the continuous spectrum of the radial Schrödinger operators

$$H_l R_l \equiv \left[-\frac{d^2}{dr^2} + \frac{l(l+1)}{r^2} + V(r) \right] R_l(r, s) = s^2 R_l(r, s),$$

the functions $f_l(s)$ are connected with the asymptotes of these solutions for large r:

$$R_l(r, s) = \frac{1}{2is} \left[e^{isr} - (-1)^l e^{-isr} \right] + f_l(s) e^{isr} + o(1).$$

The condition of being unitary (see no. 2) appears in the following form in terms of the coefficients f_l:

$$f_l(s) - \overline{f_l(s)} = 2is f_l(s) \overline{f_l(s)} = 2is |f_l(s)|^2.$$

Consequently $f_l(s)$ can be written in the form

$$f_l(s) = \frac{1}{2is} \left[e^{2i\delta_l(s)} - 1 \right],$$

where the $\delta_l(s)$ are real functions which are called *asymptotic phases*. This terminology is connected with the fact that the asymptote $R_l(r, s)$ can be rewritten in the form

$$R_l(r, s) = C(s) \sin\left(sr - \frac{l\pi}{2} + \delta_l(s) \right) + o(1),$$

$$C(s) = \frac{i^l \exp i\delta_l(s)}{s}.$$

For the solutions $R_l(r, s)$ we can write out the integral equations which are analogous to the equation for $u(x, k)$. However it is more convenient to deal with other solutions. If we introduce to the case under consideration the solution $g_l(r, s)$ of the equation $H_l R_l = s^2 R_l(r, s)$ with the condition

$$g_l(r, s) = e^{isr} + o(1) \quad (r \to \infty),$$

then the integral equation for $g_l(r, s)$ assumes the form

$$g_l(r, s) = i^{l+1} h_l^{(1)}(sr) - \int_r^\infty J^{(l)}(s; r, t) \, V(t) \, g_l(t, s) \, dt,$$

$$J^{(l)}(s; r, t) = \frac{(-1)^{l+1}}{is} [j_l(sr) h_l^{(1)}(-st) - j_l(st) h_l^{(1)}(-sr)].$$

Here $j_l(t)$ and $h_l^{(1)}(t)$ are Bessel and Hankel functions. In the case $l = 0$,

$$J^{(0)}(s; r, t) = \frac{\sin s(r - t)}{s}.$$

The phases $\delta_l(s)$ are defined by means of the functions

$$M_l(s) = 1 + \int_0^\infty j_l(sr) \, V(r) \, g_l(r, s) \, dr$$

by the formula

$$S_l(s) = \exp\{ -2i\delta_l(s) \} = \frac{M_l(s)}{M_l(-s)}.$$

The basic convenience of the choice of the solutions $g_l(r, s)$ consists in the Volterra-ness of the corresponding integral equation as a result of which the solvability of the problem can be investigated very easily. Another method, also connected with an equation of Volterra type, consists in finding a solution $\varphi_l(r, s)$ which satisfies the condition

$$\lim_{r \to 0} \frac{(2l+1)!!}{r^{l+1}} \varphi_l(r, s) = 1 \,.$$

The corresponding integral equation has the form

$$\varphi_l(r, s) = \frac{j_l(sr)}{s^{l+1}} + \int_0^r J^{(l)}(s; r, t) \, V(t) \, \varphi_l(t, s) \, dt$$

The representation

$$M_l(s) = 1 + \int_0^\infty h_l^{(1)}(s, r) \, V(r) \, \varphi_l(r, s) \, dr$$

is valid for the functions $M_l(s)$. The Volterra-ness of the equations for g_l and φ_l guarantees the convergence of the method of successive approximations for them if $V(r)$ satisfies the single condition

$$\int_0^\infty r \, |V(r)| \, dr < \infty \,.$$

5. *General case.* In the general case of a non-symmetric potential, a simplification of the problem, connected with the Volterra-ness of the integral equations, apparently does not exist. The integral equation of the theory of scattering (see no. 3) itself constitutes a basic method for finding $f(|k|; \alpha, \beta)$. The convergence of the successive approximations for this equation for all k can be indicated only for the condition that the potential $V(x)$ satisfies some condition of smallness, for example:

$$\max_x \int |V(y)| \frac{1}{|x - y|} \, dy < 4\pi \,.$$

Under this condition the operator H does not have a completely discrete spectrum. If a discrete spectrum is present, then the successive approximations do not automatically converge for all k. At the same time, for $|k|$ sufficiently large, the successive approximations always converge independently of the magnitude of the potential. This fact is a justification of the so-called Born formula for the amplitude of scattering which is obtained if in the expression for $f(|k|; \alpha, \beta)$ (see no. 3) we substitute in place of $u(x, k)$ its zero approximation, the corresponding plane wave

$$f_B(|k|; \alpha, \beta) = -\frac{1}{4\pi} \int e^{-i|k|(x,\alpha)} V(x) e^{i|k|(x,\beta)} dx,$$

i.e. a Born amplitude is simply the Fourier transformation of the potential. A precise statement is that for large $|k|$

$$f(|k|; \alpha, \beta) - f_B(|k|; \alpha, \beta) = o(1)$$

uniformly with respect to α and β where if $V(x)$ is a differentiable function and $|\text{grad } V(x)| = O\left(\dfrac{1}{|x|^{2+\varepsilon}}\right)$ then instead of $o(1)$ we can substitute $O\left(\dfrac{1}{|k|}\right)$ here. Sometimes $f_B(|k|; \alpha, \beta)$ is called the *first Born approxima-tion* and the sequence which is obtained if we substitute in the expression for $f(|k|; \alpha, \beta)$ the sequence of successive approximations for the solution $u(k, k)$ a *Born sequence*. The successive approximations of the integral equation for $t(k, l)$ (see no. 3) give the expressions for higher Born approximations over the first, i.e. over the Fourier transformation of the potential $\tilde{V}(k)$.

The integral equation of the theory of scattering allows us to apply also other approximate methods for the determination of the amplitude $f(|k|; \alpha, \beta)$. For example, the scheme of the Galerkin method for this equation is strongly connected with the so-called variational method of Schwinger for the problem of scattering.

6. *Inverse problem of the theory of scattering.* The above described problem of finding the operator of scattering with respect to a potential can be called the direct problem of the theory of scattering. We can pose the inverse problem, i.e., the problem of the reconstruction of the potential with respect to an S-matrix. At the present time this problem is solved only for the case of a spherical symmetric potential $V(x) = V(r)$ which satisfies the condition $\int_0^\infty r|V(r)|dr < \infty$. The potential is determined by prescribing one of the phases $\delta_1(s)$ for all s. The set of phases corresponding to the potential with this condition is characterized completely. If the corresponding radial Schrödinger operator H_l has a discrete spectrum, then the potential is defined with respect to the phase $\delta_l(s)$ ambiguously.

The operator H_l does not have a discret spectrum for sufficiently large l; thus the potential is defined uniquely with respect to the phase $\delta_l(s)$ for l sufficiently large.

Here we mention the scheme of one of the methods for the solution of the inverse problem for the case $l=0$. The function $S_0(s)=\exp\{2i\delta_0(s)\}$ has the following properties:

1) $|S_0(s)| = S_0(\infty) = S_0(0) = 1$;

2) $S_0(-s) = \overline{S_0(s)} = S_0^{-1}(s)$;

3) $S_0(s) = 1 + \displaystyle\int_{-\infty}^{\infty} F(t)\,e^{-ist}\,dt$,

where $\displaystyle\int_{-\infty}^{\infty} |F(t)|\,dt < \infty,\qquad \int_0^{\infty} t\,|F'(t)|\,dt < \infty$;

4) $\arg S_0(s)\,|_{-\infty}^{\infty} = -4\pi m$, where $m \geq 0$ is the number of discrete eigenvalues of the corresponding operator H_0 whose potential satisfies the condition indicated above. Now let the function $S_0(s)$ having the listed properties be given. We form the function

$$F_1(t) = F(t) + \sum_{n=1}^{m} b_n e^{-\kappa_n t},$$

where b_n and κ_n are arbitrary positive numbers, where among the κ_n there are none which coincide. The integral equation

$$A(x, y) = F_1(x + y) + \int_x^{\infty} A(x, t)\,F_1(t + y)\,dt = 0 \quad (x < y)$$

is solvable for all $x \geq 0$.

The function

$$V(r) = -2\,\frac{d}{dr}\,A(r, r)$$

satisfies the condition $\int_0^{\infty} r|V(r)|dr < \infty$ and $S_0(s)$ is an S-function for the operator H_0 with $V(r)$ as a potential. The solution $g_0(r, s)$ is given by the formula

$$g_0(r, s) = e^{irs} + \int_r^\infty A(r, t) e^{its} \, dt.$$

For the case $l > 0$ an analogue of this scheme has been worked out. It is interesting that the necessary and sufficient conditions 1)–4), which the phase $\delta_0(s)$ must satisfy, remain valid for the case $l > 0$.

The integral equation for $A(x, y)$ allows an explicit solution in the case when $S_0(s)$ is a rational function. The solutions and potential are obtained in this case in the form of rational functions of trigonmetric and hyperbolic functions. The following is the simplest example:

$$S_0(s) = \frac{s + i\alpha}{s + i\beta} \frac{s - i\beta}{s - i\alpha}.$$

The corresponding potential has the form

$$V(r) = 2 \frac{\beta^2 (\beta^2 - \alpha^2)}{(\beta \operatorname{ch} \beta^2 + \alpha \operatorname{sh} \beta^2)^2}.$$

The inverse problem has also been studied for systems of equations with radial operators and for the one-dimensional Schrödinger equation.

CHAPTER VIII

GENERALIZED FUNCTIONS

§ 1. Generalized functions and operations on them

1. *Introductory remarks.* Some physical quantities (for example, the density of a concentrated load) cannot be expressed by means of ordinary functions. Therefore, in physics and technology, *generalized* or *singular* functions*), expressing such magnitudes, have long been employed. An example of a singular function is the so-called *δ-function*, defined by the following property:

For an arbitrary continuous function $\varphi(x)$ *the equality*

$$\int_{-\infty}^{\infty} \delta(x)\, \varphi(x)\, dx = \varphi(0)$$

is satisfied. It is evident that the δ-function is not an ordinary function. In fact it follows from the definition that $\delta(x)=0$ for $x\neq0$ and $\int_{-\infty}^{\infty} \delta(x)\, dx=1$. No classical function has such properties. We can only construct a sequence of ordinary functions $f_n(x)$ such that we have

$$\lim_{n\to\infty} \int_{-\infty}^{\infty} f_n(x)\, \varphi(x)\, dx = \varphi(0)$$

for an arbitrary continuous function $\varphi(x)$. For example, we can set

$$f_n(x) = \begin{cases} \dfrac{n}{2} & \text{for} \quad |x| \leqslant \dfrac{1}{n}, \\[2mm] 0 & \text{for} \quad |x| > \dfrac{1}{n} \end{cases}$$

---

*) The term *distribution* is also used [Editor].

or

$$f_n(x) = \frac{n}{\sqrt{2\pi}} e^{-\frac{n^2 x^2}{2}}.$$

Such sequences of functions are called *δ-shaped sequences*. In many cases in which the δ-function was spoken of, (for example, in questions connected with point sources and sinks, with Green's function, and so on), instead of the δ-function, δ-shaped sequences were used and a passage to the limit was made. This complicated mathematical physics to the extent that mathematical analysis would be complicated by the systematic replacement of all derivatives by limits of difference quotients and of all integrals by limits of approximating sums.

The elimination of these difficulties was made possible only after the construction of a rigorous theory of generalized functions, the establishment of rules of operations on them, and the creation of a sufficiently developed algorithmic apparatus. Such a construction was carried out on the basis of the investigation of continuous linear functionals in certain linear topological spaces.

2. *Notation.* Since in the sequel we consider functions both of one and of several variables, for brevity in writing we shall adopt the following notation:

1) $x = (x_1, \ldots, x_n)$;
2) $|x|^2 = x_1^2 + \cdots + x_n^2$;
3) $q = (q_1, \ldots, q_n)$, $q_k \geq 0$, q_k integers;
4) $|q| = q_1 + \cdots + q_n$, $q_k \geq 0$;
5) $q! = q_1! \ldots q_n!$;
6) $x^q = x_1^{q_1} \ldots x_n^{q_n}$;
7) $\varphi(x) = \varphi(x_1, \ldots, x_n)$;
8) $\varphi^{(q)}(x) = \dfrac{d^q \varphi}{dx^q} = \dfrac{\partial^{q_1 + \cdots + q_n} \varphi(x_1, \ldots, x_n)}{\partial x_1^{q_1} \ldots \partial x_n^{q_n}}$.

With this notation, for example, the Taylor series for functions of several variables is written the same as for functions of one variable:

$$\varphi(x + h) = \sum_{|q|=0}^{\infty} \frac{\varphi^{(q)}(x) h^q}{q!}.$$

Moreover, for integration over the space R_n, notation for the domain of integration is omitted:

$$\int \varphi(x)\,dx = \int_{R_n} \varphi(x_1, \ldots, x_n)\,dx_1 \ldots dx_n.$$

3. *Generalized functions.* A continuous linear functional (f, φ) on the space K is called a *generalized function.* K consists of the infinitely differentiable functions which have compact support and assume complex values. A function $\varphi(x)$ is said to have *compact support* if an a can be found such that $\varphi(x)=0$ for $|x|\geq a$. The topology on the space K is defined as follows: a sequence of functions $\{\varphi_n(x)\}$ of the space K is called *convergent to zero* if:

a) all the functions $\varphi_n(x)$ become zero outside some fixed ball $|x|\leq a$;

b) for arbitrary q, the equality

$$\lim_{n\to\infty}\sup_x |\varphi_n^{(q)}(x)| = 0$$

holds.

Thus a generalized function f is considered to be defined if a number (f, φ) is associated to each function $\varphi(x)$ of the space K where the following conditions are satisfied:

a) $(f, \varphi_1+\varphi_2)=(f, \varphi_1)+(f, \varphi_2)$;

b) $(f, \alpha\,\varphi)=\alpha\,(f, \varphi)$ for an arbitrary complex number α;

c) $\lim_{n\to\infty}(f, \varphi_n)=0$, if the functions $\varphi_n(x)$ are convergent to zero in the topology of the space K.

A generalized function f is called *real* if for all real functions $\varphi(x)$ of the space K the numbers (f, φ) are real.

To every continuous function $f(x)$ there corresponds a generalized function (f, φ) given by the equality

$$(f, \varphi) = \int f(x)\,\varphi(x)\,dx.$$

In fact, the integral on the right defines a continuous linear functional on the space K. In an analogous manner we can define the generalized function (f, φ) corresponding to an arbitrary locally summable function $f(x)$ (a function $f(x)$ is called *locally summable* if it is summable on every ball $|x| \leqslant a$). If $f(x)$ is a locally summable function and $\varphi(x)$ is an infinitely differentiable function with compact support, then the integral written above is convergent and defines a continuous linear functional on K.

Thus a generalized function (f, φ) corresponds to every locally summable function $f(x)$. This defines an imbedding of the space of locally summable functions in the space K' of all generalized functions. In this connection *different generalized functions correspond to different locally summable functions*: if $f_1(x)$ and $f_2(x)$ are locally summable functions and the equality $(f_1, \varphi) = (f_2, \varphi)$ is satisfied for all functions $\varphi(x)$ of the space K, then the equality $f_1(x) = f_2(x)$ holds for almost all values of x.

The generalized functions (f, φ) which correspond to locally summable functions $f(x)$ are called *regular generalized functions*. The *jump function* (θ, φ), given by the formula

$$(\theta, \varphi) = \int_0^\infty \varphi(x)\, dx$$

can serve as an example of a regular generalized function. It corresponds to the function $\theta(x)$ equal to zero for $x < 0$ and equal to unity for $x > 0$.

Continuous linear functionals on the space K which are not representable in integral form with a locally summable function $f(x)$ are called *singular* generalized functions. The δ-function, mentioned above, is an example of a singular generalized function. The corresponding functional is given by the equality

$$(\delta, \varphi) = \varphi(0).$$

A wide class of singular generalized functions is given by formulas of the form

$$(\mu, \varphi) = \int \varphi(x)\, d\mu(x),$$

where μ is a measure on the space R_n having finite variation in every ball $|x| \leqslant a$, and the integral is understood in the sense of a Stieltjes integral (in particular, μ can be an arbitrary positive measure such that the μ-measure of an arbitrary ball $|x| \leqslant a$ is finite).

The δ-function, corresponding to unit measure concentrated at the point $x = 0$, belongs to the class indicated.

Singular generalized functions are often denoted by the same symbol $f(x)$ as used for ordinary functions and are written

$$(f, \varphi) = \int f(x) \varphi(x) \, dx.$$

In this connection, one should keep in mind that generalized functions, generally speaking, do not have values at individual points.

4. *Operations on generalized functions.* The sum of generalized functions is defined by the equality

$$(f_1 + f_2, \varphi) = (f_1, \varphi) + (f_2, \varphi),$$

and the product of a generalized function with a complex number α is defined by the equality

$$(\alpha f, \varphi) = \alpha(f, \varphi).$$

If the generalized functions are regular, then these definitions coincide with the usual definitions of the sum of functions and the product of a function by a number.

Generally, in the definition of operations on generalized functions, we require that these definitions coincide with the usual definitions for regular generalized functions. For example, from the identity

$$\int (f(x) \alpha(x)) \varphi(x) \, dx = \int f(x) (\alpha(x) \varphi(x)) \, dx$$

it follows that the product of a generalized function $f(x)$ and an infinitely differentiable function $\alpha(x)$ is given by the formula

$$(\alpha f, \varphi) = (f, \alpha\varphi).$$

The product of two generalized functions, generally speaking, is not defined so that, for example, the generalized function $\delta^2(x)$ does not have meaning.

In some cases we can carry out a substitution for the variable in a generalized function. We call the generalized function given by the formula

$$(f(x - h), \varphi) = (f(x), \varphi(x + h))$$

the *translation of the generalized function $f(x)$ by the vector h.* For example,

$$(\delta(x - h), \varphi) = (\delta(x), \varphi(x + h)) = \varphi(h).$$

If U is a linear transformation in n-dimensional space, then we set

$$(f(Ux), \varphi(x)) = |U|(f, \varphi(U^{-1}x)),$$

where $|U|$ is the determinant of the transformation. Thus a *similarity transformation* for $\alpha > 0$ is defined by the formula

$$(f(\alpha x), \varphi) = \alpha^n \left(f, \varphi\left(\frac{x}{\alpha}\right) \right).$$

If $\alpha(x)$ is an infinitely differentiable function, all zeros of which are simple, then we set

$$\delta[\alpha(x)] = \sum_n \frac{\delta(x - x_n)}{|\alpha'(x_n)|},$$

where the summation is extended over all the zeros of the function $\alpha(x)$.
For example,

$$\delta(x^2 - 1) = \frac{\delta(x-1)}{2} + \frac{\delta(x+1)}{2},$$

$$\delta(\sin x) = \sum_{n=-\infty}^{\infty} \delta(x - \pi n).$$

5. *Differentiation and integration of generalized functions.* Corresponding to the equality

$$\int f^{(q)}(x)\,\varphi(x)\,dx = (-1)^q \int f(x)\,\varphi^{(q)}(x)\,dx$$

we define the *q-th derivative of a generalized function f* of one variable by the formula

$$(f^{(q)}, \varphi) = (-1)^q (f, \varphi^{(q)}).$$

For functions of several variables, the analogous formula holds:

$$(f^{(q)}, \varphi) = (-1)^{|q|} (f, \varphi^{(q)}).$$

For example,

$$(\delta^{(q)}, \varphi) = (-1)^{|q|} (\delta, \varphi^{(q)}) = (-1)^{|q|} \varphi^{(q)}(0).$$

All generalized functions are infinitely differentiable since the functions $\varphi(x)$ of the space K are infinitely differentiable. In particular, an arbitrary locally summable function is infinitely differentiable in the generalized sense. However, we must keep in mind that if the function $f(x)$ has almost everywhere an ordinary derivative, then the functional defined by this derivative might not coincide with the derivative of $f(x)$ as a generalized function.

It is important that derivatives of higher order of generalized functions do not depend on the order of differentiation.

EXAMPLE. The generalized function (θ', φ) is given by the formula

$$(\theta', \varphi) = -(\theta, \varphi') = -\int_0^\infty \varphi'(x)\, dx = \varphi(0) = (\delta, \varphi).$$

Therefore

$$\theta'(x) = \delta(x).$$

Using this formula we can differentiate in the generalized sense an arbitrary function $f(x)$ having discontinuities of the first kind and a locally summable derivative at points of continuity. Namely, if the discontinuities of the function $f(x)$ are located at the points $x_1, \ldots, x_n$ and the jumps at these points are equal to h_k, then

$$(f', \varphi) = \sum_{k=1}^n h_k \varphi(x_k) + \int_{-\infty}^\infty f'(x)\, \varphi(x)\, dx.$$

If $\alpha(x)$ is an infinitely differentiable function having simple zeros, then

$$\delta^{(q)}[\alpha(x)] = \sum_n \frac{1}{|\alpha'(x_n)|} \left(\frac{1}{\alpha'(x)}\frac{d}{dx}\right)^q \delta(x - x_n),$$

where the sum is extended over all the zeros of the function $\alpha(x)$.

For every generalized function f of one variable, there exists an anti-derivative generalized function f_1 defined to within a constant term, i.e. a function such that $f_1' = f$. It is defined by the equality

$$(f_1, \varphi') = -(f, \varphi)$$

on all functions which are derivatives of functions of K. These functions form a subspace which differs from K by one dimension. Therefore we can set

$$(f_1, \psi_0) = C$$

for a fixed function $\psi_0(x)$ of K such that $\int\limits_{-\infty}^{\infty} \psi_0(x)\, dx \neq 0$, after which the generalized function f_1 will be uniquely defined.

6. *Limit of a sequence of generalized functions.* A sequence $\{f_k\}$ of generalized functions is said to *converge to the generalized function f* if the equality

$$\lim_{k \to \infty} (f_k, \varphi) = (f, \varphi)$$

holds for an arbitrary function $\varphi(x)$ of the space K.

For every generalized function f we can construct a sequence of functions $\{\psi_k(x)\}$ of the space K which converges to it, i.e. a sequence such that

$$\lim_{k \to \infty} \int \psi_k(x)\, \varphi(x)\, dx = (f, \varphi)$$

for all the functions $\varphi(x)$ in K.

If a sequence $\{f_k(x)\}$ of locally summable functions is such that

$$\lim_{k \to \infty} \int |f(x) - f_k(x)|\, dx = 0,$$

then the generalized functions (f_k, φ) converge to the generalized function (f, φ). However, if the equality $\lim\limits_{k \to \infty} f_k(x) = f(x)$ is satisfied at an arbitrary point x, it does not follow that $\lim\limits_{k \to \infty} (f_k, \varphi) = (f, \varphi)$.

For example, for all values of x

$$\lim_{k \to \infty} \frac{2k^3 x^2}{\pi(1 + k^2 x^2)^2} = 0.$$

However, for an arbitrary function $\varphi(x)$ of the space K we have

$$\lim_{k \to \infty} \frac{2k^3}{\pi} \int\limits_{-\infty}^{\infty} \frac{x^2 \varphi(x)\, dx}{(1 + k^2 x^2)^2} = \varphi(0),$$

and therefore in the sense of generalized functions

$$\lim_{k \to \infty} \frac{2k^3 x^2}{(1 + k^2 x^2)^2} = \delta(x).$$

A sequence of regular generalized functions which converges to the

δ-function is called a δ-*shaped sequence*. The following are examples of δ-shaped sequences for functions of one variable:

a) $f_m(x) = \dfrac{m}{\pi(1 + m^2 x^2)}$,

b) $f_m(x) = \dfrac{1}{\pi} \dfrac{\sin mx}{x}$,

c) $f_m(x) = \dfrac{m}{\pi} \dfrac{\sin^2 mx}{x^2}$.

Two other examples of δ-shaped sequences are indicated in no. 1.

A series $\sum\limits_{k=1}^{\infty} f_k$ of generalized functions is said to *converge to the generalized function f* if

$$\lim_{j \to \infty} \sum_{k=1}^{j} f_k = f.$$

For example, the series

$$1 + \sum_{k=1}^{\infty} \left[\cos kx - \cos(k-1)x \right]$$

converges in the generalized sense to zero since

$$\lim_{j \to \infty} \left(1 + \sum_{k=1}^{j} \left[\cos kx - \cos(k-1)x \right], \varphi\right) =$$

$$= \lim_{j \to \infty} (\cos jx, \varphi) = \lim_{j \to \infty} \int_{-\infty}^{\infty} \varphi(x) \cos jx \, dx = 0$$

for an arbitrary function $\varphi(x)$ of the space K.

A convergent series of generalized functions can be termwise differentiated. In other words, if $\sum\limits_{k=1}^{\infty} f_k = f$, then we have

$$\sum_{k=1}^{\infty} f_k^{(q)} = f^{(q)}$$

for arbitrary q.

EXAMPLE. The series $\sum\limits_{k=-\infty}^{\infty} e^{ikx}$ is convergent in the generalized sense to the generalized function $2\pi \sum\limits_{k=-\infty}^{\infty} \delta(x - 2\pi k)$. If we apply this equality to a function $\varphi(x)$ of the space K, then we obtain the *Poisson formula*:

$$\sum_{k=-\infty}^{\infty} \psi(k) = 2\pi \sum_{k=-\infty}^{\infty} \varphi(2\pi k),$$

where

$$\psi(\lambda) = \int_{-\infty}^{\infty} \varphi(x) \, e^{i\lambda x} \, dx$$

is the Fourier transformation of the function $\varphi(x)$.

Moreover, it follows from the equality

$$\sum_{k=-\infty}^{\infty} e^{ikx} = 2\pi \sum_{k=-\infty}^{\infty} \delta(x - 2\pi k)$$

that

$$\tfrac{1}{2} + \sum_{k=1}^{\infty} \cos kx = \pi \sum_{k=-\infty}^{\infty} \delta(x - 2\pi k).$$

Differentiating this equation gives

$$\sum_{k=1}^{\infty} k^q \cos\left(kx + \frac{q\pi}{2}\right) = \pi \sum_{k=-\infty}^{\infty} \delta^{(q)}(x - 2\pi k).$$

In exactly the same way, from the equation

$$\sum_{k=1}^{\infty} \frac{\cos kx}{k} = -\ln\left|2 \sin \frac{x}{2}\right|$$

it follows that

$$\sum_{k=1}^{\infty} k^{q-1} \cos\left(kx + \frac{q\pi}{2}\right) = -\left\{\ln\left|2 \sin \frac{x}{2}\right|\right\}^{(q)},$$

where the derivative on the right hand side of this equation is to be understood in the generalized sense.

7. *Local properties of generalized functions.* We say that the generalized function $f(x)$ is *equal to zero in the domain* Ω if $(f, \varphi) = 0$ for every function $\varphi(x)$ of the space K which is equal to zero outside of a closed set A contained in Ω. For example, the generalized function $\delta(x)$ is equal to zero in the domain Ω obtained from the space R_n by deleting the point $x = 0$.

The generalized function $f(x)$ is said to be *concentrated on the closed set B* if it is equal to zero on the complement of this set. The smallest closed set in which the generalized function $f(x)$ is concentrated is called the *support* of this function. For example, the support of the generalized function $\delta(x)$ and of all of its derivatives is the point $x=0$. The support of a regular function $f(x)$ is the closure of the set of points on which this function is different from zero.

A generalized function is said to have *compact support* if it is concentrated in some ball $|x| \leqslant a$.

The generalized functions $f_1(x)$ and $f_2(x)$ are said to *coincide in the open domain* Ω if $f_1 - f_2 = 0$ in this domain. In particular, the generalized function $f(x)$ is said to be *regular in the open domain* Ω if in this domain it coincides with some ordinary locally summable function. In this case, we can speak of the values of the generalized function $f(x)$ at points of the set Ω. For example, the generalized function $\delta(x)$ is regular in the complement of the point $x=0$ and is equal to zero on the complement.

EXAMPLE. For an arbitrary function $\varphi(x)$ in the space K, let

$$(|x|^{-\frac{3}{2}}, \varphi) = \int_0^\infty \frac{\varphi(x) + \varphi(-x) - 2\varphi(0)}{x^{\frac{3}{2}}}\, dx.$$

This equation defines a continuous linear functional on the space K, that is, a generalized function. Outside the point $x=0$, this generalized function coincides with the regular generalized function described by the function $|x|^{\frac{3}{2}}$. In other words, if the function $\varphi(x)$ of the space K vanishes in some neighborhood of the point $x=0$, then we have the equality

$$(|x|^{-\frac{3}{2}}, \varphi) = \int_{-\infty}^\infty |x|^{-\frac{3}{2}} \varphi(x)\, dx.$$

8. *Direct product of generalized functions.* Let $f(x)$ be a generalized function on the space K_x of functions $\varphi(x)$ of m variables, and let $g(y)$ be a generalized function on the space K_y of functions $\psi(y)$ of n variables. By $f(x) \times g(y)$ we denote the generalized function on the space $K_{x,y}$ of functions $\chi(x, y)$ of $m+n$ variables defined by the formula

$$(f \times g, \chi) = (f, (g, \chi(x, y))).$$

This generalized function is called the *direct product* of the generalized functions $f(x)$ and $g(y)$. If the function $\chi(x, y)$ of the space $K_{x, y}$ has the form $\chi(x, y) = \varphi(x)\psi(y)$ where $\varphi(x) \in K_x$, $\psi(y) \in K_y$, then

$$(f \times g, \chi) = (f, \varphi)(g, \psi).$$

The following formulas for the direct product of generalized functions hold:

$$f(x) \times g(y) = g(y) \times f(x),$$

$$f(x) \times \{g(y) \times h(z)\} = \{f(x) \times g(y)\} \times h(z).$$

If a generalized function $f(x, y)$ is invariant with respect to translations in the variable x (i.e. if $(f, \varphi(x+h, y)) = (f, \varphi(x, y))$ for arbitrary h), then it has the form

$$f(x, y) = 1_x \times g(y),$$

where $g(y)$ is a generalized function in the space K_y and 1_x is the generalized function in the space K_x defined by the formula

$$\left(1_x, \varphi(x)\right) = \int \varphi(x)\, dx.$$

9. *Convolution of generalized functions.* Let $f(x)$ and $g(x)$ be generalized functions of one variable where either one of the following conditions is satisfied:

a) one of the functions $f(x)$, $g(x)$ has bounded support;

b) the supports of the generalized functions $f(x)$ and $g(x)$ are bounded on the same side (for example, $f(x) = 0$ for $x < a$, $g(x) = 0$ for $x < b$).

Then the expression

$$(f(x) \times g(y), \varphi(x + y)),$$

which is denoted by $(f*g, \varphi)$, is defined for an arbitrary function $\varphi(x)$ of the space K. We call the generalized function $f*g$ the *convolution* of the generalized functions $f(x)$ and $g(x)$.

If the generalized functions $f(x)$ and $g(x)$ are regular and satisfy one of the conditions a), b), then the generalized function $f*g$ is also regular and is defined by the formula

$$f*g(x) = \int\limits_{-\infty}^{\infty} f(x - y)\, g(y)\, dy.$$

EXAMPLES. 1. If $f(x)$ is an arbitrary generalized function, then $\delta * f(x) = f(x)$. Thus the δ-function assumes the role of the identity element with respect to the operation of convolution. In particular, $\delta * \delta(x) = \delta(x)$.

2. The convolution of the generalized function $f(x)$ with $\delta(x-h)$ is equivalent to the translation of $f(x)$ by h:

$$\delta(x - h) * f(x) = f(x - h).$$

The equations

$$f * g(x) = g * f(x)$$

and

$$(f * g) * h(x) = f * (g * h)(x),$$

which express the commutativity and associativity of the convolution of generalized functions, are valid.

The formula for differentiation of the convolution has the form

$$\frac{d}{dx}(f * g) = \frac{df}{dx} * g = f * \frac{dg}{dx}.$$

If $\lim_{v \to \infty} f_v = f$, then $\lim_{v \to \infty} f_v * g = f * g$ under each of the following assumptions:

a) all the generalized functions $f_v(x)$ are concentrated on a single bounded set;

b) the generalized function g is concentrated on a bounded set;

c) the supports of the generalized functions $f_v(x)$ and $g(x)$ are bounded on the same side by a constant, not depending on v.

Hence it follows that if the generalized function $f_t(x)$ depends on the parameter t and is differentiable with respect to this parameter, then the formula

$$\frac{\partial}{\partial t}(f_t * g(x)) = \frac{\partial f_t}{\partial t} * g(x)$$

is valid if $f_t(x)$ and $g(x)$ satisfy any one of the assumptions a)–c).

The convolution of generalized functions $f(x)$ and $g(x)$ of several variables is defined exactly as for functions of one variable. In this connection, it is required that one of the factors, for example $f(x)$, have the property that for an arbitrary function $\varphi(x)$ of the space K, the function

$$\psi(y) = (f, \varphi(x + y))$$

belongs to this same space.

The *convolution of $f(x)$ with the function* $\varphi(x)$ of the space K is defined by the formula

$$(f * g, \varphi) = (g, f * \varphi).$$

If $f(x)$ is a regular function, then this formula assumes the following form:

$$f * \varphi(x) = \int f(y - x)\,\varphi(y)\,dy.$$

10. *General form of generalized functions.*   Let the generalized function $f(x)$ have compact support. Then a parallelepiped

$$a_j \leqslant x_j \leqslant b_j,$$

$$1 \leqslant j \leqslant n,$$

can be found on which this generalized function is concentrated. It is possible to show that for an arbitrary $\varepsilon > 0$ an integer $p > 0$ and continuous functions $f_{q\varepsilon}(x)$, $0 \leqslant |q| \leqslant p$ becoming zero for $a_j - \varepsilon \leqslant x_j \leqslant b_j + \varepsilon$ can be found which satisfy the relation

$$f(x) = \sum_{|q|=0}^{p} f_{q\varepsilon}^{(q)}(x).$$

Thus *every generalized function with compact support is a linear combination of derivatives of continuous functions with compact support* (where it is understood that the derivatives are regarded in a generalized sense).

Analogously, every linear functional on the space $K(a)$ of infinitely differentiable functions becoming zero for $|x| \leqslant a$ has the form

$$(f, \varphi) = (F^{(q)}, \varphi) \equiv (-1)^{|q|} \int F(x)\,\varphi^{(q)}(x)\,dx,$$

where $F(x)$ is a continuous function on the ball $|x| \leqslant a$.

If $f(x)$ is an arbitrary generalized function, then we can construct a sequence of generalized functions with compact support $f_1(x), \ldots, f_n(x), \ldots$ such that

1) $\lim\limits_{n \to \infty} f_n(x) = f(x)$;

2) for every $a > 0$, an N can be found such that we have $f_n(x) = f_m(x)$ in the ball $|x| \leqslant a$ for $n \geqslant N$, $m \geqslant N$.

The generalized functions concentrated at one point have a particularly simple structure. For example, all generalized functions concentrated at the point $x = 0$ are finite linear combinations of the δ-function and its derivatives, i.e. have the form

$$f(x) = \sum_{|q|=0}^{n} c_q \delta^{(q)}(x).$$

11. *Kernel Theorem.* In many applications of generalized functions the following theorem turns out to be useful:

KERNEL THEOREM. *Let $B(\varphi, \psi)$ be a bilinear functional, where $\varphi(x)$ runs through the space K_x of infinitely differentiable functions of m variables with compact support and $\psi(y)$ runs through the space K_y of infinitely differentiable functions of n variables with compact support. If the functional $B(\varphi, \psi)$ is continuous with respect to each of the variables $\varphi(x)$ and $\psi(y)$, then a generalized function $f(x, y)$ exists on the space $K_{x, y}$ of infinitely differentiable functions of $m+n$ variables with compact support such that*

$$B(\varphi, \psi) = (f, \varphi(x)\psi(y)).$$

§ 2. Generalized functions and divergent integrals

1. *Regularization of divergent integrals.* In several problems of mathematical physics, divergent integrals occur. By means of the apparatus of generalized functions we can obtain an algorithm which allows the assigning of a numerical value to some divergent integrals, and, using this value, we obtain solutions to these problems. This algorithm is called the *regularization of a divergent integral.*

Let $f(x)$ be some function. We call the point x_0 a *point of local summability for the function* $f(x)$ if a neighborhood $v(x_0)$ of this point exists in which the function $f(x)$ is summable. Points which are not points of local summability are called *singular points.* Here we consider functions which have only a finite set of singular points on every interval.

Let K_f be the subspace of K consisting of functions $\varphi(x) \in K$ vanishing in some neighborhood of each singular point of the function $f(x)$. A sequence of functions $\{\varphi_m(x)\}$ of the space K_f *converges to zero* if all the functions $\varphi_m(x)$ are concentrated on a single compact set which does not contain singular points of the function $f(x)$, and if the equality

$$\lim_{m \to \infty} \sup_x |\varphi_m^{(q)}(x)| = 0$$

is satisfied for arbitrary q.

The integral $\int f(x)\,\varphi(x)\,dx$ is convergent for an arbitrary function $\varphi(x)$ of the space K_f, and the equality

$$(f, \varphi) = \int f(x)\,\varphi(x)\,dx$$

defines a linear functional on the space K_f. This functional can be extended to the entire space K.*) The value (f, φ) of this functional at a function $\varphi(x)$ of the space K is called a *regularized value of the integral* $\int f(x)\,\varphi(x)\,dx$ (if the function $\varphi(x)$ does not belong to the subspace K_f, then this integral can, generally speaking, be divergent). We call the generalized function (f, φ), obtained from the extension, a *regularization* of the function $f(x)$. This regularization of the function $f(x)$ coincides with $f(x)$ on the set of points complementary to the set of singular points.

EXAMPLE. The equality

$$(|x|^{-\frac{3}{2}}, \varphi) = \int\limits_{0}^{\infty} \frac{\varphi(x) + \varphi(-x) - 2\varphi(0)}{|x|^{\frac{3}{2}}}\,dx$$

gives a regularization of the generalized function $|x|^{-\frac{3}{2}}$.

Generally speaking, a function can have different regularizations. In this connection, regularizations of different functions might not be in agreement with one another, so that, for example, the equality $(f_1 + f_2, \varphi) = (f_1, \varphi) + (f_2, \varphi)$ can be violated. We introduce the concept of *canonical regularization*. Let L be a linear space consisting of functions $f(x)$ (generally speaking, not locally summable), each of which has a discrete set of singular points and is infinitely differentiable on the complement of this set. We assume that the space L contains, along with the functions $f(x)$, all of their derivatives (on the complements of the sets of singular points) and all the functions $\alpha(x)f(x)$ where the $\alpha(x)$ are infinitely differentiable functions.

Let a linear functional (f, φ), a regularization of the function, be associated with every function $f(x)$ of the space L. The regularization is called *canonical*, and the functional is denoted by c.r. $f(x)$, if the following

*) The extension of a continuous linear functional from a subspace to the whole space is possible in a locally convex linear topological space (see ch. 1, § 4, no. 2).

conditions are satisfied:

1) c.r. $[\lambda_1 f_1(x) + \lambda_2 f_2(x)] = \lambda_1$ c.r. $f_1(x) + \lambda_2$ c.r. $f_2(x)$;

2) c.r. $\left(\dfrac{df}{dx}\right) = \dfrac{d}{dx}(\text{c.r.}\, f(x))$;

here, on the left $\dfrac{d}{dx}$ is the derivative of the function in the usual sense, and on the right it is the derivative of the generalized function;

3) $\qquad\qquad$ c.r. $(\alpha(x) f(x)) = \alpha(x)$ c.r. $f(x)$

for an arbitrary infinitely differentiable function $\alpha(x)$.

The *set of functions with algebraic singularities* serves as an example of a space of functions for which a canonical regularization exists. The point x_0 is called an algebraic singular point of the function $f(x)$ if in a neighborhood of this point the function $f(x)$ is representable in the form

$$f(x) = \sum_{j=1}^{m} \alpha_j(x) h_j(x),$$

where the $\alpha_j(x)$ are infinitely differentiable functions and the $h_j(x)$ one of the following functions:

$$(x - x_0)^{\lambda}_{+}, \quad (x - x_0)^{\lambda}_{-}, \quad (x - x_0)^{-n}, \quad \lambda \neq -1, -2, \ldots$$

The function x^{λ}_{+} is defined by the equality

$$x^{\lambda}_{+} = \begin{cases} x^{\lambda} & \text{if } x > 0, \\ 0 & \text{if } x < 0, \end{cases}$$

and the function x^{λ}_{-} by the equality

$$x^{\lambda}_{-} = \begin{cases} 0 & \text{if } x > 0, \\ |x|^{\lambda} & \text{if } x < 0. \end{cases}$$

The function $f(x)$ is called a *function with algebraic singularities* if it has a finite set of algebraic singular points on each interval.

In order for a canonical regularization to be defined for the space of functions with algebraic singularities, we first solve the problem concerning the regularization of the functions $x^{\lambda}_{+}, x^{\lambda}_{-}, x^{-n}$.

2. *Regularization of the functions* $x_+^\lambda, x_-^\lambda, x^{-n}$ *and their linear combinations.*

1) If $\operatorname{Re}\lambda > -1$, then the integral

$$(x_+^\lambda, \varphi) = \int_0^\infty x^\lambda \varphi(x)\, dx$$

is convergent for an arbitrary function $\varphi(x)$ of the space K and gives a linear functional on this space. In order to define the functional (x_+^λ, φ) for $\operatorname{Re}\lambda < -1$ we use condition 2) of a canonical regularization. Let $\operatorname{Re}\lambda > -n-1$, $\lambda \neq -1, -2, \ldots, -n, \ldots$; then $\operatorname{Re}(\lambda+n) > -1$ and therefore the functional $(x_+^{\lambda+n}, \varphi)$ is defined by the equality

$$(x_+^{\lambda+n}, \varphi) = \int_0^\infty x^{\lambda+n} \varphi(x)\, dx.$$

But $x_+^\lambda = \dfrac{\Gamma(\lambda+1)}{\Gamma(\lambda+n+1)}(x_+^{\lambda+n})^{(n)}$. Therefore by virtue of condition 2) the equality

$$(x_+^\lambda, \varphi) = \frac{\Gamma(\lambda+1)}{\Gamma(\lambda+n+1)}[(x_+^{\lambda+n})^{(n)}, \varphi] =$$

$$= (-1)^n \frac{\Gamma(\lambda+1)}{\Gamma(\lambda+n+1)}[x_+^{\lambda+n}, \varphi^{(n)}] =$$

$$= (-1)^n \frac{\Gamma(\lambda+1)}{\Gamma(\lambda+n+1)} \int_0^\infty x^{\lambda+n} \varphi^{(n)}(x)\, dx$$

must hold. Thus the functional (x_+^λ, φ) is given by the formula

$$(x_+^\lambda, \varphi) = (-1)^n \frac{\Gamma(\lambda+1)}{\Gamma(\lambda+n+1)} \int_0^\infty x^{\lambda+n} \varphi^{(n)}(x)\, dx$$

for $\operatorname{Re}\lambda > -n-1$.

We can verify, integrating by parts, that this formula is equivalent to the following:

$$(x_+^\lambda, \varphi) = \int_0^1 x^\lambda \left[\varphi(x) - \varphi(0) - x\varphi'(0) - \cdots - \right.$$

$$\left. - \frac{x^{n-1}}{(n-1)!} \varphi^{(n-1)}(0) \right] dx + \int_1^\infty x^\lambda \varphi(x) \, dx +$$

$$+ \sum_{k=1}^n \frac{\varphi^{(k-1)}(0)}{(k-1)! \, (\lambda + k)}.$$

For $-n-1 < \mathrm{Re}\,\lambda < -n$, we can use a simpler formula:

$$(x_+^\lambda, \varphi) = \int_0^\infty x^\lambda \left[\varphi(x) - \varphi(0) - x\varphi'(0) - \cdots - \frac{x^{n-1}}{(n-1)!} \varphi^{(n-1)}(0) \right] dx.$$

The formulas mentioned give the regularization of the functions x^λ for $\mathrm{Re}\,\lambda < -1$. It remains to remark that this generalized function is not defined for $\lambda = -1, -2, \ldots, -n, \ldots$.

2) The generalized function (x_-^λ, φ) is given for $\mathrm{Re}\,\lambda > -n-1$, $\lambda \neq -1, -2, \ldots, -n, \ldots$ by the formula

$$(x_-^\lambda, \varphi) = \int_0^1 x^\lambda \left[\varphi(-x) - \varphi(0) + x\varphi'(0) - \cdots - \right.$$

$$\left. - (-1)^{n-1} \frac{x^{n-1}}{(n-1)!} \varphi^{(n-1)}(0) \right] dx + \int_1^\infty x^\lambda \varphi(-x) \, dx +$$

$$+ \sum_{k=1}^n \frac{(-1)^{k-1} \varphi^{(k-1)}(0)}{(k-1)! \, (\lambda + k)},$$

and for $-n-1 < \mathrm{Re}\,\lambda < -n$ by the simpler formula:

$$(x_-^\lambda, \varphi) = \int_0^\infty x^\lambda \left[\varphi(-x) - \varphi(0) + x\varphi'(0) - \cdots - \right.$$

$$\left. - (-1)^{n-1} \frac{x^{n-1}}{(n-1)!} \varphi^{(n-1)}(0) \right] dx.$$

3) Along with the generalized functions x_+^λ and x_-^λ it is sometimes useful to consider their linear combinations (for $\lambda \neq -1, -2, \ldots$):

$$|x|^\lambda = x_+^\lambda + x_-^\lambda, \quad |x|^\lambda \operatorname{sign} x = x_+^\lambda - x_-^\lambda,$$

$$(x + i0)^\lambda = \lim_{\varepsilon \to +0} (x + i\varepsilon)^\lambda = x_+^\lambda + e^{i\lambda\pi} x_-^\lambda,$$

$$(x - i0)^\lambda = \lim_{\varepsilon \to +0} (x - i\varepsilon)^\lambda = x_+^\lambda + e^{-i\lambda\pi} x_-^\lambda.$$

We can give more convenient formulas for these linear combinations. Thus, for $-2m-1 < \operatorname{Re}\lambda < -2m+1$,

$$(|x|^\lambda, \varphi) = \int_0^\infty x^\lambda \left\{ \varphi(x) + \varphi(-x) - \right.$$

$$\left. - 2\left[\varphi(0) + \frac{x^2}{2!} \varphi''(0) + \cdots + \frac{x^{2m-2}}{(2m-2)!} \varphi^{(2m-2)}(0) \right] \right\} dx.$$

For $\lambda = -2m$, we write $|x|^{-2m} = x^{-2m}$ so that

$$(x^{-2m}, \varphi) = \int_0^\infty x^{-2m} \left\{ \varphi(x) + \varphi(-x) - \right.$$

$$\left. - 2\left[\varphi(0) + \frac{x^2}{2!} \varphi''(0) + \cdots + \frac{x^{2m-2}}{(2m-2)!} \varphi^{(2m-2)}(0) \right] \right\} dx.$$

In particular,

$$(x^{-2}, \varphi) = \int_0^\infty \frac{\varphi(x) + \varphi(-x) - 2\varphi(0)}{x^2} dx.$$

For $-2m-2 < \operatorname{Re}\lambda < -2m$, the formula

$$(|x|^\lambda \operatorname{sign} x, \varphi) = \int_0^\infty x^\lambda \left\{ \varphi(x) - \varphi(-x) - \right.$$

$$\left. - 2\left[x\varphi'(0) + \frac{x^3}{3!} \varphi^{(3)}(0) + \cdots + \frac{x^{2m-1}}{(2m-1)!} \varphi^{(2m-1)}(0) \right] \right\} dx$$

holds. We write $|x|^{-2m-1} \operatorname{sign} x = x^{-2m-1}$ for $\lambda = -2m-1$, so that

$$(x^{-2m-1}, \varphi) = \int_0^\infty x^{-2m-1} \left\{ \varphi(x) - \varphi(-x) - \right.$$

$$\left. - 2\left[x\varphi'(0) + \frac{x^3}{3!} \varphi^{(3)}(0) + \cdots + \frac{x^{2m-1}}{(2m-1)!} \varphi^{(2m-1)}(0) \right] \right\} dx.$$

In particular,

$$(x^{-1}, \varphi) = \int_0^\infty \frac{\varphi(x) - \varphi(-x)}{x} \, dx,$$

$$(x^{-3}, \varphi) = \int_0^\infty \frac{\varphi(x) - \varphi(-x) - 2x\varphi'(0)}{x^3} \, dx.$$

The expression (x^{-1}, φ) is called the *principal value of the integral*

$$\int_{-\infty}^\infty \frac{\varphi(x)}{x} \, dx \text{ in the Cauchy sense.}$$

For $\lambda = -n$ the generalized function $(x+i0)^\lambda$ is defined by the equality

$$(x + i0)^{-n} = x^{-n} - \frac{i\pi(-1)^{n-1}}{(n-1)!} \delta^{(n-1)}(x),$$

and the generalized function $(x-i0)^\lambda$ by the equality

$$(x - i0)^{-n} = x^{-n} + \frac{i\pi(-1)^{n-1}}{(n-1)!} \delta^{(n-1)}(x),$$

where (x^{-n}, φ) is defined above for even and odd n.

3. *Regularization of functions with algebraic singularities.* Suppose the function $f(x)$ has the single singular point $x=0$ and can be written in the form

$$f(x) = \sum_{j=1}^n \alpha_j(x) h_j(x),$$

where the $\alpha_j(x)$ are infinitely differentiable functions, and the $h_j(x)$ are taken from the functions $x_+^\lambda, x_-^\lambda, x^{-n}, \lambda \neq -1, -2, \ldots$. Then

$$(f, \varphi) = \sum_{j=1}^n (h_j, \alpha_j\varphi).$$

Since the $\alpha_j(x) \varphi(x)$ are functions of the space K, then all the terms on the right hand side make sense. This defines the regularization of the function $f(x)$.

Analogously, we define the regularization of a function $f(x)$ having an algebraic singular point $x=x_0$. For an arbitrary function $f(x)$ with algebraic singularities, we can construct functions $f_k(x)$, $1 \leqslant k < \infty$, such that $f = \sum f_k$, each of the functions $f_k(x)$ has one singular point, and only a finite number of the functions are different from zero on every segment $|x| \leqslant a$. Then we set

$$(f, \varphi) = \sum_{k=1}^{\infty} (f_k, \varphi).$$

It is possible to show that the value (f, φ) does not depend on the method of partitioning and that the formula mentioned defines a canonical regularization in the space of functions with algebraic singularities.

Thus an algorithm is given for the regularization of integrals of the form $\int f(x) \varphi(x) \, dx$, where $f(x)$ is an arbitrary function with algebraic singularities and $\varphi(x)$ is a function of the space K. The formulas obtained are applicable to a wider class of functions. For example, if the function $f(x)$ has *exponential growth* (i.e. if a, p and C exist such that $|f(x)| \leqslant C(1+|x|^2)^p$ for $|x| \geqslant a$), then the formulas for regularization are applicable to the infinitely differentiable functions which are decreasing, together with all derivatives, as $|x| \to \infty$ faster than an arbitrary power of $|x|$.

EXAMPLE 1. Let the infinitely differentiable function $\varphi(x)$ vanish outside the segment $\left[-\dfrac{\pi}{2}, \dfrac{\pi}{2} \right]$. It follows from the formulas of regularization that

$$\left(\cot \frac{x}{2}, \varphi(x) \right) = \int_0^{\infty} [\varphi(x) - \varphi(-x)] \cot \frac{x}{2} \, dx.$$

Analogously,

$$\left(\frac{1}{\sin^2 \dfrac{x}{2}}, \varphi \right) = \int_0^{\infty} \left[\frac{\varphi(x) + \varphi(-x)}{\sin^2 \dfrac{x}{2}} - \frac{8\varphi(0)}{x^2} \right] dx.$$

EXAMPLE 2. The well-known integral expression

$$\Gamma(\lambda) = \int_0^{\infty} x^{\lambda-1} e^{-x} \, dx$$

for the Γ-function is convergent only for $\mathrm{Re}\,\lambda > 0$.

However, it remains valid for arbitrary values of λ, $\lambda \neq -1, -2,\ldots$ if we understand the integral in the generalized sense. In this connection, if $-n > \operatorname{Re}\lambda > -n-1$, then the formula in expanded form is written as follows:

$$\Gamma(\lambda) = \int_0^\infty x^{\lambda-1}\left[e^{-x} - \sum_{k=0}^{n-1} \frac{(-1)^k x^k}{k!}\right]dx.$$

4. *Regularization on a finite segment.* Let

$$x_{[0,b]}^\lambda = \begin{cases} x^\lambda & \text{if } x \in [0, b], \\ 0 & \text{if } x \notin [0, b]. \end{cases}$$

Since the function $\varphi(x)$ might not be equal to zero at the point $x=b$, the formulas in no. 2 are not applicable in this case. The formula in the form

$$(x_{[0,b]}^\lambda, \varphi) = \int_0^b x^\lambda\left[\varphi(x) - \varphi(0) - \cdots - \frac{x^{n-1}}{(n-1)!}\varphi^{(n-1)}(0)\right]dx +$$
$$+ \varphi(0)\frac{b^{\lambda+1}}{\lambda+1} + \cdots + \varphi^{(n-1)}(0)\frac{b^{\lambda+n}}{n!(\lambda+n)}$$

holds for $\operatorname{Re}\lambda > -n-1$, $\lambda \neq -1, -2,\ldots$. We can consider this formula as a regularization of the integral $\int_0^b x^\lambda \varphi(x)\, dx$. If $-n-1 < \operatorname{Re}\lambda < -n$, then as $b \to \infty$, the formula passes to one of the formulas in no. 2.

EXAMPLE. For arbitrary value of λ different from $-1, -2,\ldots, -n, \ldots$,

$$\int_0^b x^\lambda\, dx = \frac{b^{\lambda+1}}{\lambda+1}$$

(this integral is divergent for $\operatorname{Re}\lambda < -1$, and the expression on the right hand side gives a regularized value of the integral).

It is useful to remark that, for $0 < c < b$, the equality

$$\int_0^b x^\lambda\varphi(x)\, dx = \int_0^c x^\lambda\varphi(x)\, dx + \int_c^b x^\lambda\varphi(x)\, dx$$

is satisfied for arbitrary functions $\varphi(x)$ of the space K if we understand by the first two integrals the regularized value indicated above.

As with the preceding, we can regularize integrals of the form

$$\int_a^b (x - a)^\lambda \, \alpha(x) \, \varphi(x) \, dx$$

and

$$\int_a^b (b - x)^\lambda \, \alpha(x) \, \varphi(x) \, dx,$$

where $\alpha(x)$ is an infinitely differentiable function and $\varphi(x)$ is a function coinciding with a function of the space K on the segment $[a, b]$.

If the points a and b are singular points for the function $f(x)$, then we set

$$\int_a^b f(x) \, \varphi(x) \, dx = \int_a^c f(x) \, \varphi(x) \, dx + \int_c^b f(x) \, \varphi(x) \, dx,$$

regularizing the terms on the right by the above indicated method. The result obtained does not depend on the choice of the point c.

Examples

1. The equality

$$B(\lambda, \mu) = \int_0^1 x^{\lambda - 1} (1 - x)^{\mu - 1} \, dx$$

is valid in the classical sense for $\operatorname{Re} \lambda > 0$, $\operatorname{Re} \mu > 0$; it remains valid for all λ and μ except the values $-1, -2, \ldots$ if we understand the integral in the sense of a regularized value. However, the formula in expanded form is cumbersome for $\operatorname{Re} \lambda > -k$, $\operatorname{Re} \mu > -s$:

$$B(\lambda, \mu) = \int_0^{\frac{1}{2}} x^{\lambda - 1} \left[(1 - x)^{\mu - 1} - \sum_{r=0}^{k-1} (-1)^r \frac{\Gamma(\mu) \, x^r}{r! \, \Gamma(\mu - r)} \right] dx +$$

$$+ \int_{\frac{1}{2}}^1 (1 - x)^{\mu - 1} \left[x^{\lambda - 1} - \sum_{r=0}^{s-1} (-1)^r \frac{\Gamma(\lambda) \, (1 - x)^r}{r! \, \Gamma(\lambda - r)} \right] dx +$$

$$+ \sum_{r=0}^{k-1} \frac{(-1)^r \, \Gamma(\mu)}{2^{r+\lambda} \, r! \, \Gamma(\mu - r) \, (r + \lambda)} + \sum_{r=0}^{s-1} \frac{(-1)^r \, \Gamma(\lambda)}{2^{r+\mu} \, r! \, \Gamma(\lambda - r) \, (r + \mu)}.$$

2. The integral representation of the spherical function

$$\mathfrak{P}_p^{-q}(x) = \frac{(x^2 - 1)^{q/2}}{2^q \sqrt{\pi}\, \Gamma(q + \tfrac{1}{2})} \int_{-1}^{1} \frac{(1 - t^2)^{q - \frac{1}{2}}}{(x + t\sqrt{x^2 - 1})^{q - p}}\, dt$$

$$|x| > 1,$$

is valid in the classical sense for $\operatorname{Re} q > -\tfrac{1}{2}$; it remains valid for all q, $q \neq -\tfrac{1}{2}$, $-\tfrac{3}{2}, \ldots$ if we consider regularized values of the integrals.

5. *Regularization at infinity.* Let $b > 0$ and $K(b, \infty)$ be the class of all functions $\varphi(x)$ which are defined and infinitely differentiable for all $x > b$ and such that the inversion transformation $\varphi(x) \to \varphi\left(\frac{1}{x}\right)$ takes them onto functions $\psi(x)$ coinciding on the interval $(0, \frac{1}{b})$ with functions of the space K. For $\lambda \neq -1, 0, 1, \ldots$, according to the definition,

$$\int_b^\infty x^\lambda \varphi(x)\, dx = \int_0^{1/b} y^{-\lambda - 2}\, \varphi\left(\frac{1}{y}\right) dy,$$

where the integral in the right side is understood in the sense indicated in no. 4.

If the function $f(x)$ has the form $x^\lambda g(x)$, where $g(x)$ is a function of the class $K(b, \infty)$, then we set

$$\int_b^\infty f(x)\, \varphi(x)\, dx = \int_b^\infty x^\lambda g(x)\, \varphi(x)\, dx.$$

In an analogous manner, functions on the interval $(-\infty, -b)$ are regularized. For the regularization of a function on the entire axis, we set

$$\int_{-\infty}^\infty f(x)\, \varphi(x)\, dx = \int_{-\infty}^{-b} f(x)\, \varphi(x)\, dx +$$

$$+ \int_{-b}^b f(x)\, \varphi(x)\, dx + \int_b^\infty f(x)\, \varphi(x)\, dx,$$

applying the above indicated formulas to separate terms.

EXAMPLES

1. The substitution $y = \dfrac{1}{x}$ shows that

$$\int\limits_{b}^{\infty} x^{\lambda}\, dx = \int\limits_{0}^{1/b} y^{-\lambda-2}\, dy = -\frac{b^{\lambda+1}}{\lambda+1}$$

for $\lambda \neq -1$ (compare with the example in no. 4). Therefore

$$\int\limits_{0}^{\infty} x^{\lambda}\, dx = \int\limits_{0}^{b} x^{\lambda}\, dx + \int\limits_{b}^{\infty} x^{\lambda}\, dx = 0$$

for $\lambda \neq -1$.

2. The equality

$$\mathrm{B}(\lambda, \mu) = \int\limits_{0}^{\infty} x^{\lambda-1} (1+x)^{-\lambda-\mu}\, dx,$$

valid in the classical sense for $\mathrm{Re}\,\lambda > 0$, $\mathrm{Re}\,\mu > 0$, remains valid for all values of λ and μ (except λ, $\mu = -1, -2, \ldots$) if we understand the integral in the sense of a regularized value.

3. The integral representation of the MacDonald function

$$K_p(x) = \frac{\left(\dfrac{x}{2}\right)^p \Gamma\left(\dfrac{1}{2}\right)}{\Gamma\left(p + \dfrac{1}{2}\right)} \int\limits_{1}^{\infty} e^{-xt} (t^2 - 1)^{p-\frac{1}{2}}\, dt,$$

valid in the classical sense only for $\mathrm{Re}\,p > -\frac{1}{2}$, remains valid for all $p\,(p \neq -\frac{1}{2}, -\frac{3}{2}, \ldots)$ if we understand the integral in the sense of a regularized value.

6. *Non-canonical regularizations.* In some cases, non-canonical regularizations of divergent integrals turn out to be useful.

1) Let x_+^{-n} be the function defined by the equalities

$$x_+^{-n} = \begin{cases} x^{-n} & \text{if } x > 0, \\ 0 & \text{if } x < 0. \end{cases}$$

To it corresponds the functional (x_+^{-n}, φ) of the form

$$(x_+^{-n}, \varphi) = \int_0^\infty x^{-n} \left[\varphi(x) - \varphi(0) - x\varphi'(0) - \cdots - \right.$$

$$\left. - \frac{x^{n-2}}{(n-2)!} \varphi^{(n-2)}(0) - \frac{x^{n-1}}{(n-1)!} \varphi^{(n-1)}(0) \theta(1-x) \right] dx,$$

where $\theta(x)=0$ for $x<0$ and $\theta(x)=1$ for $x>0$. This functional is not a value of the functional x_+^λ for $\lambda=-n$.

2) The generalized function

$$(x_-^{-n}, \varphi) = \int_0^\infty x^{-n} \left[\varphi(-x) - \varphi(0) + x\varphi'(0) - \cdots - \right.$$

$$\left. - (-1)^{n-1} \frac{x^{n-1}}{(n-1)!} \varphi^{(n-1)}(0) \theta(1-x) \right] dx$$

corresponds to the function

$$x_-^{-n} = \begin{cases} 0 & \text{if } x > 0, \\ |x|^{-n} & \text{if } x < 0. \end{cases}$$

It also is not a value of the generalized function x_-^λ for $\lambda=-n$.

3) The generalized function

$$(|x|^{-2m-1}, \varphi) = \int_0^\infty x^{-2m-1} \left\{ \varphi(x) + \varphi(-x) - \right.$$

$$\left. - 2\left[\varphi(0) + \frac{x^2}{2!} \varphi''(0) + \cdots + \frac{x^{2m}}{(2m)!} \varphi^{(2m)}(0) \theta(1-x) \right] \right\} dx$$

corresponds to the function $|x|^{-2m-1}$. This function is not a value of the generalized function $|x|^\lambda$ for $\lambda=-2m-1$.

4) The generalized function

$$(|x|^{-2m} \operatorname{sign} x, \varphi) = \int_0^\infty x^{-2m} \left\{ \varphi(x) - \varphi(-x) - \right.$$

$$- 2\left[x\varphi'(0) + \frac{x^3}{3!} \varphi'''(0) + \cdots + \right.$$

$$\left. \left. + \frac{x^{2m-1}}{(2m-1)!} \varphi^{(2m-1)}(0) \theta(1-x) \right] \right\} dx$$

corresponds to the function $|x|^{-2m} \operatorname{sign} x$.

This function is not a value of the generalized function $|x|^\lambda \operatorname{sign} x$ for $\lambda = -2m$.

5) The generalized function

$$(x_+^\lambda \ln^m x_+, \varphi) = \int_0^1 x^\lambda \ln^m x \left[\varphi(x) - \varphi(0) - x\varphi'(0) - \cdots - \right.$$

$$\left. - \frac{x^{n-1}}{(n-1)!} \varphi^{(n-1)}(0) \right] dx + \int_1^\infty x^\lambda \ln^m x \varphi(x) \, dx +$$

$$+ \sum_{k=1}^n \frac{(-1)^m \, m! \, \varphi^{(k-1)}(0)}{(k-1)! \, (\lambda + k)^{m+1}}$$

corresponds to the function

$$x_+^\lambda \ln^m x_+ = \begin{cases} x^\lambda \ln^m x & \text{if } x > 0, \\ 0 & \text{if } x < 0 \end{cases}$$

for $\operatorname{Re}\lambda > -n-1$, $\lambda \neq -1, -2, \ldots$. We can give a simpler formula for $-n-1 < \operatorname{Re}\lambda < -n$:

$$(x_+^\lambda \ln^m x_+, \varphi) = \int_0^\infty x^\lambda \ln^m x \left[\varphi(x) - \varphi(0) - \right.$$

$$\left. - x\varphi'(0) - \cdots - \frac{x^{n-1}}{(n-1)!} \varphi^{(n-1)}(0) \right] dx.$$

The last formula is obtained from the formula for x_+^λ (see no. 2) by replacing x^λ by $x^\lambda \ln^m x$. In the same way, the generalized function $x_-^\lambda \ln^m x_-$ is given by the formula obtained from the formula for x_-^λ by replacing x^λ by $x^\lambda \ln^m x$.

An analogous remark is valid for the generalized functions

$$x_+^{-n} \ln^m x_+, \quad x_-^{-n} \ln^m x_-, \quad |x|^\lambda \cdot \ln^m |x|, \quad |x|^\lambda \operatorname{sign} x \ln^m |x|.$$

They are defined by the formulas obtained from the formulas for x_+^{-n}, x_-^{-n}, $|x|^\lambda$, $|x|^\lambda \operatorname{sign} x$ by replacing x^{-n} (or x^λ) by $x^{-n} \ln^m x$ (or $x^\lambda \ln^m x$) respectively.

6) The generalized functions $\ln(x+i0)$ and $\ln(x-i0)$ are defined by

the formulas:

$$\ln(x + i0) = \lim_{\varepsilon \to +0} \ln(x + i\varepsilon) = \ln|x| + i\pi\theta(-x),$$

$$\ln(x - i0) = \lim_{\varepsilon \to +0} \ln(x - i\varepsilon) = \ln|x| - i\pi\theta(-x),$$

where, as above, $\theta(x)=0$ for $x<0$ and $\theta(x)=1$ for $x>0$.

7) The generalized functions $(x+i0)^\lambda \ln(x+i0)$ and $(x-i0)^\lambda \ln(x-i0)$ are given by the formulas:

$$(x + i0)^\lambda \ln(x + i0) =$$
$$= \begin{cases} x_+^\lambda \ln x_+ + i\pi e^{i\lambda\pi}x_-^\lambda + e^{i\lambda\pi}x_-^\lambda \ln x_- & \text{if } \lambda \neq -n, \\ (-1)^n i\pi x_-^{-n} + (-1)^{n-1} \dfrac{\pi^2}{2} \dfrac{\delta^{(n-1)}(x)}{(n-1)!} + x^{-n}\ln|x| & \text{if } \lambda = -n; \end{cases}$$

$$(x - i0)^\lambda \ln(x - i0) =$$
$$= \begin{cases} x_+^\lambda \ln x_+ - i\pi e^{-i\lambda\pi}x_-^\lambda + e^{-i\lambda\pi}x_-^\lambda \ln x_- & \text{if } \lambda \neq -n, \\ (-1)^{n-1} i\pi x_-^{-n} + (-1)^{n-1} \dfrac{\pi^2}{2} \dfrac{\delta^{(n-1)}(x)}{(n-1)!} + x^{-n}\ln|x| & \text{if } \lambda = -n. \end{cases}$$

7. Generalized functions x_+^λ, x_-^λ and functions which are analogous to them as function of the parameter λ.

1) The regularization of the function x_+^λ was based on the equality $(x_+^\lambda)' = \lambda x_+^{\lambda-1}$. Another method for the regularization of this function exists (which gives the same result) which is based on the idea of analytic continuation. If $\varphi(x)$ is a fixed function of the space K, then the expression (x_+^λ, φ) is an analytic function of λ in the half-plane $\operatorname{Re}\lambda > -1$. Continuing this function analytically to the entire λ-plane, we obtain values for the expression (x_+^λ, φ) also for $\operatorname{Re}\lambda \leqslant -1$. It turns out that the value of this analytic continuation is given by the formula in no. 2 for $\operatorname{Re}\lambda > -n-1$, $\lambda \neq -1, -2, \ldots, -n$ and by the simpler formula mentioned there for $-n-1 < \operatorname{Re}\lambda < -n$. The analytic function (x_+^λ, φ) has simple poles for $\lambda = -1, -2, \ldots, -k, \ldots$. The residues at these poles are given by the formulas

$$\operatorname*{Res}_{\lambda=-k} (x_+^\lambda, \varphi) = \frac{\varphi^{(k-1)}(0)}{(k-1)!}.$$

We can say that the generalized function x_+^λ is an analytic function of λ with poles at the points $\lambda = -1, \ldots, -k, \ldots$, where

$$\operatorname*{Res}_{\lambda=-k} x_+^\lambda = \frac{(-1)^{k-1}}{(k-1)!} \delta^{(k-1)}(x), \quad k = 1, 2, \ldots$$

It is convenient to normalize this generalized function, considering instead the function $x_+^\lambda/\Gamma(\lambda+1)$. This function is an entire analytic function of λ assuming, at $\lambda=-1,\ldots,-k,\ldots$, the values $\delta^{(k-1)}(x)$ (since the function $\Gamma(\lambda+1)$ has poles at the same points as the function x_+^λ).

The expansion of the function x_+^λ in powers of $\lambda-\lambda_0$, $\lambda_0\neq-1,\ldots,-k,\ldots$ has the form

$$x_+^\lambda = x_+^{\lambda_0} + (\lambda-\lambda_0)\,x_+^{\lambda_0}\ln x_+ + \cdots + \frac{(\lambda-\lambda_0)^m}{m!}\,x_+^{\lambda_0}\ln^m x_+ + \cdots$$

The expansion of the same function x_+^λ in powers of $\lambda+k$ has the form

$$x_+^\lambda = \frac{(-1)^{k-1}\,\delta^{(k-1)}(x)}{(k-1)!}\,\frac{1}{\lambda+k} + x_+^{-k} + (\lambda+k)\,x_+^{-k}\ln x_+ + \cdots +$$
$$+ \frac{(\lambda+k)^m}{m!}\,x_+^{-k}\ln^m x_+ + \cdots$$

Analogous statements are valid for the function x_-^λ. It has poles at $\lambda=-1,\ldots,-k,\ldots$ with residues $\dfrac{\delta^{(k-1)}(x)}{(k-1)!}$. The normalized function $\dfrac{x_-^\lambda}{\Gamma(\lambda+1)}$ is an entire analytic function of λ, assuming the values $(-1)^{k-1}\,\delta^{(k-1)}(x)$ for $\lambda=-1,\ldots,-k,\ldots$. The formulas

$$x_-^\lambda = x_-^{\lambda_0} + (\lambda-\lambda_0)\,x_-^{\lambda_0}\ln x_- + \cdots + \frac{(\lambda-\lambda_0)^m}{m!}\,x_-^{\lambda_0}\ln^m x_- + \cdots,$$
$$\lambda_0 \neq -1,\,-2,\ldots,\,-k,\ldots,$$

and

$$x_-^\lambda = \frac{\delta^{(k-1)}(x)}{(k-1)!}\,\frac{1}{\lambda+k} + x_-^{-k} + (\lambda+k)\,x_-^{-k}\ln x_- + \cdots +$$
$$+ \frac{(\lambda+k)^m}{m!}\,x_-^{-k}\ln^m x_- + \cdots$$

hold.

2) The generalized function $|x|^\lambda$ is an analytic function of λ having simple poles at $\lambda=-1,\,-3,\ldots,\,-2k-1,\ldots$ with residues

$$\operatorname*{Res}_{\lambda=-2k-1}\,|x|^\lambda = \frac{2\delta^{(2k)}(x)}{(2k)!}.$$

The normalized function $|x|^\lambda/\Gamma\left(\dfrac{\lambda+1}{2}\right)$ is an entire analytic function of λ, assuming the values

$$\frac{(-1)^k\,k!\,\delta^{(2k)}(x)}{(2k)!}$$

at $\lambda=-2k-1$.

The formulas

$$|x|^\lambda = |x|^{\lambda_0} + (\lambda-\lambda_0)\,|x|^{\lambda_0}\ln|x| +\cdots+ \frac{1}{m!}(\lambda-\lambda_0)^m\,|x|^{\lambda_0}\ln^m|x| +\cdots,$$

$$\lambda \neq -1,-3,\ldots,-2k-1,\ldots,$$

and

$$|x|^\lambda = \frac{2\delta^{(2k)}(x)}{(2k)!}\,\frac{1}{\lambda+2k+1} + |x|^{-2k-1} +$$

$$+ (\lambda+2k+1)\,|x|^{-2k-1}\ln|x| +\cdots+$$

$$+ \frac{(\lambda+2k+1)^m}{m!}\,|x|^{-2k-1}\ln^m|x| +\cdots$$

are valid.

3) The generalized function $|x|^\lambda\operatorname{sign} x$ is also analytic and has simple poles at $\lambda=-2,-4,\ldots,-2k,\ldots$ with residues

$$\operatorname*{Res}_{\lambda=-2k}\ |x|^\lambda\operatorname{sign} x = -2\,\frac{\delta^{(2k-1)}(x)}{(2k-1)!}.$$

The normalized function $|x|^\lambda\operatorname{sign} x/\Gamma\left(\dfrac{\lambda+2}{2}\right)$ is an entire analytic function of λ, assuming the values

$$(-1)^k\frac{(k-1)!\,\delta^{(2k-1)}(x)}{(2k-1)!}$$

for $\lambda=-2,-4,\ldots,-2k,\ldots$.

The following formulas are valid:

$$|x|^\lambda\operatorname{sign} x = |x|^{\lambda_0}\operatorname{sign} x + (\lambda-\lambda_0)\,|x|^{\lambda_0}\ln|x|\operatorname{sign} x +\cdots+$$

$$+ \frac{1}{m!}\,|x|^{\lambda_0}\ln^m|x|\operatorname{sign} x +\cdots,$$

$$\lambda \neq -2,-4,\ldots,-2k,\ldots,$$

and

$$|x|^\lambda \operatorname{sign} x = -2 \frac{\delta^{(2k-1)}(x)}{(2k-1)!} \frac{1}{\lambda+2k} + |x|^{-2k} \operatorname{sign} x +$$

$$+ (\lambda+2k) |x|^{-2k} \ln|x| \operatorname{sign} x + \cdots +$$

$$+ \frac{(\lambda+2k)^m}{m!} |x|^{-2k} \ln^m|x| \operatorname{sign} x + \cdots$$

4) The generalized functions $(x+i0)^\lambda$ and $(x-i0)^\lambda$ are entire analytic functions of λ. They assume the values

$$(x + i0)^{-k} = x^{-k} - \frac{i\pi(-1)^{k-1}}{(k-1)!} \delta^{(k-1)}(x)$$

and

$$(x - i0)^{-k} = x^{-k} + \frac{i\pi(-1)^{k-1}}{(k-1)!} \delta^{(k-1)}(x)$$

for $\lambda = -1, -2, \ldots, -k, \ldots$, respectively.

8. *Homogeneous generalized functions.* A generalized function $f(x)$ of one variable is called a *homogeneous function of degree* λ if the equality $f(ax) = a^\lambda f(x)$ is satisfied for arbitrary $a > 0$. For every λ, the homogeneous generalized functions of degree λ are linear combinations of two linearly independent homogeneous functions. We can take $(x+i0)^\lambda$ and $(x-i0)^\lambda$ as these functions for example. For $\lambda \neq -1, -2, \ldots, -k, \ldots$ we can take the functions x_+^λ and x_-^λ, and for $\lambda = -k$ we can take the functions x^{-k} and $\delta^{(k-1)}(x)$.

A generalized function $f(x)$ is called an *associated homogeneous function of degree* λ *and order* m if the equality

$$f(ax) = a^\lambda f(x) + a^\lambda \ln^m a f_0(x)$$

holds, where $f_0(x)$ is an associated homogeneous generalized function of degree λ and order $m-1$ (the homogeneous generalized functions of degree λ are the associated homogeneous functions of the same degree and zero order).

For an arbitrary value of λ, there exist two linearly independent associated homogeneous generalized functions of degree λ and order m of which all such functions are linear combinations. For arbitrary λ, we can choose $(x+i0)^\lambda \ln^m(x+i0)$ and $(x-i0)^\lambda \ln^m(x-i0)$. For $\lambda \neq -1, \ldots,$

$-k,...,$ we can take the functions $x_+^\lambda \ln^m x_+$ and $x_-^\lambda \ln^m x_-$, and for $\lambda = -1, ..., -k, ...$ we can take the functions $x_+^{-k} \ln^m x_+$ and $x^{-k} \ln^m x_-$.

9. Table of derivatives of some generalized functions.

	$f(x)$	$f'(x)$
1	x_+^λ	$\lambda x_+^{\lambda-1}, \quad \lambda \neq -1, -2, ..., -k, ...$
2	x_-^λ	$\lambda x_-^{\lambda-1}, \quad \lambda \neq -1, -2, ..., -k, ...$
3	$|x|^\lambda$	$\lambda |x|^{\lambda-1}, \lambda \neq -1, -3, ..., -2k-1, ...$
4	$|x|^\lambda \operatorname{sign} x$	$\lambda |x|^{\lambda-1} \operatorname{sign} x, \lambda \neq -2, -4, ..., -2k, ...$
5	$(x + i0)^\lambda$	$\lambda (x + i0)^{\lambda-1}$
6	$(x - i0)^\lambda$	$\lambda (x - i0)^{\lambda-1}$
7	x_+^{-k}	$-kx_+^{-k-1} + \dfrac{(-1)^k \delta^{(k)}(x)}{k!}$
8	x_-^{-k}	$-kx_-^{-k-1} - \dfrac{\delta^{(k)}(x)}{k!}$
9	$|x|^{-2k-1}$	$-(2k+1) x^{-2k-2} - \dfrac{2\delta^{(2k+1)}(x)}{(2k+1)!}$
10	$|x|^{-2k} \operatorname{sign} x$	$-2kx^{-2k-1} + \dfrac{2\delta^{(2k)}(x)}{2k!}$
11	$\ln(x + i0)$	$(x + i0)^{-1} = \dfrac{1}{x} - i\pi\delta(x)$
12	$\ln(x - i0)$	$(x - i0)^{-1} = \dfrac{1}{x} + i\pi\delta(x)$
13	$\ln x_+$	x_+^{-1}
14	$\ln x_-$	x_-^{-1}
15	$\ln|x|$	x^{-1}
16	$\theta(x)$	$\delta(x)$

It remains to remark that a *q-th* order anti-derivative of the function $|x|^\lambda$ is given by the formula

$$\underbrace{\int \cdots \int}_{q} |x|^\lambda \, dx = \frac{|x|^{\lambda+q} (\operatorname{sign} x)^q}{(\lambda+1)...(\lambda+q)} - \sum_{k=1}^{\left[\frac{q}{2}\right]} \frac{x^{q-2k}}{(2k-1)!(q-2k)!} \frac{1}{\lambda+2k}.$$

10. *Differentiation and integration of arbitrary order.* Let $f(x)$ be a function which is equal to zero on the half-axis $(-\infty, 0)$ and integrable on an arbitrary finite segment of the half-axis $(0, \infty)$. Then its *q-th* order anti-derivative which is zero on the half-axis $(-\infty, 0)$ is given by the *Cauchy formula*:

$$f_q(x) = \frac{1}{\Gamma(q)} \int_0^x f(t)(x-t)^{q-1}\, dt.$$

We can write this formula in the form

$$f_q(x) = f(x) * \frac{x_+^{q-1}}{\Gamma(q)}.$$

In analogy with this equality, an anti-derivative of order λ of a generalized function $f(x)$ which is equal to zero on the half-axis $(-\infty, 0)$ is defined by the following formula:

$$f_\lambda(x) = f(x) * \frac{x_+^{\lambda-1}}{\Gamma(\lambda)}.$$

The generalized function $x_+^{\lambda-1}/\Gamma(\lambda)$ is denoted by $\Phi_\lambda(x)$. In no. 7 it was indicated that

$$\Phi_{-k}(x) = \lim_{\lambda \to -k} \frac{x_+^{\lambda-1}}{\Gamma(\lambda)} = \delta^{(k)}(x) \quad (k = 0, 1, \ldots).$$

Therefore

$$f_{-k}(x) = f(x) * \Phi_{-k}(x) = f(x) * \delta^{(k)}(x) = f^{(k)}(x).$$

Thus the anti-derivative of order $-k$ is nothing other than the derivative of order k of the generalized function $f(x)$. In accordance with this, we set

$$\frac{d^\lambda f}{dx^\lambda} = f_{-\lambda} = f(x) * \Phi_{-\lambda}(x)$$

for arbitrary λ. The formula

$$\frac{d^\beta}{dx^\beta}\left(\frac{d^\gamma f}{dx^\gamma}\right) = \frac{d^{\beta+\gamma} f}{dx^{\beta+\gamma}},$$

implied by the equality

$$\Phi_\beta * \Phi_\gamma = \Phi_{\beta+\gamma},$$

is valid. For the function $\Phi_\mu(x)$,

$$\frac{d^\lambda}{dx^\lambda}\left(\frac{x_+^{\mu-1}}{\Gamma(\mu)}\right) = \frac{x_+^{\mu-\lambda-1}}{\Gamma(\mu-\lambda)}.$$

In particular,

$$\frac{d^\lambda\theta(x)}{dx^\lambda} = \frac{x_+^{-\lambda}}{\Gamma(-\lambda+1)},$$

$$\frac{d^\lambda}{dx^\lambda}\left(\delta^{(k)}(x)\right) = \frac{x_+^{-k-\lambda-1}}{\Gamma(-k-\lambda)},$$

$$\frac{d^\lambda}{dx^\lambda}\left(\frac{x_+^{\lambda-k-1}}{\Gamma(\lambda-k)}\right) = \delta^{(k)}(x).$$

The *Abel integral equation*

$$g(x) = \frac{1}{\Gamma(1-\alpha)}\int_0^x \frac{f(t)\,dt}{(x-t)^\alpha}$$

can be written in the form

$$g(x) = f(x) * \Phi_\lambda(x),$$

where $\lambda = -\alpha+1$.

By virtue of the formula for the convolution of the functions $\Phi_\lambda(x)$,

$$g(x) * \Phi_{-\lambda}(x) = f(x) * \Phi_\lambda(x) * \Phi_{-\lambda}(x) = f(x) * \delta(x) = f(x).$$

The formula

$$f(x) = \frac{1}{\Gamma(\alpha-1)}\int_0^x (x-t)^{\alpha-2} g(t)\,dt$$

is obtained for $0<\alpha<1$ where we understand the integral as a regularized value.

If $g(x)$ is a differentiable function, then the last equality can be written as follows:

$$f(x) = \frac{1}{\Gamma(\alpha)}\int_0^x (x-t)^{\alpha-1} g'(t)\,dt.$$

11. *Expression of some special functions in the form of derivatives of fractional order.* Using the operations of differentiation and integration

of fractional order, we can quickly obtain integral representations of some special functions. For example, the following equalities are valid for the hypergeometric function:

$$\frac{x^{\gamma-1}}{\Gamma(\gamma)} F(\alpha, \beta, \gamma; x) = \frac{d^{\beta-\gamma}}{dx^{\beta-\gamma}} \left(\frac{x_+^{\beta-1}(1-x)_+^{-\alpha}}{\Gamma(\beta)} \right), \qquad 0 < x < 1,$$

$$\frac{x^{\gamma-1}(1-x)^{\alpha+\beta-\gamma}}{\Gamma(\gamma)} F(\alpha, \beta, \gamma; x) =$$

$$= \frac{d^{-\beta}}{dx^{-\beta}} \left[\frac{x_+^{\gamma-\beta-1}(1-x)^{\alpha-\gamma}}{\Gamma(\gamma-\beta)} \right], \qquad 0 < x < 1.$$

For a Bessel function the equality

$$2^p \sqrt{\pi} \, x^{p/2} J_p(\sqrt{x}) = \frac{d^{-p-\frac{1}{2}}}{dx^{-p-\frac{1}{2}}} \left[\frac{\cos\sqrt{x}}{\sqrt{x}} \right]$$

is valid.

Using formulas for differentiation and integration, we can obtain several relations for hypergeometric and Bessel functions from these expressions. For example, taking the derivative of order $-q-1$ of both sides of the last equality, we obtain

$$\frac{d^{-q-1}}{dx^{-q-1}} [x^{\frac{p}{2}} J_p(x)] = 2^{q+1} x^{\frac{p+q+1}{2}} J_{p+q+1}(\sqrt{x}).$$

In integral form this equality means the following:

$$2^{q+1} x^{\frac{p+q+1}{2}} J_{p+q+1}(\sqrt{x}) = \int_0^x t^{\frac{p}{2}} J_p(\sqrt{t}) \frac{(x-t)^q}{\Gamma(q+1)} \, dt.$$

§ 3. Some generalized functions of several variables

1. *The generalized function r^λ.* An analogue of the generalized function $|x|^\lambda$, for functions of several variables, is the generalized function r^λ, $r = \sqrt{x_1^2 + \cdots + x_n^2}$. It is given by the formula

$$(r^\lambda, \varphi) = \int r^\lambda \varphi(x) \, dx =$$

$$= \int (x_1^2 + \cdots + x_n^2)^{\lambda/2} \, \varphi(x_1, \ldots, x_n) \, dx_1 \ldots dx_n$$

for $\operatorname{Re}\lambda> -n$. The expression (r^λ, φ) is an analytic function of λ in the domain $\operatorname{Re}\lambda> -n$. Continuing this function analytically, we obtain

$$(r^\lambda, \varphi) = \Omega_n \int_0^\infty r^{\lambda+n-1} S_\varphi(r)\, dr,$$

where $\Omega_n = 2(\sqrt{\pi})^n / \Gamma\left(\dfrac{n}{2}\right)$ is the area of the surface of the unit ball in n-dimensional space, $S_\varphi(r)$ is the mean value of the function $\varphi(x)$ on the ball of radius r, and the integral on the right hand side is thought of in the sense of a regularized value (as $(r_+^{\lambda+n-1}, S_\varphi(r))$). The generalized function r^λ has simple poles at the points $\lambda = -n, -n-2,\ldots, -n-2k,\ldots$. An expansion in a Laurent series in a neighborhood of the pole $\lambda = -n-2k$ has the form

$$r^\lambda = \Omega_n\left[\frac{\delta^{(2k)}(r)}{(2k)!}\frac{1}{\lambda+n+2k} + r^{-n-2k} + (\lambda+n+2k)\, r^{-n-2k}\ln r +\cdots\right].$$

The functionals $\delta^{(2k)}(r)$, r^{-n-2k}, $r^{-n-2k}\ln^m r$ are applied to the function $S_\varphi(r)$ and understood as

$$(\delta^{(2k)}(r), S_\varphi(r)),\ (r_+^{-n-2k}, S_\varphi(r)),\ (r_+^{-n-2k}\ln^m r_+, S_\varphi(r)).$$

The quantity $(\delta^{(2k)}, S_\varphi)$ is expressed in terms of the function $\varphi(x)$ and its derivatives by the formula

$$(\delta^{(2k)}, S_\varphi) = S_\varphi^{(2k)}(0) = \frac{(2k)!\, \Delta^k\varphi(0)}{2^k k!\, n(n+2)\ldots(n+2k-2)},$$

where Δ is the Laplace operator.

The generalized function $2r^\lambda / \Omega_n \Gamma\left(\dfrac{\lambda+n}{2}\right)$ is an entire analytic function of λ, and the value of this function at the point $\lambda = -n-2k$ is equal to

$$\frac{(-1)^k \Delta^k\delta(x)}{2^k k!\, n(n+2)\ldots(n+2k-2)}.$$

The formula

$$\Delta r^\lambda = \lambda(\lambda+n-2)\, r^{\lambda-2},$$

valid in the classical sense only for $\operatorname{Re}\lambda> -n+2$, retains its sense for all values of λ, $\lambda\neq -n-2k$, $k=0, 1,\ldots$, if we regard both sides of the equality as generalized functions.

In many problems, it is useful to expand the function r^λ in terms of functions which assume fixed values on planes. This expansion has the form

$$2 \, \frac{r^\lambda}{\Gamma\left(\dfrac{\lambda + n}{2}\right)} = \frac{1}{\pi^{\frac{n-1}{2}} \, \Gamma\left(\dfrac{\lambda + 1}{2}\right)} \int_\Omega |\omega_1 x_1 + \cdots + \omega_n x_n|^\lambda \, d\Omega,$$

where the point $(\omega_1, \ldots, \omega_n)$ ranges over the unit ball Ω.

In particular, for odd n

$$\delta(x) = \frac{(-1)^{\frac{n-1}{2}}}{2(2\pi)^{n-1}} \int_\Omega \delta^{(n-1)}(\omega_1 x_1 + \cdots + \omega_n x_n) \, d\Omega,$$

and for even n

$$\delta(x) = \frac{(-1)^{n/2} (n-1)!}{(2\pi)^n} \int_\Omega (\omega_1 x_1 + \cdots + \omega_n x_n)^{-n} \, d\Omega.$$

For arbitrary n, the formula

$$\delta(x) = \frac{(n-1)!}{(2\pi i)^n} \int_\Omega (\omega_1 x_1 + \cdots + \omega_n x_n - i0)^{-n} \, d\Omega$$

is valid.

It follows from the first formula for $\delta(x)$ that

$$\varphi(0) = \frac{(-1)^{\frac{n-1}{2}}}{2(2\pi)^{n-1}} \int_\Omega d\omega \int_{\Sigma \omega_k x_k = 0} \frac{\partial^{n-1} \varphi(x)}{\partial v^{n-1}} \, d\sigma_0,$$

where $d\sigma_0$ is an area element in the plane $\Sigma \omega_k x_k = 0$ and $\dfrac{\partial}{\partial v}$ is differentiation in the direction of the vector ω orthogonal to it.

Analogously, it follows from the second formula that

$$\varphi(0) = \frac{(-1)^{n/2} (n-1)!}{(2\pi)^n} \int_\Omega ((\omega_1 x_1 + \cdots + \omega_n x_n)^{-n},$$

$$\psi(\omega_1 x_1 + \cdots + \omega_n x_n)) \, d\omega,$$

where we set

$$\psi(t) = \int_{\Sigma \omega_k x_k = t} \varphi(x) \, d\sigma_t,$$

$d\sigma_t$ is an area element in the plane $\sum \omega_k x_k = t$.

2. *Generalized functions connected with quadratic forms.* Let $P = \sum\limits_{\alpha, \beta = 1}^{n} g_{\alpha\beta} x_\alpha x_\beta$ be a positive definite quadratic form. By means of a non-singular linear transformation this form can be reduced to the form $P = \sum\limits_{\alpha = 1}^{n} y_\alpha^2$. Therefore, the consideration of generalized functions of the form P^λ in the case of positive definite quadratic forms P reduces to the consideration of the generalized function $r^{2\lambda}$ carried out in no. 1.

The situation is more complicated for non-definite quadratic forms $P = \sum\limits_{\alpha, \beta = 1}^{n} g_{\alpha\beta} x_\alpha x_\beta$ since in this case the function P^λ is not defined uniquely. Forms with complex coefficients can be written in the form $\mathscr{P} = P_1 + iP_2$, where P_1 and P_2 are real quadratic forms. Let the "upper half-plane" be the set of all quadratic forms $P_1 + iP_2$ for which the form P_2 is positive definite, and the "lower half-plane" be the set of the forms of the form $P_1 - iP_2$ where P_2 is positive definite.

If the quadratic form $\mathscr{P}$ belongs to the upper half-plane, then we set

$$\mathscr{P}^\lambda = e^{\lambda \, [\ln |\mathscr{P}| \, + \, i \arg \mathscr{P}]},$$

where $0 < \arg \mathscr{P} < \pi$ and introduce the generalized function

$$(\mathscr{P}^\lambda, \varphi) = \int \overline{\mathscr{P}^\lambda} \varphi(x) \, dx$$

(integration is carried out over the entire space R_n). The function $\mathscr{P}^\lambda(x)$ is defined uniquely, and the integral is convergent for $\operatorname{Re} \lambda > 0$ and is an analytic function of λ. Analytically continuing this function with respect to λ, we define the functional $(\mathscr{P}^\lambda, \varphi)$ for other values of λ.

Considering the generalized function $\mathscr{P}^\lambda$ for $P_1 = 0$ we obtain the generalized function $(iP_2)^\lambda$. Inasmuch as the form P_2 is positive definite, the investigation of this generalized function reduces to the investigation, carried out above, of the generalized function $r^{2\lambda}$. Analytically continuing the expression for the residues of the form $(iP_2)^\lambda$ with respect to the co-efficients of the form P_2, we find an expression for the residues of the form $\mathscr{P}^\lambda$. The result is formulated in the following way:

If $\mathscr{P} = \sum\limits_{\alpha, \beta = 1}^{n} g_{\alpha\beta} x_\alpha x_\beta$ is an arbitrary quadratic form with a positive definite imaginary part, then the generalized function $\mathscr{P}^\lambda$ is a regular *analytic function of λ everywhere except at the points*

$$\lambda = -\frac{n}{2}, \ -\frac{n}{2} - 1, \ldots, \ -\frac{n}{2} - k, \ldots,$$

at which the function has simple poles. In this connection

$$\operatorname*{Res}_{\lambda = -\frac{n}{2} - k} \mathscr{P}^{\lambda} = \frac{e^{-\frac{\pi n i}{4}} \pi^{\frac{n}{2}}}{4^{k} k! \ \Gamma\left(\frac{n}{2} + k\right) \sqrt{(-i)^{n} |g|}} L^{k} \delta(x).$$

Here, L denotes the differential operator

$$L = \sum_{\alpha, \beta = 1}^{n} g^{\alpha\beta} \frac{\partial^{2}}{\partial x_{\alpha} \partial x_{\beta}},$$

whose coefficients are connected with the coefficients of the quadratic form $\mathscr{P}$ by the relations

$$\sum_{\beta = 1}^{n} g^{\alpha\beta} g_{\beta\gamma} = \delta_{\gamma}^{\alpha} \quad (\delta_{\gamma}^{\alpha} \text{ is the Kronecker symbol}).$$

The determinant of the form $\mathscr{P}$ is denoted by $|g|$.

The function $\sqrt{(-i)^{n} |g|}$ is given by the formula

$$\sqrt{(-i)^{n} |g|} = \sqrt{|b|^{2} (1 - \lambda_{1} i)^{\frac{1}{2}} \ldots (1 - \lambda_{n} i)^{\frac{1}{2}}}.$$

Here $|b|$ is the determinant of the matrix $\|b_{\alpha\beta}\|$ of the transformation with real coefficients:

$$x_{\alpha} = \sum_{\beta = 1}^{n} b_{\alpha\beta} \gamma_{\beta},$$

reducing the forms P_1 and P_2, $\mathscr{P} = P_1 + iP_2$, to the forms

$$P_1 = \lambda_1 y_1^2 + \cdots + \lambda_n y_n^2,$$
$$P_2 = y_1^2 + \cdots + y_n^2,$$

and the values of the square roots are given by the formula

$$\sqrt{z} = |z|^{\frac{1}{2}} e^{\frac{i}{2} \arg z}, \quad -\pi < \arg z < \pi.$$

If the form $\mathscr{P} = P_1 - iP_2$ belongs to the lower half-plane, then its residues at the points $\lambda = -\frac{n}{2}, \ldots, -\frac{n}{2} - k, \ldots$ are given by the formulas with the replacement of i by $-i$.

3. *Generalized functions* $(P+i0)^\lambda$ *and* $(P-i0)^\lambda$. Let

$$P = \sum_{\alpha, \beta = 1}^{n} g_{\alpha\beta} x_\alpha x_\beta$$

be a non-degenerate quadratic form with real coefficients which has p positive and q negative terms in its canonical form. We consider quadratic forms $P_1 + i\varepsilon P_2$, where P_2 is a positive definite quadratic form with real coefficients and $\varepsilon > 0$. We set

$$(P + i0)^\lambda = \lim_{\varepsilon \to +0} (P + i\varepsilon P_2)^\lambda$$

(the generalized function $(P + i\varepsilon P_2)^\lambda$ is defined in no. 2). Analogously

$$(P - i0)^\lambda = \lim_{\varepsilon \to +0} (P - i\varepsilon P_2)^\lambda.$$

The generalized function $(P+i0)^\lambda$, considered as a function of λ, has simple poles at the points $\lambda = -\dfrac{n}{2}, \ -\dfrac{n}{2}-1, \ldots, \ -\dfrac{n}{2}-k, \ldots$ with residues

$$\operatorname*{Res}_{\lambda = -\frac{n}{2}-k} (P + i0)^\lambda = \frac{e^{-\frac{\pi}{2} qi} \pi^{\frac{n}{2}}}{4^k k! \, \Gamma\left(\dfrac{n}{2} + k\right) \sqrt{|\varDelta|}} L^k \delta(x),$$

where $\varDelta$ is the determinant of the form P.

The function $(P-i0)^\lambda$ has poles at the same points with residues obtained from the last formula by the replacement of i with $-i$.

We set

$$P_+^\lambda = \frac{e^{-\pi\lambda i}(P + i0)^\lambda - e^{\pi\lambda i}(P - i0)^\lambda}{-2i \sin \pi\lambda},$$

$$P_-^\lambda = \frac{(P + i0)^\lambda - (P - i0)^\lambda}{2i \sin \pi\lambda}.$$

For $\operatorname{Re}\lambda > -n$, the generalized function P_+^λ coincides with the regular generalized function given by the formula

$$(P_+^\lambda, \varphi) = \int_{P > 0} P^\lambda \varphi(x) \, dx,$$

and P_-^λ coincides with the regular generalized function given by the formula

$$(P_-^\lambda, \varphi) = \int\limits_{P<0} |P(x)|^\lambda \, \varphi(x) \, dx.$$

4. *Generalized functions of the form* $\mathscr{P}^\lambda f(\mathscr{P}, \lambda)$. Let $f(z, \lambda)$ be an entire function of z and λ. If $\mathscr{P}$ is a complex quadratic form with a positive definite imaginary part, then we set

$$(\mathscr{P}^\lambda f(\mathscr{P}, \lambda), \varphi(x)) = \int \mathscr{P}^\lambda f(\mathscr{P}, \lambda) \varphi(x) \, dx$$

for $\operatorname{Re}\lambda > 0$. By means of analytic continuation the generalized function $\mathscr{P}^\lambda f(\mathscr{P}, \lambda)$ is defined for other values of λ. The generalized function

$$\mathscr{P}^\lambda \ln^m \mathscr{P} f(\mathscr{P}, \lambda)$$

is defined analogously.

The generalized function $(P+i0)^\lambda f(P+i0, \lambda)$ is given by the equality

$$(P + i0)^\lambda \, f(P + i0, \lambda) = \lim_{\varepsilon \to +0} (P + i\varepsilon P_2)^\lambda \, f(P + i\varepsilon P_2, \lambda),$$

where P_2 is a positive definite quadratic form. Analogously

$$(P - i0)^\lambda \, f(P - i0, \lambda) = \lim_{\varepsilon \to +0} (P - i\varepsilon P_2)^\lambda \, f(P - i\varepsilon P_2, \lambda).$$

If the form P is positive definite, then

$$(P + i0)^\lambda \, f(P + i0, \lambda) = (P - i0)^\lambda \, f(P - i0, \lambda) = P^\lambda f(P, \lambda).$$

Moreover,

$$f(P + i0, \lambda) = f(P - i0, \lambda) = f(P, \lambda).$$

The equalities

$$(P + i0)^\lambda \, f(P, \lambda) = P_+^\lambda f(P_+, \lambda) + e^{i\lambda\pi} P_-^\lambda f(P_-, \lambda)$$

and

$$(P - i0)^\lambda \, f(P, \lambda) = P_+^\lambda f(P_+, \lambda) + e^{-i\lambda\pi} P_-^\lambda f(P_-, \lambda)$$

are valid and give expressions for our generalized functions in terms of the variables P_+, P_-.

Such functions as

$$Z_\lambda[(P + i0)^{\frac{1}{2}}], \quad (P + i0)^{-\frac{\lambda}{2}} Z_\lambda[(P + i0)^{\frac{1}{2}}],$$

where $Z_\lambda(x)$ is a cylindrical function, belong to the class of generalized functions being considered.

To every real non-degenerate quadratic form $P = \sum\limits_{\alpha, \beta = 1}^{n} g_{\alpha\beta} x_\alpha x_\beta$ there corresponds a form

$$Q = \sum_{\alpha, \beta = 1}^{n} g^{\alpha\beta} s_\alpha s_\beta,$$

conjugate to it, where $\sum\limits_{\beta=1}^{n} g_{\alpha\beta} g^{\beta\gamma} = \delta_\alpha^\gamma \, (\alpha, \gamma = 1, 2, ..., n)$. In the sequel the generalized function

$$\frac{K_{\frac{n}{2}+\lambda}\left[c(Q + i0)^{\frac{1}{2}}\right]}{(Q + i0)^{\frac{1}{2}\left(\frac{n}{2}+\lambda\right)}}$$

is used, where c is a real number. It is defined for non-integral λ by the expansion

$$\frac{K_{\frac{n}{2}+\lambda}\left[c(Q + i0)^{\frac{1}{2}}\right]}{(Q + i0)^{\frac{1}{2}\left(\frac{n}{2}+\lambda\right)}} =$$

$$= \frac{\pi\left(\dfrac{c}{2}\right)^{\frac{n}{2}+\lambda}}{2\sin\left(\dfrac{n}{2}+\lambda\right)\pi}\left[\sum_{m=0}^{\infty}\frac{\left(\dfrac{c}{2}\right)^{-2\lambda-n+2m}(Q+i0)^{-\lambda-\frac{n}{2}+m}}{m!\,\Gamma\left(-\lambda-\dfrac{n}{2}+m+1\right)} - \right.$$

$$\left. - \sum_{m=0}^{\infty}\frac{\left(\dfrac{c}{2}\right)^{2m}(Q+i0)^{m}}{m!\,\Gamma\left(\lambda+\dfrac{n}{2}+m+1\right)}\right].$$

For $\lambda = s$, $s > 0$, this function has a pole with residue

$$\operatorname*{Res}_{\lambda = s}\frac{K_{\frac{n}{2}+\lambda}\left[c(Q + i0)^{\frac{1}{2}}\right]}{(Q + i0)^{\frac{1}{2}\left(\frac{n}{2}+\lambda\right)}} =$$

$$= \frac{(-1)^s\,\pi^{\frac{n}{2}}\left(\dfrac{c}{2}\right)^{s-\frac{n}{2}}e^{-\frac{\pi q i}{2}}\sqrt{|\Delta|}}{2}\sum_{m=0}^{s}\frac{(-1)^m\left(\dfrac{c}{2}\right)^{-2m}}{4^m m!\,(s-m)!}\,L^m\delta(x),$$

where q is the number of negative terms in the canonical form of the form Q.

By

$$\frac{K_{\frac{n}{2}+s}\left[c(Q+i0)^{\frac{1}{2}}\right]}{(Q+i0)^{\frac{1}{2}\left(\frac{n}{2}+s\right)}}$$

we denote the regular part of this generalized function for $\lambda=s$.

5. *Generalized functions on smooth surfaces.* Let the surface S be defined by the equation $P(x_1,\ldots,x_n)=0$ in an n-dimensional space, where P is an infinitely differentiable function such that $\operatorname{grad}P\neq0$ for $P=0$ (i.e. there are no singular points on the surface $P=0$). We define the generalized function $\delta(P)$ in the following way. In a sufficiently small neighborhood U_x of an arbitrary point x on the surface S new coordinates are introduced by setting $u_1=P$ and choosing the remaining coordinates $u_2,\ldots,u_n$ arbitrarily with only the restriction that the jacobian $D\!\left(\dfrac{x}{u}\right)$ be different from zero in U_x. If the function $\varphi(x)$ becomes zero outside U_x, then we set

$$(\delta(P),\varphi)=\int\psi(0,u_2,\ldots,u_n)\,du_2\ldots du_n,$$

where

$$\psi(u_1,\ldots,u_n)=\varphi_1(u_1,\ldots,u_n)\,D\!\left(\frac{x}{u}\right),$$

$$\varphi_1(u_1,\ldots,u_n)=\varphi(x_1,\ldots,x_n).$$

In the general case, we expand the function $\varphi(x)$ in terms of $\varphi_k(x)$ which become zero outside the neighborhoods U_{x_k}.

We can show that the generalized function $\delta(P)$ depends only on the function P and does not depend on the choice of the coordinates $u_1,\ldots,u_n$. We can also define this function in the following way. Let $\theta(P)$ be the generalized function

$$(\theta(P),\varphi)=\int\limits_{P\geqslant0}\varphi(x)\,dx.$$

Then $\delta(P) = \theta'(P)$ in the sense that

$$\frac{\partial \theta(P)}{\partial x_j} = \frac{\partial P}{\partial x_j} \delta(P)$$

for arbitrary j.

We set $(\delta^{(k)}(P), \varphi) = (-1)^k \int \psi_{u_1}^{(k)}(0, u_2, \ldots, u_n) du_2 \ldots du_n$ for the generalized functions $\delta^{(k)}(P)$. Every functional f of the form

$$(f, \varphi) = \sum_{i_1 + \cdots + i_n \leqslant k} \int_{P=0} a_{i_1 i_2 \ldots i_n}(x) \frac{\partial^{i_1 + \cdots + i_n} \varphi(x)}{\partial x_1^{i_1} \ldots \partial x_n^{i_n}} dx$$

is expressed in terms of $\delta(P), \ldots, \delta^{(k)}(P)$ in the following manner:

$$(f, \varphi) = \sum_{j=0}^{k} \int b_j(x) \delta^{(j)}(P) dx.$$

In this connection, the expression is unique: if $f = 0$, then

$$b_j(x) = 0, \quad 1 \leqslant j \leqslant k.$$

The formula for the differentiation of the composite function

$$\frac{\partial}{\partial x_j} \delta^{(k)}(P) = \frac{\partial P}{\partial x_j} \delta^{(k+1)}(P)$$

is valid. Moreover the equalities

$$P\delta(P) = 0,$$
$$P\delta^{(k)}(P) + k\delta^{(k-1)}(P) = 0, \quad k = 1, 2, \ldots$$

hold.

If the surfaces $P = 0$ and $Q = 0$ do not intersect and Q has the same properties as P, then

$$\delta(PQ) = P^{-1}\delta(Q) + Q^{-1}\delta(P).$$

If the function $a(x)$ does not become zero, then

$$\delta^{(k)}(aP) = \frac{\delta^{(k)}(P)}{a^k(x) |a(x)|}.$$

Now let the surface S have dimension $n-k$ in n-dimensional space and be defined by the k equations

$$P_1(x_1, \ldots, x_n) = 0, \ldots, P_k(x_1, \ldots, x_n) = 0,$$

where the $P_j(x_1, \ldots, x_n)$ are infinitely differentiable functions and where the surfaces $P_1 = 0, \ldots, P_k = 0$ form a regular network. In other words, it is assumed that in a neighborhood of each point x of the surface S we can introduce a system of coordinates $u_1, \ldots, u_n$ such that $u_j = P_j$, $1 \leqslant j \leqslant k$, and the jacobian $D\left(\dfrac{x}{u}\right)$ is different from zero. Then we set

$$(\delta(P_1, \ldots, P_k), \varphi) = \int \psi(0, \ldots, 0, u_{k+1}, \ldots, u_n)\, du_{k+1} \ldots du_n$$

and

$$\left(\frac{\partial^{i_1 + \cdots + i_k}\delta}{\partial P_1^{i_1} \ldots \partial P_k^{i_k}}, \varphi\right) = (-1)^{i_1 + \cdots + i_k} \times$$

$$\times \int \frac{\partial^{i_1 + \cdots + i_k}\psi(0, \ldots, 0, u_{k+1}, \ldots, u_n)}{\partial u_1^{i_1} \ldots \partial u_k^{i_k}}\, du_{k+1} \ldots du_n.$$

We can show that these generalized functions do not depend on the choice of the coordinates $u_1, \ldots, u_n$.

The following equalities are valid for the functions which are introduced:

$$\frac{\partial}{\partial x_j}\delta(P_1, \ldots, P_k) = \sum_{i=1}^{k} \frac{\partial\delta(P_1, \ldots, P_k)}{\partial P_i} \cdot \frac{\partial P_i}{\partial x_j},$$

$$P_i\delta(P_1, \ldots, P_k) = 0,$$

$$P_iP_j\delta(P_1, \ldots, P_k) = 0,$$

$$\cdots \cdots \cdots \cdots \cdots$$

$$P_1P_2 \ldots P_k\delta(P_1, \ldots, P_k) = 0.$$

Identities obtained by formal differentiation of these last equalities are also valid.

EXAMPLES.

1. The generalized function $\delta(\alpha_1 x_1 + \cdots + \alpha_n x_n)$ is given by the equality

$$(\delta(\alpha_1 x_1 + \cdots + \alpha_n x_n), \varphi) = \int_{\Sigma \, \alpha_k x_k = 0} \varphi\, d\sigma,$$

where $d\sigma$ is an area element of the plane $\displaystyle\sum_{k=1}^{n} \alpha_k x_k = 0$.

2. The generalized function $\delta(xy-c)$ is given by the formula

$$(\delta(xy - c), \varphi) = -\int \varphi\left(x, \frac{c}{x}\right)\frac{dx}{x}$$

for $c \neq 0$.

3. The generalized function $\delta(r-c)$, where $r^2 = \sum\limits_{k=1}^{n} x_k^2$, $c>0$, is given by the formula

$$(\delta(r - c), \varphi) = \int\limits_{\Omega_c} \varphi \, d\Omega_c,$$

where $d\Omega_c$ is an area element of the surface of the ball Ω_c of radius c. This ball can be given by the equation $r^2 = c^2$. Then

$$(\delta(r^2 - c^2), \varphi) = \frac{1}{2c}\int\limits_{\Omega_c} \varphi \, d\Omega_c.$$

4. If $P = x_1 - f(x_2, ..., x_n)$, then the equality

$$(\delta(P), \varphi) = \int \varphi\left[f(x_2, ..., x_n), x_2, ..., x_n\right] dx_2 ... dx_n$$

holds.

§ 4. Fourier transformation of generalized functions

1. *The space S and generalized functions of exponential growth.* In order to define the Fourier transformation for generalized functions, we introduce a new function space. A function $\varphi(x)$ is called *rapidly decreasing* as $|x| \to \infty$ if the equality $\lim\limits_{|x| \to \infty} |x|^m \varphi(x) = 0$ is satisfied for an arbitrary m.

The space S consists of functions $\varphi(x)$ which are rapidly decreasing as $|x| \to \infty$ together with derivatives of every order (e^{-x^2} and also functions of the form $P_k(x) e^{-x^2}$ where $P_k(x)$ is an arbitrary polynomial can serve as examples of such functions). A sequence of functions $\{\varphi_k(x)\}$ of the space S is said to be *convergent to zero* if the equality

$$\lim\limits_{k \to \infty} \sup\limits_{x} \left[(1 + |x|^2)^m |\varphi_k^{(q)}(x)|\right] = 0, \quad 0 \leqslant |q| \leqslant m,$$

is satisfied for an arbitrary m.

Every infinitely differentiable function with compact support belongs to the space S. Hence a continuous one-to-one mapping of the space K

onto an everywhere dense subset of the space S is defined. Therefore, every continuous linear functional on the space S defines a continuous linear functional on the space K, i.e., some generalized function. Such generalized functions are called *generalized functions of exponential growth*. For example, $|x|^\lambda$, $(x+i0)^\lambda$, and so on, are generalized functions of exponential growth. An arbitrary generalized function with compact support has exponential growth. Locally summable functions $f(x)$ for which the integral $\int |f(x)| (1+|x|^2)^{-m} dx$ is convergent for some m define regular generalized functions of exponential growth.

2. *Fourier transformation of generalized functions of exponential growth.* We call the function $\tilde{\varphi}(x)$, defined by the equality

$$\tilde{\varphi}(s) = \int \varphi(x)\, e^{i(x,\, s)}\, dx,$$

the *Fourier transformation of the summable function* $\varphi(x)$ where we set

$$(x,\, s) = x_1 s_1 + \cdots + x_n s_n.$$

The Fourier transformation of an arbitrary function $\varphi(x)$ of the space S belongs to the same space, and the mapping $\varphi(x) \to \tilde{\varphi}(s)$ is a one-to-one bicontinuous mapping of the space S onto itself.

If $f(x)$ is a summable function having a summable square, then the equality (Plancherel equality)

$$\int f(x)\, \varphi(-x)\, dx = \frac{1}{(2\pi)^n} \int \tilde{f}(s)\, \tilde{\varphi}(s)\, ds$$

is satisfied for an arbitrary function $\varphi(x)$ of the space S. In connection with this we define the *Fourier transform of the generalized function $f(x)$ of exponential growth* as the generalized function $\tilde{f}(s)$ such that the equality

$$(f,\, \varphi(-x)) = \frac{1}{(2\pi)^n} (\tilde{f},\, \tilde{\varphi})$$

holds for an arbitary function $\varphi(x)$ of the space S. *The generalized function $\tilde{f}$ also has exponential growth.* The indicated definition coincides with the usual definition for regular generalized functions $f(x)$. The Fourier transform of the generalized function $f(x)$ is also denoted by $F(f)$.

For the Fourier transformation of generalized functions, the following formulas for the differentiation of ordinary Fourier transforms are retained:

$$F\left[P\left(\frac{\partial}{\partial x_1}, ..., \frac{\partial}{\partial x_n}\right) f\right] = P(-is_1, ..., -is_n)\, \tilde{f},$$

$$F\left[P(ix_1, ..., ix_n)\, f\right] = P\left(\frac{\partial}{\partial s_1}, ..., \frac{\partial}{\partial s_n}\right) \tilde{f},$$

where P is an arbitrary polynomial. It remains to remark that

$$F[F(f(x))] = (2\pi)^n f(-x).$$

EXAMPLE. It follows from the Fourier inversion formula that

$$\varphi(x) = \frac{1}{(2\pi)^n} \int \tilde{\varphi}(s)\, e^{-i(s,\, x)}\, ds.$$

Setting $x=0$, we obtain

$$(\delta, \varphi) = \varphi(0) = \frac{1}{(2\pi)^n} \int \tilde{\varphi}(s)\, ds = \frac{1}{(2\pi)^n} (1, \tilde{\varphi}).$$

Therefore $\tilde{\delta} = 1$.

Applying the formula for differentiation, we find the formula for the Fourier transform of a polynomial

$$F[P(x_1, ..., x_n)] = (2\pi)^n P\left(-i\frac{\partial}{\partial s_1}, ..., -i\frac{\partial}{\partial s_n}\right) \delta(s).$$

For example,

$$F(|x|^2) = F(x_1^2 + \cdots + x_n^2) = -(2\pi)^n \Delta\delta(s),$$

where Δ is the Laplace operator:

$$\Delta = \frac{\partial^2}{\partial s_1^2} + \cdots + \frac{\partial^2}{\partial s_n^2}.$$

If $f(x)$ is an arbitrary locally summable function of exponential growth, then it can be written in the form $f(x) = (1+|x|^2)^p h(x)$, where $h(x)$ is a summable function and $p>0$ is an integer. If the Fourier transform of the function $h(x)$ is equal to $\tilde{h}(s)$, then the Fourier transform of the function $f(x)$ has the form

$$\tilde{f}(s) = (2\pi)^n (1 - \Delta)^p \tilde{h}(s),$$

where differentiation is thought of in the sense of generalized functions.

If $f(x)$ is a periodic locally summable function with period $a = (a_1, \ldots, a_n)$, then it can be expanded in a Fourier series:

$$f(x) = \sum c_m e^{2\pi i \left(\frac{m}{a}, x \right)},$$

where $\left(\dfrac{m}{a}, x \right) = \displaystyle\sum_{k=1}^{n} \frac{m_k x_k}{a_k}$. The Fourier transform of $f(x)$ has the form

$$\tilde{f}(s) = \sum_m c_m \delta \left(s + \frac{2\pi m}{a} \right).$$

Thus the functional $\tilde{f}(s)$ is concentrated on a denumerable set of points of the form

$$2\pi \left(\frac{m_1}{a_1}, \ldots, \frac{m_n}{a_n} \right),$$

where $m_1, \ldots, m_n$ are integers.

3. *Fourier transformation of arbitrary generalized functions.* In order to define the Fourier transform of an arbitrary generalized function, we introduce still another space. We denote by Z the space consisting of entire analytic functions $\varphi(z)$ satisfying inequalities of the form

$$|z^m \varphi(z)| \leqslant C e^{a|y|}$$

for $a > 0$, $z = x + iy$ where the constants C and a depend on $\varphi(z)$ and m. A sequence of functions $\{\varphi_k(z)\}$ of the space Z is said to be *convergent to zero* if the functions $\varphi_k(z)$ are uniformly convergent to zero on every bounded set and if there exist constants C_m and a constant $a > 0$ such that

$$|z^m \varphi_k(z)| \leqslant C_m e^{a|y|}$$

for all values of k.

The space Z is dual to the space K with respect to the Fourier transformation: if $\varphi(x)$ is some function of the space K, then the function

$$\tilde{\varphi}(z) = \int \varphi(x) e^{i(x, z)} \, dx$$

belongs to the space Z and the mapping $\varphi(x) \to \tilde{\varphi}(z)$ is a one-to-one bicontinuous mapping of the space K onto the space Z.

Let $f(x)$ be an arbitrary generalized function. *Its Fourier transform is defined to be the continuous linear functional $\tilde{f}$ on the space Z such that* *the relation*

$$(\ , \tilde{\varphi}) = (2\pi)^n (f, \varphi)$$

is satisfied for an arbitrary function $\varphi(x)$ *of the space K.* The continuous linear functionals on the space Z are henceforth called *generalized functions* on this space. Thus *the Fourier transform of a generalized function on the space K is a generalized function on the space Z.*

EXAMPLES.

1. The generalized function $(2\pi)^n \delta(s-ib)$ on the space Z is the Fourier transformation of the regular generalized function e^{bx}. It is impossible to extend the functional $\delta(s-ib)$ from the space Z to the space S inasmuch as there are non-analytic functions in the space S which are defined only for real values of the argument.

2. The generalized function

$$(\tilde{f}, \tilde{\varphi}) = \int\limits_{-i\infty}^{i\infty} e^{\frac{s^2}{2}} \tilde{\varphi}(s) \, ds$$

is the Fourier transform of the regular generalized function of one variable

$$f(x) = -\frac{i}{\sqrt{2\pi}} e^{\frac{x^2}{2}}.$$

4. *Table of Fourier transforms of generalized functions of one variable.*
The values of the functions $A(\lambda)$, $B(\lambda)$, $C(\lambda)$, $D(\lambda)$ and the coefficients $a_{-1}^{(n)}$ and so on are given at the end of the table.

Entry no.	Generalized function $f(x)$	Its Fourier transform $\tilde{f}(s)$
0	Usual integrable function $f(x)$	$F[f] = \int\limits_{-\infty}^{\infty} f(x) e^{ixs} \, dx$
1	$\delta(x)$	1
2	$x_+^\lambda \ (\lambda \neq -1, -2, \ldots)$	$A(\lambda)(s+i0)^{-\lambda-1} = A(\lambda) s_+^{-\lambda-1} + B(\lambda) s_-^{-\lambda-1}$
3	x_+^n	$i^{n+1} n! \, s^{-n-1} + (-i)^n \pi \delta^{(n)}(s)$

Entry no.	Generalized function $f(x)$	Its Fourier transform $\tilde{f}(s)$
4	$\theta(x)$	$is^{-1} + \pi\delta(s)$
5	x_-^λ $(\lambda \neq -1, -2, \ldots)$	$B(\lambda)(s - i0)^{-\lambda-1} = A(\lambda)s_-^{-\lambda-1} + $ $+ B(\lambda)s_+^{-\lambda-1}$
6	x_-^n	$(-i)^{n+1}n!\,s^{-n-1} + i^n\pi\delta^{(n)}(s)$
7	$(x + i0)^\lambda$	$\dfrac{2\pi e^{i\lambda\frac{\pi}{2}}}{\Gamma(-\lambda)}s_-^{-\lambda-1}$
8	$(x - i0)^\lambda$	$\dfrac{2\pi e^{-i\lambda\frac{\pi}{2}}}{\Gamma(-\lambda)}s_+^{-\lambda-1}$
9	$\lvert x\rvert^\lambda\,(\lambda \neq -1, -3, \ldots)$	$C(\lambda)\lvert s\rvert^{-\lambda-1}$
10	$\lvert x\rvert^\lambda\,\mathrm{sign}\,x$ $(\lambda \neq -2, -4, \ldots)$	$iD(\lambda)\lvert s\rvert^{-\lambda-1}\,\mathrm{sign}\,s$
11	x^m	$2(-i)^m\,\pi\delta^{(m)}(s)$
12	x^{-m}	$i^m\,\dfrac{\pi}{(m-1)!}\,s^{m-1}\,\mathrm{sign}\,s$
13	x^{-1}	$i\pi\,\mathrm{sign}\,s$
14	x^{-2}	$-\pi\lvert s\rvert$
15	$\lvert x\rvert^{-2m-1}$	$c_0^{(2m+1)}s^{2m} - c_{-1}^{(2m+1)}s^{2m}\ln\lvert s\rvert$
16	$x^{-2m}\,\mathrm{sign}\,x$	$id_0^{(2m)}s^{2m-1} - id_{-1}^{(2m)}s^{2m-1}\ln\lvert s\rvert$
17	$x_+^\lambda\ln x_+$ $(\lambda \neq -1, -2, \ldots)$	$ie^{i\lambda\frac{\pi}{2}}\Big\{\Big[\Gamma'(\lambda+1) + $ $+ i\,\dfrac{\pi}{2}\Gamma(\lambda+1)\Big](s+i0)^{-\lambda-1} - $ $- \Gamma(\lambda+1)(s+i0)^{-\lambda-1}\ln(s+i0)\Big\}$
18	$x_-^\lambda\ln x_-$ $(\lambda \neq -1, -2, \ldots)$	$-ie^{-i\lambda\frac{\pi}{2}}\Big\{\Big[\Gamma'(\lambda+1) - $ $- i\,\dfrac{\pi}{2}\Gamma(\lambda+1)\Big](s-i0)^{-\lambda-1} - $ $- \Gamma(\lambda+1)(s-i0)^{-\lambda-1}\ln(s-i0)\Big\}$

Entry no.	Generalized function $f(x)$	Its Fourier transform $\tilde{f}(s)$
19	$\ln x_+$	$i\left\{\left(\Gamma'(1) + i\dfrac{\pi}{2}\right)[(s + i0)^{-1} - (s + i0)^{-1}\ln(s + i0)]\right\}$
20	$\ln x_-$	$-i\left\{\left(\Gamma'(1) - i\dfrac{\pi}{2}\right)[(s - i0)^{-1} - (s - i0)^{-1}\ln(s - i0)]\right\}$
21	$\|x\|^\lambda \ln\|x\|$ $(\lambda \neq -1, -2, \ldots)$	$ie^{i\lambda\frac{\pi}{2}}\left\{\left[\Gamma'(\lambda + 1) + i\dfrac{\pi}{2}\Gamma(\lambda + 1)\right](s + i0)^{-\lambda-1} - \Gamma(\lambda + 1)(s + i0)^{-\lambda-1}\ln(s + i0)\right\} - ie^{-i\lambda\frac{\pi}{2}}\left\{\left[\Gamma'(\lambda + 1) - i\dfrac{\pi}{2}\Gamma(\lambda + 1)\right](s - i0)^{-\lambda-1} - \Gamma(\lambda + 1)(s - i0)^{-\lambda-1}\ln(s - i0)\right\}$
22	$\|x\|^\lambda \ln\|x\| \operatorname{sign} x$ $(\lambda \neq -1, -2, \ldots)$	$ie^{i\lambda\frac{\pi}{2}}\left\{\left[\Gamma'(\lambda + 1) + i\dfrac{\pi}{2}\Gamma(\lambda + 1)\right](s + i0)^{-\lambda-1} - \Gamma(\lambda + 1)(s + i0)^{-\lambda-1}\ln(s + i0)\right\} + ie^{-i\lambda\frac{\pi}{2}}\left\{\left[\Gamma'(\lambda + 1) - i\dfrac{\pi}{2}\Gamma(\lambda + 1)\right](s - i0)^{-\lambda-1} - \Gamma(\lambda + 1)(s - i0)^{-\lambda-1}\ln(s - i0)\right\}$
23	$x^{-2m}\ln\|x\|$	$c_1^{(2m)}\|s\|^{2m-1} - c_0^{(2m)}\|s\|^{2m-1}\ln\|s\|$
24	$x^{-2m-1}\ln\|x\|$	$id_1^{(2m+1)}s^{2m}\operatorname{sign} s - id_0^{(2m+1)}s^{2m}\ln\|s\|\operatorname{sign} s$

Entry no.	Generalized function $f(x)$	Its Fourier transform $\tilde{f}(s)$
25	$|x|^{-2m-1}\ln|x|$	$c_1^{(2m+1)}s^{2m} - c_0^{(2m+1)}s^{2m}\ln|s| + \dfrac{1}{2}c_{-1}^{(2m+1)}s^{2m}\ln^2|s|$
26	$|x|^{-2m}\ln|x|\,\mathrm{sign}\,x$	$id_1^{(2m)}s^{2m-1} - id_0^{(2m)}s^{2m-1}\ln|s| + \dfrac{i}{2}d_{-1}^{(2m)}s^{2m-1}\ln^2|s|$
27	$(1-x^2)_+^{\lambda}$ $\quad(\lambda \neq -1, -2, \ldots)$	$\sqrt{\pi}\,\Gamma(\lambda+1)\left(\dfrac{s}{2}\right)^{-\lambda-\frac{1}{2}}J_{\lambda+\frac{1}{2}}(s)$
28	$\delta^{(n-1)}(1-x^2)$	$\sqrt{\pi}\left(\dfrac{s}{2}\right)^{n-\frac{1}{2}}J_{-n+\frac{1}{2}}(s)$
29	$(1+x^2)^{\lambda}$	$\dfrac{2\sqrt{\pi}}{\Gamma(-\lambda)}\left(\dfrac{|s|}{2}\right)^{-\lambda-\frac{1}{2}}K_{-\lambda-\frac{1}{2}}(|s|)$
30	$(x^2-1)_+^{\lambda}$ $\quad(\lambda \neq -1, -2, \ldots)$	$-\Gamma(\lambda+1)\sqrt{\pi}\left|\dfrac{s}{2}\right|^{-\lambda-\frac{1}{2}}N_{-\lambda-\frac{1}{2}}(|s|)$
31	$(x^2-1)_+^{n}$	$2\pi(-1)^n\left(1+\dfrac{d^2}{ds^2}\right)^n\delta(s) + (-1)^{n+1}\sqrt{\pi}\left(\dfrac{s}{2}\right)^{-n-\frac{1}{2}}J_{n+\frac{1}{2}}(s)$
32	e^{bx}	$2\pi\delta(s-ib)$
33	$e^{\frac{x^2}{2}}$	Functional $(\tilde{f}, \tilde{\varphi}) = -2\pi\displaystyle\int_{-i\infty}^{i\infty} e^{\frac{s^2}{2}}\tilde{\varphi}(s)\,ds$
34	$\delta^{(2m)}(x)$	$(-1)^m s^{2m}$
35	$\delta^{(2m+1)}(x)$	$(-1)^m is^{2m+1}$
36	$\sin bx$	$-i\pi[\delta(s+b)-\delta(s-b)]$
37	$\cos bx$	$\pi[\delta(s+b)+\delta(s-b)]$
38	Polynomial $P(x)$	$2\pi P\left(-i\dfrac{d}{ds}\right)\delta(s)$

Here and in the sequel, the following standard notations are used:

$$N_\lambda(z) = \frac{1}{\sin \lambda \pi} \left[\cos \lambda \pi J_\lambda(z) - J_{-\lambda}(z) \right],$$

$$K_\lambda(z) = \frac{\pi}{2 \sin \lambda \pi} \left[I_{-\lambda}(z) - I_\lambda(z) \right],$$

where

$$I_\lambda(z) = e^{\frac{-\pi \lambda i}{2}} J_\lambda(iz),$$

$$H_\lambda^{(1)}(z) = J_\lambda(z) + iN_\lambda(z),$$

$$H_\lambda^{(2)}(z) = J_\lambda(z) - iN_\lambda(z), \; \lambda \text{ is not an integer}.$$

The functions $A(\lambda)$, $B(\lambda)$, $C(\lambda)$, $D(\lambda)$ are given by the equalities:

$$A(\lambda) = ie^{i\lambda\frac{\pi}{2}} \Gamma(\lambda + 1) = \frac{a_{-1}^{(n)}}{\lambda + n} + a_0^{(n)} + a_1^{(n)}(\lambda + n) + \cdots,$$

$$B(\lambda) = -ie^{-i\lambda\frac{\pi}{2}} \Gamma(\lambda + 1) = \frac{b_{-1}^{(n)}}{\lambda + n} + b_0^{(n)} + b_1^{(n)}(\lambda + n) + \cdots,$$

$$C(\lambda) = -2 \sin \frac{\lambda\pi}{2} \Gamma(\lambda + 1) = \frac{c_{-1}^{(n)}}{\lambda + n} + c_0^{(n)} + c_1^{(n)}(\lambda + n) + \cdots,$$

$$D(\lambda) = 2 \cos \frac{\lambda\pi}{2} \Gamma(\lambda + 1) = \frac{d_{-1}^{(n)}}{\lambda + n} + d_0^{(n)} + d_1^{(n)}(\lambda + n) + \cdots$$

The coefficients $a_{-1}^{(n)}$, $a_0^{(n)}$ and $a_1^{(n)}$ have the form:

$$a_{-1}^{(n)} = \frac{i^{n-1}}{(n-1)!},$$

$$a_0^{(n)} = \frac{i^{n-1}}{(n-1)!} \left[1 + \frac{1}{2} + \cdots + \frac{1}{n-1} + \Gamma'(1) + i\frac{\pi}{2} \right],$$

$$a_1^{(n)} = \frac{i^{n-1}}{(n-1)!} \left\{ \sum_{j=1}^{n-1} \frac{1}{j^2} + \sum_{\substack{j \neq k \\ 1 \leqslant j, k \leqslant n-1}} \frac{1}{jk} - \frac{\pi^2}{8} + \right.$$

$$+ \left(1 + \frac{1}{2} + \cdots + \frac{1}{n-1} \right) \Gamma'(1) + \Gamma''(1) +$$

$$\left. + i\frac{\pi}{2} \left[1 + \frac{1}{2} + \cdots + \frac{1}{n-1} + \Gamma'(1) \right] \right\}.$$

The coefficients $b_{-1}^{(n)}$, and so on, are expressed in terms of the $a_k^{(n)}$ by the formulas:

$$b_k^{(n)} = \overline{a_k^{(n)}},$$
$$c_k^{(n)} = 2\operatorname{Re} a_k^{(n)},$$
$$d_k^{(n)} = 2\operatorname{Im} a_k^{(n)}.$$

In particular,

$$b_{-1}^{(n)} = \frac{(-i)^{n-1}}{(n-1)!}, \quad c_{-1}^{(n)} = \frac{2(-1)^{n-1}}{(n-1)!}\cos(n-1)\frac{\pi}{2},$$

$$d_{-1}^{(n)} = \frac{2(-1)^{n}}{(n-1)!}\sin(n-1)\frac{\pi}{2}.$$

(Table 5 see pp. 344–348)

Regarding notation concerning quadratic forms, see § 3; in particular, Q is the form associate to P.

The coefficients $c_{-1}^{(n+2m)}$, $c_0^{(n+2m)}$, and so on, are the coefficients in the Laurent expansion of the function

$$C_\lambda = 2^{\lambda+n}\pi^{\frac{n}{2}}\frac{\Gamma\left(\dfrac{\lambda+n}{2}\right)}{\Gamma\left(-\dfrac{\lambda}{2}\right)}$$

in a neighborhood of the point $\lambda = -n-2m$.

6. *Positive definite generalized functions.* The generalized function $f(x)$ is said to be *positive definite* if the inequality $(f, \varphi*\varphi^*) \geqslant 0$ is satisfied for an arbitrary function $\varphi(x)$ of the space K, where we set $\varphi^*(x) = \overline{\varphi(-x)}$.

In order for the generalized function $f(x)$ to be positive definite, it is necessary and sufficient that it be the Fourier transform of a positive measure μ of exponential growth (i.e. a measure μ such that the integral $\int (1+|x|^2)^{-p}\,d\mu(x)$ is convergent for some $p>0$).

An arbitary positive definite generalized function can be represented in the form $f(x) = (1-\varDelta)^m f_1(x)$, where $f_1(x)$ is a positive definite continuous function. The convolution of two positive definite generalized functions is positive definite.

5. Table of Fourier transforms of generalized functions of several variables

Entry no.	Generalized function $f(x)$	Its Fourier transform $f(\varrho)$
1	r^λ $(\lambda \neq -n, -n-2, \ldots)$	$2^{\lambda+n}\pi^{\frac{n}{2}}\dfrac{\Gamma\left(\dfrac{\lambda+n}{2}\right)}{\Gamma\left(-\dfrac{\lambda}{2}\right)}\varrho^{-\lambda-n}, \quad \varrho^2 = s_1^2 + \cdots + s_n^2$
2	$r^\lambda \ln r \quad (\lambda \neq -n, -n-2, \ldots)$	$C'_\lambda \varrho^{-\lambda-n} + C_\lambda \varrho^{-\lambda-n}\ln\varrho$
3	$r^\lambda \ln^2 r \quad (\lambda \neq -n, -n-2, \ldots)$	$C''_\lambda \varrho^{-\lambda-n} + 2C'_\lambda \varrho^{-\lambda-n}\ln\varrho + C_\lambda \varrho^{-\lambda-n}\ln^2\varrho$
4	$\delta^{(2m)}(r)$	$\dfrac{(2m)!}{\Omega_n}c_{-1}^{(n+2m)}\varrho^{2m}$
5	r^{-2m-n}	$\dfrac{1}{\Omega_n}\left[c_{-1}^{(n+2m)}\varrho^{2m}\ln\varrho + c_0^{(n+2m)}\varrho^{2m}\right]$
6	$r^{-2m-n}\ln r$	$\dfrac{1}{\Omega_n}\left[\tfrac{1}{2}c_{-1}^{(n+2m)}\varrho^{2m}\ln^2\varrho + c_0^{(n+2m)}\varrho^{2m}\ln\varrho + c_1^{(n+2m)}\varrho^{2m}\right]$
7	$\delta(r-a)$	$\Omega_{n-1}a^{\frac{n}{2}}\varrho^{-\frac{n}{2}+1}J_{\frac{n}{2}-1}(a\varrho)$
8	$\left(\dfrac{d}{a\,da}\right)^m \dfrac{\delta(r-a)}{a}$	$\Omega_{n-1}\sqrt{\dfrac{2}{\pi}}\cdot\dfrac{\sin a\varrho}{\varrho}, \quad n = 2m+3$

Entry no.	Generalized function $f(x)$	Its Fourier transform $\tilde{f}(Q)$
9	$(P + i0)^\lambda$	$\dfrac{e^{-\frac{\pi}{2}qi} 2^{n+2\lambda} \pi^{\frac{n}{2}} \Gamma\left(\lambda + \frac{n}{2}\right)}{\sqrt{|\Delta|}\, \Gamma(-\lambda)} (Q - i0)^{-\lambda - \frac{n}{2}}$
10	$(P - i0)^\lambda$	$\dfrac{e^{\frac{\pi}{2}qi} 2^{n+2\lambda} \pi^{\frac{n}{2}} \Gamma\left(\lambda + \frac{n}{2}\right)}{\sqrt{|\Delta|}\, \Gamma(-\lambda)} (Q + i0)^{-\lambda - \frac{n}{2}}$
11	P_+^λ	$\dfrac{2^{n+2\lambda} \pi^{\frac{n}{2}-1} \Gamma(\lambda+1) \Gamma\left(\lambda + \frac{n}{2}\right)}{\sqrt{|\Delta|}} \cdot \dfrac{1}{2i} \left[e^{-i\left(\frac{q}{2}+\lambda\right)\pi} (Q - i0)^{-\lambda - \frac{n}{2}} - e^{i\left(\frac{q}{2}+\lambda\right)\pi} (Q + i0)^{-\lambda - \frac{n}{2}} \right]$
12	P_-^λ	$-\dfrac{2^{n+2\lambda} \pi^{\frac{n}{2}-1} \Gamma(\lambda+1) \Gamma\left(\lambda + \frac{n}{2}\right)}{\sqrt{|\Delta|}} \cdot \dfrac{1}{2i} \left[e^{-\frac{\pi}{2}qi} (Q - i0)^{-\lambda - \frac{n}{2}} - e^{\frac{\pi}{2}qi} (Q + i0)^{-\lambda - \frac{n}{2}} \right]$

Entry no.	Generalized function $f(x)$	Its Fourier transform $\tilde{f}(\sigma)$
13	$(c^2 + P + i0)^\lambda$	$$\frac{2^{\lambda+1}(\sqrt{2\pi})^n c^{\frac{n}{2}+\lambda} K_{\frac{n}{2}+\lambda}[c(Q-i0)^{\frac{1}{2}}]}{\Gamma(-\lambda)\sqrt{|\Delta|}\quad (Q-i0)^{\frac{1}{2}(\frac{n}{2}+\lambda)}} =$$ $$= \frac{2^{\lambda+\frac{n}{2}+1}\pi^{\frac{n}{2}}e^{-\frac{q\pi i}{2}}c^{\lambda+\frac{n}{2}}}{\Gamma(-\lambda)\sqrt{|\Delta|}}\left[\frac{K_{\lambda+\frac{n}{2}}(cQ_+^{\frac{1}{2}})}{Q_+^{\frac{1}{2}(\lambda+\frac{n}{2})}} + \frac{\pi i}{2}\cdot\frac{H^{(1)}_{-\lambda-\frac{n}{2}}(cQ_-^{\frac{1}{2}})}{Q_-^{\frac{1}{2}(\lambda+\frac{n}{2})}}\right]$$
14	$(c^2 + P - i0)^\lambda$	$$\frac{2^{\lambda+1}(\sqrt{2\pi})^n c^{\frac{n}{2}+\lambda} K_{\frac{n}{2}+\lambda}[c(Q+i0)^{\frac{1}{2}}]}{\Gamma(-\lambda)\sqrt{|\Delta|}\quad (Q+i0)^{\frac{1}{2}(\frac{n}{2}+\lambda)}} =$$ $$= \frac{2^{\lambda+\frac{n}{2}+1}\pi^{\frac{n}{2}}e^{\frac{q\pi i}{2}}c^{\lambda+\frac{n}{2}}}{\Gamma(-\lambda)\sqrt{|\Delta|}}\left[\frac{K_{\lambda+\frac{n}{2}}(cQ_+^{\frac{1}{2}})}{Q_+^{\frac{1}{2}(\lambda+\frac{n}{2})}} - \frac{\pi i}{2}\cdot\frac{H^{(2)}_{-\lambda-\frac{n}{2}}(cQ_-^{\frac{1}{2}})}{Q_-^{\frac{1}{2}(\lambda+\frac{n}{2})}}\right]$$

Entry no.	Generalized function $f(x)$	Its Fourier transform $\tilde{f}(\omega)$
15	$\dfrac{(c^2+P)_+^\lambda}{\Gamma(\lambda+1)}$	$2^{\lambda+\frac{n}{2}} i\pi^{\frac{n}{2}-1} c^{\frac{n}{2}+\lambda} \dfrac{1}{\sqrt{|\Delta|}} \left\{ e^{-i(\lambda+\frac{q}{2})\pi} \dfrac{K_{\frac{n}{2}+\lambda}\left[c(Q-i0)^{\frac{1}{2}}\right]}{(Q-i0)^{\frac{1}{2}(\lambda+\frac{n}{2})}} - e^{i(\lambda+\frac{q}{2})\pi} \dfrac{K_{\frac{n}{2}+\lambda}\left[c(Q+i0)^{\frac{1}{2}}\right]}{(Q+i0)^{\frac{1}{2}(\lambda+\frac{n}{2})}} \right\}$
16	$\dfrac{(c^2+P)_-^\lambda}{\Gamma(\lambda+1)}$	$2^{\lambda+\frac{n}{2}} i\pi^{\frac{n}{2}-1} c^{\frac{n}{2}+\lambda} \dfrac{1}{\sqrt{|\Delta|}} \left\{ e^{-\frac{q\pi i}{2}} \dfrac{K_{\frac{n}{2}+\lambda}\left[c(Q-i0)^{\frac{1}{2}}\right]}{(Q-i0)^{\frac{1}{2}(\lambda+\frac{n}{2})}} - e^{\frac{q\pi i}{2}} \dfrac{K_{\frac{n}{2}+\lambda}\left[c(Q+i0)^{\frac{1}{2}}\right]}{(Q+i0)^{\frac{1}{2}(\lambda+\frac{n}{2})}} \right\}$
17	$\delta^{(s-1)}(c^2+P)$	$(-1)^{s+1} \dfrac{i}{\sqrt{|\Delta|}} 2^{\frac{n-s}{2}} \pi^{\frac{n-1}{2}} c^{\frac{n-s}{2}} \times$ $\times \left[e^{-\frac{\pi i q}{2}} \dfrac{K_{\frac{n}{2}-s}\left[c(Q-i0)^{\frac{1}{2}}\right]}{(Q-i0)^{\frac{1}{2}(\frac{n}{2}-s)}} - e^{\frac{\pi i q}{2}} \dfrac{K_{\frac{n}{2}-s}\left[c(Q+i0)^{\frac{1}{2}}\right]}{(Q+i0)^{\frac{1}{2}(\frac{n}{2}-s)}} \right]$

Entry no.	Generalized function $f(x)$	Its Fourier transform $\tilde{f}(\varrho)$
18	$\dfrac{(c^2 + P)^s}{\Gamma(s+1)}$	$(2\pi)^n \displaystyle\sum_{m=0}^{n} \dfrac{(-1)^m \left(\dfrac{c}{2}\right)^{2s-2m}}{4^m m!\,(s-m)!}\, L^m \delta(\varrho)\,.$
19	Polynomial $P(x_1, \ldots, x_n)$	$(2\pi)^n P\left(-i\dfrac{\partial}{\partial s_1}, \ldots, -i\dfrac{\partial}{\partial s_n}\right)$

§ 5. Radon transformation

1. *Radon transformation of test functions and its properties.* The Fourier transformation of functions of several variables decomposes into two transformations: to the integration of the function to be transformed over planes in n-dimensional space and to a one-dimensional Fourier transformation.

Namely, if

$$\tilde{f}(\xi) = \int f(x)\, e^{i(\xi, x)}\, dx,$$

where $(\xi, x) = \xi_1 x_1 + \cdots + \xi_n x_n$, then

$$\tilde{f}(\xi) = \int_{-\infty}^{\infty} \check{f}(\xi, p)\, e^{ip}\, dp,$$

where we set

$$\check{f}(\xi, p) = \int f(x)\, \delta(p - (\xi, x))\, dx$$

(concerning the meaning of the symbol $\delta(p - (\xi, x))$, see § 3, no. 5).

We call the function $\check{f}(\xi, p)$ the *Radon transform* of the function $f(x)$. Basic properties of the Radon transform are expressed by the following formulas:

1) $\check{f}(\alpha\xi, \alpha p) = |\alpha|^{-1}\check{f}(\xi, p)$ for arbitrary $\alpha \neq 0$;

2) $\{f(x+a)\}^v = \check{f}(\xi, p + (\xi, a))$;

3) $\{f(A^{-1}x)\}^v = |\det A|\,\check{f}(A'\xi, p)$, where A is a non-degenerate linear transformation and A' is the transformation conjugate to it;

4) $\left\{\left(a, \dfrac{\partial}{\partial x}\right) f(x)\right\}^v = (a, \xi)\, \dfrac{\partial \check{f}(\xi, p)}{\partial p}$;

5) $\dfrac{\partial}{\partial p}\{(a, x)\, f(x)\}^v = -\left(a, \dfrac{\partial}{\partial \xi}\right)\check{f}(\xi, p)$;

6) $(f_1 * f_2)^v = \displaystyle\int_{-\infty}^{\infty} \check{f}_1(\xi, t)\check{f}_2(\xi, p - t)\, dt$.

The function $f(x)$ is expressed in terms of $\check{f}(\xi, p)$ by the formula

$$f(x) = \frac{(-1)^{\frac{n-1}{2}}}{2(2\pi)^{n-1}} \int_\Gamma \check{f}_p^{(n-1)}(\xi, (\xi, x)) \, d\omega,$$

if n is odd, and by the formula

$$f(x) = \frac{(-1)^{\frac{n}{2}} (n-1)!}{(2\pi)^n} \int_\Gamma \left(\int_{-\infty}^{\infty} \check{f}(\xi, p) (p - (\xi, x))^{-n} \, dp \right) d\omega,$$

if n is even. Here we denote by Γ an arbitrary surface enclosing the origin and by $d\omega$ the differential form on this surface defined by the equality

$$d\omega = \sum (-1)^{k-1} \xi_k \, d\xi_1 \dots d\xi_{k-1} \, d\xi_{k+1} \dots d\xi_n.$$

The integral with respect to p must be understood in the sense of a regularized value (see § 2, no. 2).

The analog of the Plancherel formula for the Radon transform has the following form:

$$\int f(x) \overline{g(x)} \, dx = \frac{1}{2(2\pi)^{n-1}} \int_\Gamma \left(\int_{-\infty}^{\infty} \check{f}_p^{\left(\frac{n-1}{2}\right)}(\xi, p) \overline{\check{g}_p^{\left(\frac{n-1}{2}\right)}(\xi, p)} \, dp \right) d\omega$$

for a space of odd dimension and

$$\int f(x) \overline{g(x)} \, dx = \frac{(-1)^{\frac{n}{2}} (n-1)!}{(2\pi)^n} \int_\Gamma \left(\int_{-\infty}^{\infty} \int_{-\infty}^{\infty} \check{f}(\xi, p_1) \overline{\check{g}(\xi, p_2)} \times \right.$$

$$\left. \times (p_1 - p_2)^{-n} \, dp_1 \, dp_2 \right) d\omega$$

for a space of even dimension (the integral with respect to p_1, p_2 also must be understood in the sense of a regularized value).

2. *Radon transformation of generalized functions.* Let S be the space of infinitely differentiable functions rapidly decreasing together with all derivatives, and let the number of variables n be odd. We denote by $\check{S}$ the space of functions $\psi(\xi, p)$ of the form $\psi(\xi, p) = \check{f}_p^{(n-1)}(\xi, p)$, where $f(x) \in S$. This space consists of functions $\psi(\xi, p)$ such that:

1) $\psi(\alpha\xi, \alpha p) = |\alpha|^{-n} \psi(\xi, p)$ for arbitrary $\alpha \neq 0$;

2) the functions $\psi(\xi, p)$ are infinitely differentiable with respect to ξ and with respect to p for $\xi \neq 0$;

3) the estimate $|\psi(\xi, p)| = 0(p^{-k})$ holds for arbitrary $k > 0$ as $|p| \to \infty$ and is uniform with respect to ξ when ξ runs through an arbitrary bounded closed domain not containing the point $\xi = 0$; this same estimate holds for derivatives of the function ψ;

4) for an arbitrary integer $k \geqslant 0$, the integral

$$\int_{-\infty}^{\infty} \psi(\xi, p) \, p^k \, dp$$

is a homogeneous polynomial in ξ of degree $k - n + 1$ (for $k < n - 1$ the integral is equal to zero).

To every generalized function (F, f) on the space S there is assigned a functional $\breve{F}$ on $\breve{S}$ such that

$$(F, f) = \frac{(-1)^{\frac{n-1}{2}}}{2(2\pi)^{n-1}} (\breve{F}, \psi).$$

This functional can be extended to the space of all functions $\psi(\xi, p)$ satisfying conditions 1)–3); however, in this connection, it will not be uniquely defined, but only to within linear combinations of functions of the form $p^k a_{-k-1}(\xi)$, where for $k < n - 1$, $a_{-k-1}(\xi)$ is an arbitrary function satisfying the condition of homogeneity

$$a_{-k-1}(\alpha\xi) = \alpha^{-k} |\alpha|^{-1} a_{-k-1}(\xi);$$

for $k \geqslant n - 1$, besides the condition of homogeneity, the condition

$$\int_{\Gamma} a_{-k-1}(\xi) \, P_{k-n+1}(\xi) \, d\omega = 0$$

must be satisfied for an arbitrary homogeneous polynomial $P_{k-n+1}(\xi)$ of degree $k - n + 1$.

The Radon transforms of characteristic functions of non-bounded domains are of basic interest. It turns out that *for bounded domains, the Radon transform $S(\xi, p)$ of the characteristic function of the domain V gives the area of the cross section of the domain by the plane $(\xi, x) = p$.* Therefore the Radon transform of the characteristic function of a non-bounded domain can be considered as a regularized value of the area of cross section of the domain.

3. *Table of Radon transforms of some generalized functions in an odd-dimensional space.* In the table we set

$$\theta(t) = \begin{cases} 1, \text{ when } t > 0, \\ 0, \text{ when } t < 0, \end{cases}$$

$P(x)$ is a non-degenerate quadratic form, and $Q(x)$ is the quadratic form conjugate to it.

(Table see pp. 353–357)

§ 6. Generalized functions and differential equations

1. *Fundamental solutions.* Let $P\left(\dfrac{\partial}{\partial x}\right) = P\left(\dfrac{\partial}{\partial x_1}, ..., \dfrac{\partial}{\partial x_n}\right)$ be a linear differential operator with constant coefficients. A generalized function $E(x)$ satisfying the equation $P\left(\dfrac{\partial}{\partial x}\right) E(x) = \delta(x)$ is called a *fundamental solution* corresponding to this operator. This function is defined to within a solution of the homogeneous equation $P\left(\dfrac{\partial}{\partial x}\right) u(x) = 0$. *If $\mu(x)$ is a generalized function such that the convolution $E * \mu(x)$ makes sense, then this convolution is a solution of the differential equation $P\left(\dfrac{\partial}{\partial x}\right) u(x) = \mu(x)$.*

For example, the fundamental solution $-\dfrac{r^{2-n}}{(n-2)\,\Omega_n}$, where $r = (x_1^2 + \cdots + x_2^2)$ and Ω_n is the area of the surface of the unit ball, corresponds to the Laplace operator $\Delta = \dfrac{\partial^2}{\partial x_1^2} + \cdots + \dfrac{\partial^2}{\partial x_n^2}$ for $n > 2$, and the

solution $-\dfrac{1}{2\pi} \ln \dfrac{1}{r}$ corresponds for $n = 2$. Therefore for $n > 2$, the function

$$u(x_1, ..., x_n) = -\frac{1}{(n-2)\,\Omega_n} \int \frac{\mu(\xi_1, ..., \xi_n)\, d\xi_1 ... d\xi_n}{[(x_1 - \xi_1)^2 + \cdots + (x_n - \xi_n)^2]^{\frac{n-2}{2}}}$$

serves as a solution of the Poisson equation $\Delta u = \mu$ for the condition that the masses $\mu(x)$ are concentrated in a bounded domain.

Now let the equation contain time t. Let $P\left(\dfrac{\partial}{\partial x}, \dfrac{\partial}{\partial t}\right)$ be a linear

Table of Radon Transform

Function	Radon transform
1	$p^{n-1} a(\xi)$, where $a(\xi)$ is an arbitrary even function which is homogeneous of degree $-n$ and such that $$\int_\Gamma a(\xi)\, d\omega = (-1)^{\frac{n-1}{2}} 2^n \pi^{n-1} \Gamma^{-1}(n)$$
$\theta(x_1)$	$$\tfrac{1}{2}\pi^{n-\frac{3}{2}}\Gamma\left(1-\frac{n}{2}\right)\Gamma^{-1}\left(\frac{n+1}{2}\right)\left[p_+^{n-1}(\xi_1)_+^{-1} + p_-^{n-1}(\xi_1)_-^{-1}\right]\delta(\xi_2,\ldots,\xi_n)$$
$\delta(x_1,\ldots,x_n)$	$\delta(p)$
$\delta(x_1,\ldots,x_k)$, $k-$odd $(k<n)$	$$\pi^{n-k-\frac{1}{2}}\Gamma\left(\frac{k-n+1}{2}\right)\Gamma^{-1}\left(\frac{n-k}{2}\right)|p|^{n-k-1}\delta(\xi_{k+1},\ldots,\xi_n)$$
$\delta(x_1,\ldots,x_k)$ $k-$even $(k<n)$	$$(-1)^{\frac{n-k+1}{2}} 2\pi^{n-k-\frac{1}{2}}\Gamma^{-1}\left(\frac{n-k+1}{2}\right)\Gamma^{-1}\left(\frac{n-k}{2}\right)\times$$ $$\times\, p^{n-k-1}\ln|p|\,\delta(\xi_{k+1},\ldots,\xi_n)$$
$a(x_1)\,\delta(x_2,\ldots,x_n)$,	$|\xi_1|^{-1} a\left(\dfrac{p}{\xi_1}\right)$

Function	Radon transform
where $\displaystyle\int_{-\infty}^{\infty} |a(x_1)| \, dx_1 < \infty$	
$(x_1)_+^\lambda \, \delta(x_2,\ldots,x_n)$, λ – not an integer	$p_+^\lambda (\xi_1)_+^{-1-\lambda} + p_-^\lambda (\xi_1)_-^{-1-\lambda}$
$(x_1)_+^k \, \delta(x_2,\ldots,x_n)$, $k = 0, 1, \ldots$	$p_+^k (\xi_1)_+^{-1-k} + p_-^k (\xi_1)_-^{-1-k} - \dfrac{(-1)^k}{k!} p^k \ln|p| \, \delta^{(k)}(\xi_1)$
$|x_1|^\lambda \, \delta(x_2,\ldots,x_n)$, $\lambda \neq -1, -2, \ldots,$ $\lambda \neq 2k \, (k = 0, 1, \ldots)$	$|p|^\lambda |\xi_1|^{-1-\lambda}$
$x_1^{2k} \, \delta(x_2,\ldots,x_n)$, $k = 0, 1, \ldots$	$p^{2k} |\xi_1|^{-1-2k} - \dfrac{2}{(2k)!} p^{2k} \ln|p| \, \delta^{(2k)}(\xi_1)$
$|x_1|^\lambda \, \text{sign} \, x_1 \, \delta(x_2,\ldots,x_n)$, $\lambda \neq -1, -2, \ldots,$ $\lambda \neq 2k - 1 \, (k = 1, 2, \ldots)$	$|p|^\lambda \, \text{sign} \, p \, |\xi_1|^{-1-\lambda} \, \text{sign} \, \xi_1$

Function	Radon transform
$x_1^{2k-1}\delta(x_2,\ldots,x_n)$, $k=1,2,\ldots$	$p^{2k-1}|\xi_1|^{-2k}\operatorname{sign}\xi_1 + \dfrac{2}{(2k-1)!}\,p^{2k-1}\ln|p|\,\delta^{(2k-1)}(\xi_1)$
The characteristic function of the upper half of the cone in 3-dimensional space $x_1^2 - x_2^2 - x_3^2 > 0, x_1 > 0$	$\pi[p_+^2\theta(\xi_1) + p_-^2\theta(-\xi_1)](\xi_1^2 - \xi_2^2 - \xi_3^2)_+^{-\frac{3}{2}} + p^2\ln|p|\,(\xi_1^2 - \xi_2^2 - \xi_3^2)_-^{-\frac{3}{2}}$
The characteristic function of the upper half of the hyperboloid in 3-dimensional space $x_1^2 - x_2^2 - x_3^2 > 1, x_1 > 0$	$\pi\theta(\xi,p)(\xi_1^2 - \xi_2^2 - \xi_3^2)_+^{-\frac{3}{2}}(p^2 - \xi_1^2 - \xi_2^2 - \xi_3^2)_+ +$ $+\tfrac{1}{2}(\xi_1^2 - \xi_2^2 - \xi_3^2)_-^{-\frac{3}{2}}(p^2 - \xi_1^2 + \xi_2^2 + \xi_3^2)\ln|p^2 - \xi_1^2 + \xi_2^2 + \xi_3^2|$
$P_+^\lambda(x)$	$\dfrac{(-1)^{\frac{n-1}{2}}\,\pi^{\frac{n-1}{2}}\,|p|^{2\lambda+n-1}}{(\lambda+1)\ldots\left(\lambda+\dfrac{n-1}{2}\right)\sqrt{|\varDelta|}\,\sin\pi\lambda} \times$ $\times\left[\sin\pi\left(\dfrac{l}{2}+\lambda\right)Q_+^{-\lambda-\frac{n}{2}}(\xi) - \sin\dfrac{\pi k}{2}\,Q_-^{-\lambda-\frac{n}{2}}(\xi)\right],$

where k, l correspond to the number of positive and negative squares in the canonical representation of the form $P(x)$; $\varDelta$ is the determinant of the form $P(x)$

Function	Radon transform
$\theta(x_1)(x_1^2 - x_2^2 - \cdots - x_n^2)_+^{\lambda}$	$\pi^{\frac{n-1}{2}}\dfrac{\left[p_+^{2\lambda+n-1}\theta(\xi_1) + p_-^{2\lambda+n-1}\theta(-\xi_1)\right](\xi_1^2 - \xi_2^2 - \cdots - \xi_n^2)_+^{-\lambda-\frac{n}{2}}}{(\lambda+1)\ldots\left(\lambda+\dfrac{n-1}{2}\right)} -$

$$- \frac{(-1)^{\frac{n-1}{2}}\pi^{\frac{n-1}{2}}|p|^{2\lambda+n-1}(\xi_1^2 - \xi_2^2 - \cdots - \xi_n^2)_-^{-\lambda-\frac{n}{2}}}{2(\lambda+1)\ldots\left(\lambda+\dfrac{n-1}{2}\right)\sin\pi\lambda}$$

Function	Radon transform
$[P(x) + c]_+^{\lambda}$	$\dfrac{(-1)^{\frac{n-1}{2}}\pi^{\frac{n-1}{2}}}{(\lambda+1)\ldots\left(\lambda+\dfrac{n-1}{2}\right)\sqrt{|\Delta|}\,\sin\pi\lambda} \times$

$$\times \left[\sin\pi\left(\frac{l}{2}+\lambda\right)Q_+^{-\lambda-\frac{n}{2}}(p^2+cQ)_+^{\lambda+\frac{n-1}{2}} - \right.$$

$$- \sin\frac{\pi k}{2}Q_-^{-\lambda-\frac{n}{2}}(p^2+cQ)_+^{\lambda+\frac{n-1}{2}} +$$

$$+ \sin\pi\left(\frac{l-1}{2}+\lambda\right)Q_-^{-\lambda-\frac{n}{2}}(p^2+cQ)_-^{\lambda+\frac{n-1}{2}} -$$

$$\left. - \sin\frac{\pi(k-1)}{2}Q_+^{-\lambda-\frac{n}{2}}(p^2+cQ)_-^{\lambda+\frac{n-1}{2}}\right]$$

Function	Radon transform
$\theta(x_1)(x_1^2 - x_2^2 - x_3^2 - 1)_+^\lambda$	$\dfrac{\pi}{\lambda+1}\theta(p\xi_1)(\xi_1^2 - \xi_2^2 - \xi_3^2)_+^{-\lambda-\frac{3}{2}}(p^2 - \xi_1^2 + \xi_2^2 + \xi_3^2)_+^{\lambda+1} +$ $+ \dfrac{\pi}{2(\lambda+1)\sin\pi\lambda}(\xi_1^2 - \xi_2^2 - \xi_3^2)_-^{-\lambda-\frac{3}{2}}(p^2 - \xi_1^2 + \xi_2^2 + \xi_3^2)_+^{\lambda+1}$
$\delta(x_1^2 - x_2^2 - \cdots - x_n^2)$	$(-1)^{\frac{n-1}{2}}\dfrac{n-3}{2\pi^{\frac{n-1}{2}}}\Gamma^{-1}\left(\dfrac{n-1}{2}\right)p^{n-3}\ln|p|\,(\xi_1^2 - \xi_2^2 - \cdots - \xi_n^2)_-^{-\frac{n}{2}+1}$
$\delta(x_1^2 + \cdots + x_n^2 - 1)$	$(-1)^{\frac{n-1}{2}}\pi^{\frac{n-1}{2}}\Gamma^{-1}\left(\dfrac{n-1}{2}\right)(\xi_1^2 + \cdots + \xi_n^2)_+^{-\frac{n}{2}+1}(p^2 - \xi_1^2 - \cdots - \xi_n^2)_+^{\frac{n-3}{2}}$
$\delta(x_1^2 - x_2^2 - \cdots - x_n^2 - 1)$	$\pi^{\frac{n-1}{2}}\Gamma^{-1}\left(\dfrac{n-1}{2}\right)(\xi_1^2 - \xi_2^2 - \cdots - \xi_n^2)_+^{-\frac{n}{2}+1} \times$ $\times (p^2 - \xi_1^2 + \xi_2^2 + \cdots + \xi_n^2)_+^{\frac{n-3}{2}} + (-1)^{\frac{n-1}{2}}\pi^{\frac{n-3}{2}}\Gamma^{-1}\left(\dfrac{n-1}{2}\right) \times$ $\times (\xi_1^2 - \xi_2^2 - \cdots - \xi_n^2)_-^{-\frac{n}{2}+1}(p^2 - \xi_1^2 + \xi_2^2 + \cdots + \xi_n^2)_-^{\frac{n-3}{2}} \times$ $\times \ln\left|\dfrac{p^2 - \xi_1^2 + \xi_2^2 + \cdots + \xi_n^2}{\xi_1^2 - \xi_2^2 - \cdots - \xi_n^2}\right|$

differential operator with constant coefficients, having order m with respect to t. We call the generalized function $E(x, t)$ such that $P\left(\dfrac{\partial}{\partial x}, \dfrac{\partial}{\partial t}\right) E(x, t) = 0$ and the initial conditions

$$E(x, 0) = 0, \ldots, \frac{\partial^{m-2} E(x, 0)}{\partial t^{m-2}} = 0,$$

$$\frac{\partial^{m-1} E(x, 0)}{\partial t^{m-1}} = \delta(x)$$

are satisfied the *fundamental solution of the Cauchy problem* corresponding to this operator.

The solution of the Cauchy problem for the equation $P\left(\dfrac{\partial}{\partial x}, \dfrac{\partial}{\partial t}\right) u(x, t) = 0$

with the initial conditions

$$u(x, 0) = 0, \ldots, \frac{\partial^{m-2} u(x, 0)}{\partial t^{m-2}} = 0,$$

$$\frac{\partial^{m-1} u(x, 0)}{\partial t^{m-1}} = u_{m-1}(x)$$

has the form

$$u(x, t) = E(x, t) * u_{m-1}(x)$$

(under the condition that this convolution makes sense).

The solution of the Cauchy problem with initial conditions

$$u(x, 0) = 0, \ldots, \frac{\partial^{m-2} u(x, 0)}{\partial t^{m-2}} = u_{m-2}(x), \frac{\partial^{m-1} u(x, 0)}{\partial t^{m-1}} = 0$$

can also be expressed in terms of the function $E(x, t)$. For this, we set $v(x, t) = E(x, t) * u_{m-2}(x)$ and denote $\hat{u}_{m-1}(x) = \dfrac{\partial^{m-1} v(x, 0)}{\partial t^{m-1}}$. Then the solution of the indicated Cauchy problem has the form

$$\frac{\partial v(x, t)}{\partial t} - E(x, t) * \hat{u}_{m-1}(x).$$

The solution of the Cauchy problem, if the derivative of $(m-k)$-th order is different from zero, is obtained analogously. The solution of the Cauchy problem in the general case is the sum of such solutions.

For the one-dimensional heat conduction equation $\dfrac{\partial u}{\partial t} = \dfrac{\partial^2 u}{\partial x^2}$, the fundamental solution of the Cauchy problem has the form

$$E(x, t) = \frac{1}{2\sqrt{\pi t}} e^{-\frac{x^2}{4t}}, \ t > 0.$$

Therefore the solution of the Cauchy problem for the initial condition $u(x, 0) = u_0(x)$ is given by the formula

$$u(x, t) = E(x, t) * u_0(x) = \frac{1}{2\sqrt{\pi t}} \int_{-\infty}^{\infty} e^{-\frac{\xi^2}{4t}} u_0(x - \xi) \, d\xi.$$

For the wave equation $\dfrac{\partial^2 u}{\partial t^2} = \dfrac{\partial^2 u}{\partial x^2}$, the fundamental solution of the Cauchy problem has the form

$$E(x, t) = \begin{cases} \frac{1}{2} & \text{if} \ \ |x| < t, \\ 0 & \text{if} \ \ |x| > t. \end{cases}$$

Making use of this, it is easy to obtain the solution of the Cauchy problem for this equation in d'Alembert form.

We note the following statement:

If $E(x, t)$ is the fundamental solution of the Cauchy problem for the differential equation

$$\frac{\partial u(x, t)}{\partial t} - P\left(i \frac{\partial}{\partial x} \right) u(x, t) = 0,$$

then the function

$$E_0(x, t) = \begin{cases} 0 & \text{if} \ \ t < 0, \\ E(x, t) & \text{if} \ \ t \geqslant 0 \end{cases}$$

is a fundamental solution corresponding to the operator $\dfrac{\partial}{\partial t} - P\left(i \dfrac{\partial}{\partial x} \right)$, *that is, satisfies the equation*

$$\frac{\partial E_0(x, t)}{\partial t} - P\left(i \frac{\partial}{\partial x} \right) E_0(x, t) = \delta(x, t).$$

The definition of the function $E_0(x, t)$ gives

$$(E_0(x, t), \varphi(x, t)) = \int_0^\infty (E(x, t), \varphi(x, t)) \, dt$$

for an arbitrary function $\varphi(x, t)$ of the space K.

On the left hand side of the equality, $E_0(x, t)$ is applied to the function of two variables $\varphi(x, t)$; on the right hand side, under the integral sign, $E(x, t)$ is applied, for every fixed $t > 0$, to the function of one variable $\varphi(x, t)$.

2. *Fundamental solutions for some differential equations.* For the iterated Laplace operator Δ^m, a fundamental solution has the form

$$E(x) = \begin{cases} \dfrac{(-1)^m (2\pi)^n c_{-1}^{(2m)}}{\Omega_n} r^{2m-n} \ln r, & \text{if } 2m > n \quad \text{and} \quad n \text{ is even}, \\ (-1)^m (2\pi)^n C_{-2m} r^{2m-n} & \text{in remaining cases}, \end{cases}$$

For the wave equation

$$\frac{\partial^2 u}{\partial t^2} = \frac{\partial^2 u}{\partial x_1^2} + \cdots + \frac{\partial^2 u}{\partial x_n^2}$$

in an odd-dimensional space, $n = 2m+3$, $m = 0, 1, 2, \ldots$, the fundamental solution of the Cauchy problem is given by the formula

$$E(x, t) = \sqrt{\frac{\pi}{2}} \frac{1}{\Omega_{n-1}} \left(\frac{d}{t \, dt}\right)^m \frac{\delta(r - t)}{t}$$

(Ω_{n-1} is the area of the surface of the $(n-1)$-dimensional unit ball). Therefore, for the initial conditions

$$u(x, 0) = 0, \quad \frac{\partial u(x, 0)}{\partial t} = f(x)$$

the solution of the Cauchy problem has the form

$$u(x, t) = \sqrt{\frac{\pi}{2}} \frac{1}{\Omega_{n-1}} \left(\frac{d}{t \, dt}\right)^m \frac{\delta(r - t)}{t} * f(x) =$$

$$= \sqrt{\frac{\pi}{2}} \frac{\Omega_n}{\Omega_{n-1}} \left(\frac{d}{t \, dt}\right)^m t^{n-2} M_t(f),$$

where $M_t(f)$ denotes the mean of the function $f(x-\xi)$ on the sphere $|\xi|=t$.

The fundamental solution of the Cauchy problem for the heat conduction equation in n-dimensional space is given by the formula

$$
E(x, t) = \begin{cases} \left(\dfrac{1}{2\sqrt{\pi t}}\right)^n e^{-\frac{|x|^2}{4t}} & \text{if } t > 0, \\ 0 & \text{if } t \leqslant 0. \end{cases}
$$

The fundamental solutions for a differential operator of the form L^k, where

$$
L = \sum_{\alpha, \beta = 1}^{n} g^{\alpha\beta} \frac{\partial^2}{\partial x_\alpha \partial x_\beta},
$$

have the following form:

1) If n is an odd number or if n is an even number and $k < \dfrac{n}{2}$, then

$$
E_1(x) = (-1)^k \frac{e^{\frac{\pi}{2}qt}\, \Gamma\left(\dfrac{n}{2} - k\right)\sqrt{|\Delta|}}{4^k(k-1)!\,\pi^{\frac{n}{2}}} (P + i0)^{-\frac{n}{2}+k},
$$

where

$$
P = \sum_{\alpha, \beta = 1}^{n} g_{\alpha\beta} x_\alpha x_\beta, \qquad \sum_{\beta = 1}^{n} g_{\alpha\beta} g^{\beta\gamma} = \delta_\alpha^\gamma
$$

(for the notation connected with quadratic forms P, see § 3, no. 2). A second fundamental solution has the form

$$
E_2(x) = \overline{E_1(x)}.
$$

2) If n is an even number and $k \geqslant \dfrac{n}{2}$, then

$$
E_1(x) = (-1)^{\frac{n}{2}-1} \frac{e^{\frac{\pi}{2}qt}\sqrt{|\Delta|}}{4^k\left(k - \dfrac{n}{2}\right)!\,(k-1)!} (P + i0)^{-\frac{n}{2}+k} \ln(P + i0),
$$

$$
E_2(x) = \overline{E_1(x)}.
$$

3. *Construction of fundamental solutions for elliptic equations.* Let $P\left(\dfrac{\partial}{\partial x}\right)$ be a differential operator of order $2m$ with constant coefficients,

and let P_0 be the principal part of this operator, containing only derivatives of order $2m$. The operator P is called *elliptic* if, by the substitution in P_0 of the variables $\omega_1, ..., \omega_n$ for the symbols $\dfrac{\partial}{\partial x_1}, ..., \dfrac{\partial}{\partial x_n}$, the polynomial $P_0(\omega_1, ..., \omega_n)$ which is obtained is non-zero if $\omega \neq 0$ $(\omega = (\omega_1, ..., \omega_n))$.

In order to obtain a fundamental solution $E(x_1, x_2, ..., x_n)$ corresponding to the elliptic operator P, we replace the equation

$$P\left(\frac{\partial}{\partial x}\right) E(x) = \delta(x)$$

by the equation

$$P\left(\frac{\partial}{\partial x}\right) u = \frac{2r^\lambda}{\Omega_n \Gamma\left(\dfrac{\lambda + n}{2}\right)}, \quad r = (x_1^2 + \cdots + x_n^2)^{\frac{1}{2}}.$$

After this we expand the generalized function r^λ in plane waves (see § 3, no. 1) and solve the equation

$$P\left(\frac{\partial}{\partial x}\right) v = \frac{|\omega_1 x_1 + \cdots + \omega_n x_n|^\lambda}{\Omega_n \pi^{\frac{n-1}{2}} \Gamma\left(\dfrac{\lambda + 1}{2}\right)}.$$

Integrating these solutions with respect to ω and setting $\lambda = -n$, we obtain the desired fundamental solution (since the generalized function $2r^\lambda / \Omega_n \Gamma\left(\dfrac{\lambda + n}{2}\right)$ is equal to $\delta(x)$ for $\lambda = -n$). Thus, in this way, we obtain

$$E(x_1, ..., x_n) = \int_\Omega v_\omega(\omega_1 x_1 + \cdots + \omega_n x_n, -n) \, d\Omega,$$

where Ω is the unit sphere,

$$v_\omega(\xi, \lambda) = \frac{1}{\Omega_n \pi^{\frac{n-1}{2}} \Gamma\left(\dfrac{\lambda + 1}{2}\right)} \int_{-\infty}^{\infty} G(\xi - \eta, \omega) |\eta|^\lambda \, d\eta,$$

and

$$P\left(\omega_1 \frac{d}{d\xi}, ..., \omega_n \frac{d}{d\xi}\right) G(\xi, \omega) = \delta(\xi).$$

In the case that the dimension is odd

$$E(x_1, \ldots, x_n) = C_1 \int_\Omega \frac{\partial^{n-1} G(\xi, \omega)}{\partial \xi^{n-1}} \, d\Omega,$$

where

$$C_1 = \frac{(-1)^{\frac{n-1}{2}}}{\Omega_n (2\pi)^{\frac{n-1}{2}} \, 1 \cdot 3 \ldots (n-2)}.$$

For a homogeneous elliptic differential operator $(P = P_0)$, the solution of the equation

$$P\left(\frac{\partial}{\partial x}\right) u(x) = \frac{2 r^\lambda}{\Omega_n \Gamma\left(\dfrac{\lambda + n}{2}\right)}$$

is given by the formula

$$u(x_1, \ldots, x_n) = \frac{1}{\Omega_n \pi^{\frac{n-1}{2}} \Gamma\left(\dfrac{\lambda+1}{2}\right)} \int_\Omega \left\{ \frac{|\omega_1 x_1 + \cdots + \omega_n x_n|^{\lambda + 2m}}{(\lambda + 1) \ldots (\lambda + 2m)} + \right.$$

$$\left. + Q_\lambda \left(\sum_{k=1}^n \omega_k x_k \right) \right\} \frac{d\Omega}{P(\omega_1, \ldots, \omega_n)},$$

where

$$Q_\lambda(\xi) = \sum_{k=1}^m \frac{\xi^{2m - 2k}}{(2k-1)! \, (2m - 2k)! \, (\lambda + 2k)}.$$

For $\lambda = -n$, we obtain a fundamental solution corresponding to the operator P. In the fundamental solution we can leave only those terms of the polynomial $Q_\lambda(\xi)$ which are necessary in order for the function obtained not to have a pole at $\lambda = -n$.

If n is odd and $2m \geqslant n$, then the fundamental solution has the form

$$E(x_1, \ldots, x_n) =$$

$$= \frac{(-1)^{\frac{n-1}{2}}}{4 (2\pi)^{n-1} (2m - n)!} \int_\Omega |\omega_1 x_1 + \cdots + \omega_n x_n|^{2m - n} \frac{d\Omega}{P(\omega_1, \ldots, \omega_n)}.$$

If n is even and $2m \geqslant n$, then

$$E(x_1, ..., x_n) = \frac{(-1)^{\frac{n}{2}-1}}{(2\pi)^n (2m-n)!} \int_{\Omega} |\omega_1 x_1 + \cdots + \omega_n x_n|^{2m-n} \times$$

$$\times \ln |\omega_1 x_1 + \cdots + \omega_n x_n| \frac{d\Omega}{P(\omega_1, ..., \omega_n)}.$$

If n is odd and $2m < n$, then

$$E(x_1, ..., x_n) = \frac{(-1)^{\frac{n-1}{2}}}{2(2\pi)^{\frac{n-1}{2}}} \int_{\Omega} \delta^{(n-2m-1)}(\omega_1 x_1 + \cdots + \omega_n x_n) \frac{d\Omega}{P(\omega_1, ..., \omega_n)},$$

and, if n is even and $2m < n$, then

$$E(x_1, ..., x_n) =$$

$$= \frac{(-1)^{\frac{n}{2}}(n-2m-1)!}{(2\pi)^n} \int_{\Omega} |\omega_1 x_1 + \cdots + \omega_n x_n|^{-n+2m} \frac{d\Omega}{P(\omega_1, ..., \omega_n)}.$$

The fundamental solution is an ordinary function, analytic for $x \neq 0$, and satisfies the relation

$$E(x) = \begin{cases} O(r^{2m-n} \ln r), & \text{if } n \text{ is even and } 2m \geqslant n, \\ O(r^{2m-n}) & \text{in the remaining cases} \end{cases}$$

in a neighborhood of the origin of the coordinates. The function $E(x)$ has continuous derivatives up to order $2m-n-1$ for $2m > n$ at the origin.

4. *Fundamental solutions of homogeneous regular equations.* The linear differential operator $P\left(\dfrac{\partial}{\partial x}\right)$ with constant coefficients is called *regular* if it is homogeneous (that is all terms are derivatives of the same order m) and if the gradient of the function $P(\omega_1, ..., \omega_n)$ does not become zero for $\omega \neq 0$ on the cone $P(\omega_1, ..., \omega_n) = 0$. For a regular operator,

$$E(x_1, ..., x_n) = \int_{\Omega} \frac{f_{mn}(\sum x_k \omega_k) \, d\Omega}{P(\omega_1, ..., \omega_n)}$$

is a fundamental solution, where Ω is the unit ball and the function $f_{mn}(x)$ has the following values:

1) if n is even and $m \geqslant n$, then

$$f_{mn}(x) = \frac{(-1)^{\frac{n}{2}-1}}{(2\pi)^n (m-n)!} x^{m-n} \ln|x|;$$

2) if n is even and $m < n$, then

$$f_{mn}(x) = \frac{(-1)^{\frac{n}{2}+m}(n-m-1)!}{(2\pi)^n} x^{m-n};$$

3) if n is odd and $m \geqslant n$, then

$$f_{mn}(x) = \frac{(-1)^{\frac{n-1}{2}}}{4(2\pi)^{n-1}(m-n)!} x^{m-n} \operatorname{sign} x;$$

4) if n is odd and $m < n$, then

$$f_{mn}(x) = \frac{(-1)^{\frac{n-1}{2}}}{2(2\pi)^{n-1}} \delta^{(n-m-1)}(x).$$

In this connection, the integral is understood in the sense of a regularized value:

$$E(x_1, ..., x_n) = \lim_{\varepsilon \to 0} E_\varepsilon(x_1, ..., x_n),$$

where

$$E_\varepsilon(x_1, ..., x_n) = \int_{\Omega_\varepsilon} \frac{f_{mn}(\sum x_k \omega_k) \, d\Omega}{P(\omega_1, ..., \omega_n)}.$$

We denote here by Ω_ε the set of points on the unit sphere for which $|P(\omega_1, ..., \omega_n)| > \varepsilon$.

5. *Fundamental solution of the Cauchy problem.* Let a linear differential equation with constant coefficients

$$P\left(\frac{\partial}{\partial t}, \frac{\partial}{\partial x_1}, ..., \frac{\partial}{\partial x_n}\right) u = 0$$

of order m with respect to the variable t be given. Let the differential operator

$$P_\omega\left(\frac{\partial}{\partial t}, \frac{\partial}{\partial \xi}\right) = P\left(\frac{\partial}{\partial t}, \omega_1 \frac{\partial}{\partial \xi}, \dots, \omega_n \frac{\partial}{\partial \xi}\right)$$

be such that the Cauchy problem is well posed for the equation

$$P_\omega\left(\frac{\partial}{\partial t}, \frac{\partial}{\partial \xi}\right) v = 0.$$

Then the fundamental solution of the Cauchy problem for the initial equation has the form

$$E(t, x) = \int_\Omega v_\omega(t, x_1\omega_1 + \dots + x_n\omega_n, -n)\, d\Omega,$$

where

$$v_\omega(t, \xi, \lambda) = \frac{1}{\Omega_n \pi^{\frac{n-1}{2}} \Gamma\left(\frac{\lambda+1}{2}\right)} \int_{-\infty}^{\infty} G_\omega(t, \xi - \eta) |\eta|^\lambda\, d\eta$$

and $G_\omega(t, \xi)$ is the fundamental solution of the Cauchy problem for the equation $P_\omega\left(\dfrac{\partial}{\partial t}, \dfrac{\partial}{\partial \xi}\right) v = 0$.

In the case of an odd number of variables we obtain the simpler formula:

$$E(t, x) = \frac{(-1)^{\frac{n-1}{2}} \left(\frac{n-1}{2}\right)!}{\Omega_n \pi^{\frac{n-1}{2}} (n-1)!} \int_\Omega \frac{d^{n-1}}{d\xi^{n-1}} G_\omega(t, \xi)\, d\Omega.$$

A homogeneous linear operator with constant coefficients $P\left(\dfrac{\partial}{\partial t}, \dfrac{\partial}{\partial x_1}, \dots, \dfrac{\partial}{\partial x_n}\right)$ is called *hyperbolic* if the equation of m-th order with respect to v

$$P(v, \omega_1, \dots, \omega_n) = 0$$

has m real and distinct roots for arbitrary values $\omega_1, \dots, \omega_n$, $\displaystyle\sum_{k=1}^{n} \omega_k^2 = 1$.

The solution of the hyperbolic equation

$$P\left(\frac{\partial}{\partial t}, \frac{\partial}{\partial x_1}, ..., \frac{\partial}{\partial x_n}\right) u(x, t) = 0$$

with the initial conditions

$$\frac{\partial^k u(x, 0)}{\partial t^k} = 0, \ 0 \leqslant k \leqslant m - 2,$$

$$\frac{\partial^{m-1} u(x, 0)}{\partial t^{m-1}} = \frac{2r^\lambda}{\Omega_n \Gamma\left(\dfrac{\lambda + n}{2}\right)}$$

has the form

$$u(x_1, ..., x_n, t) = -\frac{2}{\Omega_n \pi^{\frac{n-1}{2}} \Gamma\left(\dfrac{\lambda+1}{2}\right)} \times$$

$$\times \int\limits_{H=0} \left\{ \frac{|\xi|^{-\lambda-n} |\sum x_k \xi_k + t|^{\lambda+m-1} \ \mathrm{sign} \left(\sum x_k \xi_k + t\right)^{m-1}}{(\lambda+1)(\lambda+2)...(\lambda+m-1)} + \right.$$

$$\left. + Q\left(\frac{\sum x_k \xi_k + t}{|\xi|}\right) \right\} \omega,$$

where

$$Q_\lambda(\xi) = \sum_{k=1}^{\left[\frac{m-1}{2}\right]} \frac{\xi^{m-2k-1}}{(2k-1)!\,(m-2k-1)!\,(\lambda-2k)},$$

$H(\xi_1, ..., \xi_n)$ denotes the function $P(1, \xi_1, ..., \xi_n)$ and ω denotes the expression

$$\frac{d\sigma}{|\mathrm{grad}\, H|\, \mathrm{sign}\left(\displaystyle\sum_{k=1}^{n} \xi_k \frac{\partial H}{\partial \xi_k}\right)}$$

($d\sigma$ is an element of the surface $H=0$).

For $\lambda=-n$, we obtain the fundamental solution of the Cauchy problem. If n is odd, then the solution has the form

$$E(x_1, ..., x_n) = \frac{(-1)^{\frac{n+1}{2}}}{2(2\pi)^{n-1}(m-n-1)!} \times$$

$$\times \int\limits_{H=0} (\sum x_k\xi_k + t)^{m-n-1} [\text{sign}(\sum x_k\xi_k + t)]^{m-1} \omega;$$

if n is even, then the solution has the form

$$E(x_1, ..., x_n) = \frac{2(-1)^{\frac{n}{2}}}{(2\pi)^n(m-n-1)!} \times$$

$$\times \int\limits_{H=0} (\sum x_k\xi_k + t)^{m-n-1} \ln \left| \frac{\sum x_k\xi_k + t}{\sum x_k\xi_k} \right| \omega$$

(*Herglotz-Petrovskii formulas*). When the order m of the equation is less than $n-1$, the formulas for the fundamental solution of the Cauchy problem assume the forms

$$E(x_1, ..., x_n) = \frac{(-1)^{\frac{n+1}{2}}}{(2\pi)^{n-1}} \int\limits_{H=0} \delta^{(n-m)}(\sum x_k\xi_k + t) \omega$$

for odd n and

$$E(x_1, ..., x_n) = \frac{(-1)^{\frac{n}{2}}(n-m)!}{(2\pi)^n} \int\limits_{H=0} \frac{\omega}{(\sum x_k\xi_k + t)^{n-m+1}}$$

for even n.

All integrals in these formulas are understood in the sense of a regularized value.

§ 7. Generalized functions in a complex space

1. *Generalized functions of one complex variable.* In the consideration of functions of a complex variable, use is made of the operators

$$\frac{\partial}{\partial z} = \frac{1}{2}\left(\frac{\partial}{\partial x} - i\frac{\partial}{\partial y}\right),$$

$$\frac{\partial}{\partial \bar{z}} = \frac{1}{2}\left(\frac{\partial}{\partial x} + i\frac{\partial}{\partial y}\right),$$

where $z = x + iy$. For example, the Maclaurin series is written as follows:

$$f_1(x, y) \equiv f(z, \bar{z}) = \sum_{j, k=0}^{\infty} \frac{f^{(j, k)}(0, 0)}{j! \, k!} z^j \bar{z}^k,$$

where we have set $f^{(j, k)}(z, \bar{z}) \equiv \dfrac{\partial^{j+k} f(z, \bar{z})}{\partial z^j \partial \bar{z}^k}$ and $f(z, \bar{z}) = f_1\left(\dfrac{z+\bar{z}}{2}, \dfrac{z-\bar{z}}{2i}\right)$.

For analytic functions $\dfrac{\partial f(z)}{\partial \bar{z}} = 0$ (this follows from the Cauchy-Riemann conditions).

To integrate the function $f(z, \bar{z})$ we use the differential form $dz \, d\bar{z} = -2i \, dx \, dy$.

Let K be the space of infinitely differentiable functions $\varphi(z, \bar{z})$ with compact support. Let λ and μ be complex numbers such that $n = \lambda - \mu$ is an integer. Then for $\mathrm{Re}(\lambda + \mu) > -2$, the convergent integral

$$(z^\lambda \bar{z}^\mu, \varphi) = \frac{i}{2} \int z^\lambda \bar{z}^\mu \varphi(z, \bar{z}) \, dz \, d\bar{z}$$

defines a generalized function $z^\lambda \bar{z}^\mu$ on K. This function is homogeneous: the equality

$$\left(z^\lambda \bar{z}^\mu, \varphi\left(\frac{z}{a}, \frac{\bar{z}}{\bar{a}}\right)\right) = a^{\lambda+1} \bar{a}^{\mu+1}(z^\lambda \bar{z}^\mu, \varphi(z, \bar{z}))$$

is valid for an arbitrary function $\varphi(z, \bar{z})$ of K.

The generalized function $z^\lambda \bar{z}^\mu$ is defined for $\mathrm{Re}(\lambda + \mu) < -2$ by means of analytic continuation with respect to $s \equiv \lambda + \mu$:

$$(z^\lambda \bar{z}^\mu, \varphi) = \frac{i}{2} \int_{|z| \leqslant 1} z^\lambda \bar{z}^\mu \left[\varphi(z, \bar{z}) - \sum_{k+l=0}^{m-1} \frac{\varphi^{(k, l)}(0, 0)}{k! \, l!} z^k \bar{z}^l\right] dz \, d\bar{z} +$$

$$+ \frac{i}{2} \int_{|z| > 1} z^\lambda \bar{z}^\mu \varphi(z, \bar{z}) \, dz \, d\bar{z} + 2\pi \sum_{\substack{k+l=0 \\ k-l=-n}}^{m-1} \frac{\varphi^{(k, l)}(0, 0)}{k! \, l! (k + l + \lambda + \mu + 2)},$$

where $\mathrm{Re}(\lambda + \mu) > -m - 2$.

The generalized function $z^\lambda \bar{z}^\mu$ is regular everywhere except at the points $\lambda, \mu = -1, -2, \ldots$. At these points the function $z^\lambda \bar{z}^\mu$, as a function of $s = \lambda + \mu$ for fixed $n = \lambda - \mu$, has simple poles. In this connection,

$$\operatorname*{Res}_{\substack{\lambda = -k-1 \\ \mu = -l-1}} z^\lambda \bar{z}^\mu = 2\pi \frac{(-1)^{k+l}}{k!\,l!} \delta^{(k,\,l)}(z, \bar{z}),$$

where

$$(\delta, \varphi) = \varphi(0, 0) \quad \text{and} \quad \delta^{(k,\,l)}(z, \bar{z}) = \frac{\partial^{k+l} \delta(z, \bar{z})}{\partial z^k \partial \bar{z}^l}.$$

The normalized generalized function

$$\frac{z^\lambda \bar{z}^\mu}{\Gamma\left(\dfrac{s + |n| + 2}{2}\right)}$$

is an entire analytic function of $s = \lambda + \mu$ for fixed $n = \lambda - \mu$. For $\lambda = -k-1$, $\mu = -l-1$ we have

$$\frac{z^\lambda \bar{z}^\mu}{\Gamma\left(\dfrac{s + |n| + 2}{2}\right)}\Bigg|_{\substack{\lambda = -k-1 \\ \mu = -l-1}} = \frac{\pi(-1)^{k+l+j}\,j!}{k!\,l!} \delta^{(k,\,l)}(z, \bar{z}),$$

where $j = \min(k, l)$.

The generalized function z^{-k-1} is a particular case of $z^\lambda \bar{z}^\mu$. We can define it by the equality

$$z^{-k-1} = (-1)^k \frac{1}{k!} \frac{\partial^k (z^{-1})}{\partial z^k}.$$

In this connection

$$\frac{\partial(z^{-k-1})}{\partial \bar{z}} = (-1)^k \frac{\pi}{k!} \delta^{(k,\,0)}(z, \bar{z})$$

(this derivative is not equal to zero since z^{-k-1} is not analytic at the point $z = 0$).

As in the real case, the associated homogeneous functions $z^\lambda \bar{z}^\mu \ln^m |z|$ are introduced:

$$(z^\lambda \bar{z}^\mu \ln^m |z|, \varphi) = \frac{i}{2} \int z^\lambda \bar{z}^\mu \ln^m |z|\, \varphi(z, \bar{z})\, dz\, d\bar{z},$$

where the integral is understood in the sense of a regularized value. The function

$$(z^{-k-1}\bar{z}^{-l-1}, \varphi) = \frac{i}{2}\int z^{-k-1}\bar{z}^{-l-1}\left[\varphi(z, \bar{z}) - \sum_{i+j=0}^{k+l-1}\frac{\varphi^{(i,j)}(0,0)}{i!j!}z^i\bar{z}^j - \right.$$
$$\left. - \theta(1-|z|)\sum_{i+j=k+l}\frac{\varphi^{(i,j)}(0,0)}{i!j!}z^i\bar{z}^j\right]dz\,d\bar{z}$$

is an associated homogeneous function.

We call the function

$$\tilde{\varphi}(w, \bar{w}) = \frac{i}{2}\int\varphi(z, \bar{z})\,e^{i\,\mathrm{Re}\,z\bar{w}}\,dz\,d\bar{z}$$

the Fourier transform of the function $\varphi(z, \bar{z})$. The Fourier transform of the generalized function F is defined by the equality

$$(\tilde{F}, \tilde{\varphi}) = 4\pi^2(F, \varphi_1),$$

where $\varphi_1(z, \bar{z}) = \varphi(-z, -\bar{z})$.

The formula

$$\overline{\frac{z^\lambda\bar{z}^\mu}{\Gamma\left(\dfrac{s+|n|+2}{2}\right)}} = 2^{\lambda+\mu+2}\pi i^{|\lambda-\mu|}\frac{w^{-\mu-1}\bar{w}^{-\lambda-1}}{\Gamma\left(\dfrac{-s+|n|}{2}\right)}$$

holds, where $s = \lambda+\mu$, $n = \lambda-\mu$.

We also introduce generalized functions of the form $f^\lambda(z)\bar{f}^\mu(z)$, where $f(z)$ is a meromorphic function and $n = \lambda-\mu$ is an integer. If the function of compact support $\varphi(z, \bar{z})$ is concentrated in a domain containing one zero of multiplicity k of the function $f(z)$ and not containing poles of this function, then the integral

$$(f^\lambda\bar{f}^\mu, \varphi) = \frac{i}{2}\int f^\lambda(z)\,\bar{f}^\mu(z)\,\varphi(z, \bar{z})\,dz\,d\bar{z}$$

is convergent in the domain $\mathrm{Re}(\lambda+\mu) > 0$ for given $\lambda-\mu$.

For $\mathrm{Re}(\lambda+\mu) < 0$, we define its value by means of analytic continuation with respect to $s = \lambda+\mu$. Simple poles at the points $(\lambda, \mu) = \left(-\dfrac{p}{k}, -\dfrac{q}{k}\right)$, $p, q = 1, 2, \dots$ ($\lambda-\mu$ is an integer) are the only singularities of this integral considered as an analytic function of λ and μ.

In the same way, if the function $\varphi(z, \bar{z})$ is concentrated in a domain containing one pole of order l of the function $f(z)$ and not containing zeros of this function, then the integral is convergent for $\operatorname{Re}(\lambda+\mu)<0$. For $\operatorname{Re}(\lambda+\mu)>0$, its value is defined by means of analytic continuation with respect to $s = \lambda + \mu$. Simple poles at the points $(\lambda, \mu) = \left(\dfrac{p}{l}, \dfrac{q}{l}\right)$, $p, q = 1, 2, \ldots$ are the only singularities of this integral considered as a function of λ and μ.

In the general case, we define the integral by means of a partitioning of the function $\varphi(z, \bar{z})$ into a finite number of terms each of which is concentrated in a domain containing no more than one zero or pole of the function $f(z)$.

2. *Generalized functions of m complex variables.* Let S be the surface in m-dimensional complex space which is given by the equation

$$P(z) \equiv P(z_1, \ldots, z_m) = 0 \, {}^*),$$

where $P(z)$ is an infinitely differentiable function of z and $\bar{z}$. It is assumed that the differential form $dP \, d\bar{P}$ does not vanish on the surface S.

We define the differential form $d\omega$ of order $2m-2$ by the relation

$$\left(\frac{i}{2}\right)^m dz \, d\bar{z} = \frac{i}{2} dP \, d\bar{P} \, d\omega$$

and set

$$(\delta(P), \varphi) = \int_{P=0} \varphi \, d\omega .$$

Here the notations

$$\left(\frac{i}{2}\right)^m dz \, d\bar{z} = \left(\frac{i}{2}\right)^m dz_1 \, d\bar{z}_1 \ldots dz_m \, d\bar{z}_m$$

and

$$dP = \sum \frac{\partial P}{\partial z_k} dz_k + \frac{\partial P}{\partial \bar{z}_k} d\bar{z}_k$$

are used. Let

$$\delta^{(k, l)}(P) = \frac{\partial^{k+l} \delta(P)}{\partial P^k \, \partial \bar{P}^l} .$$

---

*) In order not to complicate the notation, functions of m complex variables are often denoted by $P(z)$ instead of $P(z, \bar{z})$.

For this generalized function, the properties established in the real case remain valid (see § 3, no 5):

1) $\dfrac{\partial}{\partial z_i} \delta^{(k,\,l)}(P) = \dfrac{\partial P}{\partial z_i} \delta^{(k+1,\,l)}(P) + \dfrac{\partial \bar{P}}{\partial z_i} \delta^{(k,\,l+1)}(P),$

and the analogous formula for $\dfrac{\partial}{\partial \bar{z}_i} \delta^{(k,\,l)}(P)$ holds;

2) $P\delta(P) = \bar{P}\delta(P) = 0,$
$$P\delta^{(k,\,l)}(P) + k\delta^{(k-1,\,l)}(P) = 0,$$
$$\bar{P}\delta^{(k,\,l)}(P) + l\delta^{(k,\,l-1)}(P) = 0;$$

3) if the surfaces $P=0$, $Q=0$ do not have singular points and do not intersect, then
$$\delta(PQ) = P^{-1}\bar{P}^{-1}\delta(Q) + Q^{-1}\bar{Q}^{-1}\delta(P).$$

In particular, if the function $a(z)$ does not vanish then
$$\delta(aP) = a^{-1}\bar{a}^{-1}\delta(P).$$

If the function P is analytic $\left(\dfrac{\partial P}{\partial \bar{z}_i}=0\right)$, then
$$\delta^{(k,\,l)}(aP) = a^{-k-1}\bar{a}^{-l-1}\delta^{(k,\,l)}(P)$$

for all nowhere vanishing functions $a(z)$.

Let $G(z)$ be an entire analytic function of m complex variables $z_1, ..., z_m$. If $\lambda - \mu = n$ is an integer, then we set
$$(G^\lambda \bar{G}^\mu, \varphi) = \left(\frac{i}{2}\right)^m \int G^\lambda(z)\, \bar{G}^\mu(z)\, \varphi(z, \bar{z})\, dz\, d\bar{z},$$

where
$$G^\lambda \bar{G}^\mu = |G|^{\lambda+\mu} e^{i(\lambda-\mu)\,\arg G}.$$

If the surface $G(z)=0$ does not have singular points, then the simple poles at the points
$$(\lambda, \mu) = (-k-1, -l-1), \quad k, l = 0, 1, ...,$$

with the residues
$$\operatorname*{Res}_{\substack{\lambda=-k-1 \\ \mu=-l-1}} G^\lambda \bar{G}^\mu = (-1)^{k+l}\, \frac{2\pi}{k!\, l!}\, \delta^{(k,\,l)}(G)$$

are the only singularities of the generalized function $G^\lambda \bar{G}^\mu$, considered as a function of λ, μ.

Let $P(z)$ be a non-degenerate quadratic form of m complex variables:

$$P = \sum_{i,j=1}^{m} g_{ij} z_i z_j \, .$$

Consider

$$(P^\lambda \bar{P}^\mu, \varphi) = \left(\frac{i}{2}\right)^m \int P^\lambda(z) \, \bar{P}^\mu(z) \, \varphi(z, \bar{z}) \, dz \, d\bar{z} \, .$$

This integral is convergent for $\mathrm{Re}\,(\lambda+\mu) > 0$.

In order to express $P^\lambda \bar{P}^\mu$ for $\mathrm{Re}\,(\lambda+\mu) < 0$, the differential operators

$$L_P = \sum_{i,j=1}^{m} g^{ij} \frac{\partial^2}{\partial z_i \partial z_j} \, , \quad \bar{L}_P = \sum_{i,j=1}^{m} \overline{g^{ij}} \frac{\partial^2}{\partial \bar{z}_i \partial \bar{z}_j}$$

are introduced, where

$$\sum_{j=1}^{m} g^{ij} g_{jk} = \delta_k^i \, .$$

Then, for $\mathrm{Re}\,(\lambda+\mu) > -k-l$, we define $P^\lambda \bar{P}^\mu$ by the formula

$$P^\lambda \bar{P}^\mu = C(\lambda, k) \, C(\mu, l) \, L_P^k \bar{L}_P^l P^{\lambda+k} \bar{P}^{\mu+l} \, ,$$

where

$$C(v, p) = \left\{ 4^p (v+1) \ldots (v+p) \left(v + \frac{m}{2} \right) \ldots \left(v + \frac{m}{2} + p - 1 \right) \right\}^{-1} \, .$$

The generalized function $P^\lambda \bar{P}^\mu$ has two sequences of singular points:

$$(\lambda, \mu) = (-k-1, -l-1), \quad k, l = 0, 1, 2, \ldots ,$$

and

$$(\lambda, \mu) = \left(-\frac{m}{2} - k, -\frac{m}{2} - l \right), \quad k, l = 0, 1, \ldots$$

If the point (λ, μ) belongs to only one of these sequences, then $P^\lambda \bar{P}^\mu$ has a simple pole there; if (λ, μ) belongs to both sequences, then $P^\lambda \bar{P}^\mu$ has a pole of order 2 there.

In the case when the point $\lambda = -k-1$, $\mu = -l-1$ belongs only to the first sequence, $\underset{\substack{\lambda=-k-1 \\ \mu=-l-1}}{\mathrm{Res}} \; P^\lambda \bar{P}^\mu$ is a generalized function concentrated on the surface $P=0$. We define

$$\delta^{(k,\,l)}(P) \equiv \frac{1}{2\pi}(-1)^{k+l}\,k!\,l!\;\mathop{\mathrm{Res}}_{\substack{\lambda=-k-1\\ \mu=-l-1}}\;P^\lambda\bar{P}^\mu.$$

If the point (λ, μ) belongs only to the second sequence, that is, $\lambda=-\dfrac{m}{2}-k$, $\mu=-\dfrac{m}{2}-l$, where m is an odd number, then $\mathrm{Res}\,P^\lambda\bar{P}^\mu$ at this point equals

$$\frac{(-1)^{\frac{m-1}{2}}\,2^{-m-2k-2l+1}\pi^{m+1}}{k!\,l!\,\Gamma\left(\dfrac{m}{2}+k\right)\Gamma\left(\dfrac{m}{2}+l\right)|\Delta|}\,L_P^k\,\bar{L}_P^l\,\delta(z),$$

where Δ is the discriminant of the quadratic form P.

By means of generalized functions of the form $P^\lambda\bar{P}^\mu$ we can construct fundamental solutions of the equations

$$L^k u = f(z),$$

where

$$L = \sum_{i,\,j=1}^m g^{ij}\,\frac{\partial^2}{\partial z_i\partial\bar{z}_j},$$

and the matrix $\|g^{ij}\|$ is nonsingular.

Namely, let

$$P = \sum_{i,\,j=1}^m g_{ij}z_i z_j,\quad\text{where}\quad \sum_{j=1}^m g^{ij}g_{jk} = \delta_k^i.$$

In the case of a space of odd dimensionality, the function

$$K = \frac{(-1)^{\frac{m-1}{2}+k}\,2^{m-2k}\Gamma\left(\dfrac{m}{2}\right)\Gamma\left(\dfrac{m}{2}-k\right)|\Delta|}{\pi^{m+1}(k-1)!}\,P^{-\frac{m}{2}+k}\bar{P}^{-\frac{m}{2}}$$

is a fundamental solution. If m is an even number and $k\geqslant\dfrac{m}{2}$, then there is a fundamental solution of the form

$$K = \frac{2^{m-2k} \, \Gamma\left(\dfrac{m}{2}\right) |\varDelta|}{\pi^m (k-1)! \left(k - \dfrac{m}{2}\right)!} \, P^{-\frac{m}{2}+k} \bar{P}^{-\frac{m}{2}}.$$

Finally, if m is an even number and $k < \dfrac{m}{2}$, then there is a fundamental solution of the form

$$K = \frac{(-1)^{\frac{m}{2}-1} \, 2^{m-2k} |\varDelta|}{\pi^{m-1} (k-1)!} \, \delta^{\left(\frac{m}{2}-k-1, \frac{m}{2}-1\right)}(P).$$

BIBLIOGRAPHY

[1] AGRANOVIČ, Z. S. and V. A. MARČENKO, *Inverse scattering problem*, Izd-vo Khar'kovskogo universiteta, 1960 (Russian).

[2] AHIEZER, N. I. and I. M. GLAZMAN, *Theory of linear operators in Hilbert space*, Gostehizdat, 1950 (Russian).

[2a] English edition of [2], F. Ungar Publishing Company, N. Y., in two volumes: Volume I, 1962 and Volume II, 1963.

[2b] German edition of [2], *Theorie der linearen Operatoren im Hilbert-Raum*, Akademie-Verlag, Berlin, 1954.

[3] BANACH, S., *Course in functional analysis*, Radyanska Skola, Kiev, 1948 (Ukrainian).

[4] BOURBAKI, N., *Topological vector spaces*, Fizmatgiz, 1959 (Russian).

[4a] Original French edition of [4], *Espaces vectoriels topologiques*, Hermann, Paris, 1950.

[5] BOURBAKI, N., *General topology*, Fizmatgiz, 1959 (Russian).

[5a] Original French edition of [5], *Topologie générale*, Hermann, Paris, 1958.

[6] VAINBERG, M. M., *Variational methods in the study of nonlinear operators*, Fizmatgiz, 1956 (Russian).

[7] VULIH, B. Z., *Theory of partially ordered spaces*, Fizmatgiz, 1961 (Russian).

[7a] English edition of [7] to be published soon by P. Noordhoff, Groningen.

[8] VULIH, B. Z., *Introduction to functional analysis*, Fizmatgiz, 1958 (Russian).

[8a] English edition of [8], Addison-Wesley Publishing Co., Reading, 1963.

[9] GEL'FAND, I. M., D. A. RAIKOV and G. E. ŠILOV, *Commutative normed rings* Fizmatgiz, 1960 (Russian).

[9a] English edition of [9], Chelsea Publishing Co., N. Y., 1964.

[9b] German edition of [9], VEB Deutscher Verlag der Wissenschaften, Berlin, 1964.

[10] GEL'FAND, I. M. and G. E. ŠILOV, *Generalized functions*, Issue 1: *Generalized functions and operations on them*, 2nd edition, Fizmatgiz, 1959 (Russian).

[10a] English edition of [10], Academic Press, New York and London, 1964.

[11] GEL'FAND, I. M. and G. E. ŠILOV, *Generalized functions*, Issue 2: *Spaces of test functions and generalized functions*, Fizmatgiz, 1958 (Russian).

[12] GEL'FAND, I. M. and G. E. ŠILOV, *Generalized functions*, Issue 3: *Some problems in the theory of differential equations*, Fizmatgiz, 1958 (Russian).

[13] GEL'FAND, I. M. and N. Ya. VILENKIN, *Generalized functions*, Issue 4: *Some applications of harmonic analysis. Rigged Hilbert spaces*, Fizmatgiz, 1961 (Russian).

[13a] English edition of [13], Academic Press, New York and London, 1964.

[14] GEL'FAND, I. M., M. I. GRAEV and N. Ya. VILENKIN, *Generalized functions*, Issue 5: *Integral geometry and its relations with problems of representation theory*, Fizmatgiz, 1962 (Russian).

[15] GOL'DMAN, I. I. and V. D. KRIVČENKOV, *Compendium of problems of quantum mechanics*, Gostehizdat, 1957 (Russian).

378   BIBLIOGRAPHY

[16] DUNFORD, N. and J. T. SCHWARTZ, *Linear operators*, IL, 1962 (Russian).
[16a] Original English edition of [16], Interscience Publishers, New York and London, 1958.
[17] DAY, M. M., *Normed Linear spaces*, IL, 1961 (Russian).
[17a] Original English edition of [17], Springer-Verlag, Berlin, 1960.
[18] DIRAC, P. A. M., *The principles of quantum mechanics*, Fizmatgiz, 1960 (Russian).
[18a] Original English edition of [18], Clarendon Press, Oxford, 1947.
[19] KANTOROVIČ, I. V. and G. P. AKILOV, *Functional analysis in normed spaces*, Fizmatgiz, 1959 (Russian).
[20] KANTOROVIČ, I. V., B. Z. VULIH and A. G. PINSKER, *Functional analysis in partially ordered spaces*, Gostehizdat, 1950 (Russian).
[20a] English edition of [20], to be published soon by P. Noordhoff, Groningen.
[21] KATO, T., Collection of translations "Matematika" 2:4 (1958), pp. 118–135 (Russian).
[22] KOLMOGOROV, A. N. and S. V. FOMIN, *Elements of the theory of functions and functional analysis*, Issue 1, Izd-vo Moskovskogo universiteta, 1954 (Russian).
[22a] English edition of [22], Graylock Press, 428 East Preston St., Baltimore, Maryland, 1957.
[23] KRASNOSEL'SKII, M. A., *Topological methods in the theory of nonlinear integral equations*, Gostehizdat, 1956 (Russian).
[23a] English edition of [23], The MacMillan Co., N. Y., 1964.
[24] KRASNOSEL'SKII, M. A., *Some problems of nonlinear analysis*, Uspehi matem. nauk IX: 3 (1954), pp. 57–114 (Russian).
[25] KRASNOSEL'SKII, M. A., *Positive solutions of operator equations*, Fizmatgiz, 1962 (Russian).
[25a] English edition of [25], P. Noordhoff, Groningen, 1964.
[26] KREIN, M. G. and M. A. RUTMAN, *Linear operators which leave invariant a cone in Banach space*, Uspehi matem. nauk III: 1 (1948) (Russian).
[27] LANDAU, L. D. and E. M. LIFŠIC, *Quantum mechanics*, Gostehizdat, 1948 (Russian).
[27a] English edition of [27], Pergamon Press, London, 1958.
[28] LANDAU, L. D. and E. M. LIFŠIC, *Field theory*, Fizmatgiz, 1960 (Russian).
[28a] English edition of [28], *The classical theory of fields*, Addison-Wesley Press, Cambridge, 1951.
[29] LERAY, J. and J. SCHAUDER, *Topology and functional equations*, Uspehi matem. nauk 1: 3–4 (1946), pp. 71–95 (Russian).
[30] LOOMIS, L. H., *An introduction to abstract harmonic analysis*, IL, 1956 (Russian).
[30a] Original English edition of [30], Van Nostrand, New York, 1953.
[31] LYUSTERNIK, L. A. and V. I. SOBOLEV, *Elements of functional analysis*, Gostehizdat, 1951 (Russian).
[31a] English edition of [31], F. Ungar Publishing Company, N. Y., 1960.
[31b] German edition of [31], *Elemente der Funktional analysis*, Akademie-Verlag, Berlin, 1955.
[32] MASLOV, V. P., *Quasiclassical asymptotic solutions of some problems of mathematical physics*, I, II, Žurnal vyčislitel'noi matematiki i matematičeskoi fiziki 1: 1 (1961) and 1: 4 (1961) (Russian).
[33] MIKUSIŃSKI, J. and P. SIKORSKI, *Elementary theory of generalized functions*, Issue 1, IL, 1959 (Russian).
[33a] English edition of [33], *The elementary theory of distributions* (I), PWN, Warsaw, 1957.

[34] MIKUSIŃSKI, J. and P. SIKORSKI, *Elementary theory of generalized functions*, Issue 2, IL, 1963 (Russian).

[34a] English edition of [34], *The elementary theory of distributions* (II), PWN, Warsaw, 1961.

[35] MOTT, N. F. and H. S. W. MASSEY, *The theory of atomic collisions*, IL, 1951 (Russian).

[35a] Original English edition of [35], Clarendon Press, Oxford, 1949.

[36] NAIMARK, M. A., *Normed rings*, Gostehizdat, 1956 (Russian).

[36a] English edition of [36], P. Noordhoff, Groningen, Revised edition, 1964.

[36b] German edition of [36], VEB Deutscher Verlag der Wissenschaften, Berlin, 1959.

[37] NAIMARK, M. A., *Linear differential operators*, Gostehizdat, 1954 (Russian).

[37a] German edition of [37], *Lineare Differentialoperatoren*, Akademie-Verlag, Berlin, 1960.

[38] PONTRYAGIN, L. S., *Continuous groups*, Gostehizdat, 1954 (Russian).

[38a] English edition of the first edition of [38], *Toplogical groups*, Princeton U. Press, Princeton, 1939.

[38b] German edition of the second edition of [38], *Topologische Gruppen*.

[39] RIESZ, F. and B. SZ.-NAGY, *Lectures on functional analysis*, IL, 1954 (Russian).

[39a] Original French edition of [39], *Leçons d'analyse fonctionnelle*, Budapest and Paris, 1952.

[39b] English edition of [39], F. Ungar Publishing Company, N. Y., 1955.

[39c] German edition of [39], *Vorlesungen über Funktionalanalysis*, Deutscher Verlag der Wissenschaften, Berlin, 1956.

[40] SMIRNOV, V. I., *Course in higher mathematics*, Volume V, Fizmatgiz, 1959 (Russian).

[41] SMIRNOV, N. S., *Introduction to the theory of nonlinear integral equations*, ONTI, 1936, (Russian).

[42] SOBOLEV, S. L., *Some applications of functional analysis to mathematical physics*, Izd-vo LGU, 1950 (Russian).

[43] SOBOLEVSKII, P. E., Trudy Moskovskogo matem. obscestva, Volume 10, 297–350, Fizmatgiz, 1961 (Russian).

[44] HALILOV, Z. I., *Foundations of functional analysis*, Baku, 1949 (Russian).

[45] HALMOS, P., *Finite-dimensional vector spaces*, Fizmatgiz, 1963 (Russian).

[45a] Original English edition of [45], D. Van Nostrand, N. Y., 1958.

[46] HILLE, E., *Functional analysis and semigroups*, IL, 1951 (Russian).

[46a] Original English edition of [46], AMS Colloquium Publications, Volume XXI, Providence, 1948.

[47] HILLE, E. and P. PHILLIPS, *Functional analysis and semigroups*, IL, 1962 (Russian).

[47a] Original English edition of [47], AMS Colloquium Publications, Volume XXXI. 1957.

[48] SCHWEBER, S. S., H. A. BETHE and F. DEHOFFMANN, *Mesons and fields*, I, IL, 1957 (Russian).

[48a] Original English edition of [48], 2 volumes, Row and Peterson, Evanston, 1955.

[49] ŠILOV, G. E., *Introduction to the theory of linear spaces*, 2nd edition, Fizmatgiz, 1956 (Russian).

[50] KRASNOSEL'SKII, M. A. and Ya. B. RUTICKII, *Convex functions and Orlicz spaces*, Fizmatgiz, 1958 (Russian).

[50a] English edition of [50], P. Noordhoff, Groningen, 1961.

INDEX OF LITERATURE ACCORDING TO CHAPTERS

Chapter 1

[3], [4], [5], [8], [17], [19], [22], [31], [40], [42], [44], [46], [49], [50].

Chapter 2

[2], [16], [31], [37], [39], [40], [42], [45].

Chapter 3

[21], [39], [43], [46], 47].

Chapter 4

[6], [19], [23], [24], [29], [31], [41], [46].

Chapter 5

[7], [20], [25], [26].

Chapter 6

[9], [30], [38], [47].

Chapter 7

[1], [15], [18], [27], [28], [32], [35], [48].

Chapter 8

[10], [11], [12], [13], [14], [33], [34].